Model Identification and Adaptive Control

Springer
London
Berlin
Heidelberg
New York
Barcelona
Hong Kong
Milan
Paris
Singapore
Tokyo

Graham Goodwin (Ed)

Model Identification and Adaptive Control

From Windsurfing to Telecommunications

With 70 Figures

Springer

Graham Goodwin, BSc, BE PhD, FIEEE, Hon FIE Aust., FTS, FAA
Department of Electrical and Computer Engineering, The University of Newcastle, Callaghan, NSW 2308, Australia

Springer-Verlag London Berlin Heidelberg

British Library Cataloguing in Publication Data
Model identification and adaptive control : from windsurfing to telecommunications
1.Adaptive control systems
I.Goodwin, Graham C. (Graham Clifford), 1945-
629.8'36
ISBN-13: 978-1-4471-1185-6 e-ISBN-13:978-1-4471-0711-8
DOI: 10.1007/978-1-4471-0711-8

Library of Congress Cataloging-in-Publication Data
A catalog record for this book is available from the Library of Congress

Softcover reprint of the hardcover 1st edition 2001

Typesetting: Camera ready by contributors

69/3830-543210 Printed on acid-free paper SPIN 10777845

The Centre for Integrated Dynamics and Control (A Commonwealth Special Research Centre) at *The* University *of* Newcastle decided to organize a Colloquium "Model Identification and Adaptive Control: From *Windsurfing* to *Telecommunications*" to celebrate the sixtieth birthday of the University's foundation Professor of Electrical Engineering, Professor Brian D.O. Anderson.

Brian served at *The* University *of* Newcastle between 1967 and 1981. His leadership established a world class department and his positive influence continues to this day. Beyond *The* University *of* Newcastle, Brian has had a major influence on the development of science and technology in Australia.

The colloquium received support from:

- Australian Academy of Science
- The University of Sydney, Australia
- ANUTECH, at Australian National University, Australia
- The University of Newcastle, Australia
- CIDAC, The University of Newcastle, Australia.

Acknowledgements

Many colleagues and students contributed to this book and the Colloquium. We wish to thank all of them but especially Dianne Piefke who, in reality, did most of the work and made it possible. Thanks are also due to James Welsh who put substantial work into the job of turning a collection of separate chapters into a cohesive book. Many thanks also to Heather Jauncey and Dianne Anderson who, unbeknown to Brian, provided "secret" information which appears in a serious form in this book, and in a less serious form in the speeches given at the Colloquium dinner.

BDO (Brian) Anderson is one of Australia's most distinguished scientists. He was born in Sydney Australia in January 1941 and went to Sydney Church of England Grammar School. He later undertook degrees at Sydney University in both Electrical Engineering and Mathematics. He graduated with First Class Honours in Science and University Medal in Mathematics 1962 and First Class Honours and University Medal in Electrical Engineering in 1963. He then went to Stanford University where he completed a PhD in Electrical Engineering under the direction of Professor Robert Newcomb in 1966.

He worked in industry in the United States and at Stanford before becoming the Foundation Professor of Electrical Engineering at the University of Newcastle, Australia in 1967. He was the youngest person ever to be appointed to the rank of Full Professor in Australia having received this distinction at the young age of 26. Notwithstanding his tender age, Brian proved to be one of the most energetic, forward looking and visionary professors of Electrical Engineering to have ever served in the Australian University system. Indeed the mould he set on the Department in Newcastle laid the foundations for one of the strongest Electrical Engineering departments in the world. The Department continues to benefit from Brian's legacy.

In 1981 Brian took up the newly established post of Professor and Head of the new Department of Systems Engineering in the Research School at the Australian National University in Canberra. Here, he established a new discipline area and forged exciting new research directions of benefit to Australia. He is currently Director of the Research School of Information Science and Engineering at the Australian National University.

Brian has a prolific output of research publications including 9 books and over 600 technical articles. He has also supervised 30 research students many of whom have gone on to hold senior positions in academia or industry throughout the world.

Brian has made major contributions to science and technology in Australia. He has served as a member of a number of government bodies, including the Australian Science and Technology Council and the Prime Minister's Science and Engineering Council. He is also a member of the Board of Cochlear Limited, the world's major supplier of cochlear implants. He is a Fellow of the Australian Academy of Science and Academy of Technological Sciences and Engineering, the Institute of Electrical and Electronic Engineers, and an Honorary Fellow of the Institution of Engineers, Australia. In 1989, he became a Fellow of the Royal Society, London. He holds honorary doctorates of the Catholic University of Louvain in Belgium, the Swiss Federal Institute of Technology, and the Universities of Sydney and Melbourne, and was appointed an Officer of the Order of Australia in 1993.

He was President of the International Federation of Automatic Control for the triennium 1990 to 1993, and is serving as President of the Australian Academy of Science from April 1998 until 2002.

Brian, we dedicate this to you in admiration and appreciation.

Preface

This book is based on a workshop entitled: "Model Identification and Adaptive Control: From *Windsurfing* to *Telecommunications*" held in Sydney, Australia, on December 16, 2000. The workshop was organized in honour of Professor Brian (BDO) Anderson in recognition of his seminal contributions to systems science over the past 4 decades.

The chapters in the book have been written by colleagues, friends and students of Brian Anderson. A central theme of the book is the inter relationship between identification and the use of models in real world applications. This theme has underpinned much of Brian Anderson's own contributions. The book reflects on these contributions as well as making important statements about possible future research directions.

The subtitle of the book (From *Windsurfing* to *Telecommunications*) recognizes the fact that many common life experiences, such as those we encounter when learning to ride a windsurfer are models for design methods that can be used on real world advanced technological control problems. Indeed, Brian Anderson extensively explored this link in his research work. The chapters in this book follow a similar path. The book is divided into four parts each containing three related chapters. An outline of the book is given below:

Part I addresses various issues associated with the problem of obtaining models from real data. This problem occurs in many different fields including control engineering, economics, communications, biology, etc. A common theme in these first three chapters is the issue of model quality and how it relates to the intended purpose.

Chapter 1 by Manfred Deistler, considers general aspects of the system identification problem. The chapter provides a philosophical framework for this problem and shows how developments in this field can be unified. The chapter also points to new advances which are occurring especially in the area of econometrics and creates a compelling case for greater cooperation between the different cultures who contribute to the area of system modelling using real data.

The second chapter by Goodwin and Welsh considers a specific problem in system identification, namely singularity issues which arise from the analysis of closed loop data. The first of these issues relates to the occurrence of spurious (near) pole zero cancellations which arise due to inexact satisfaction of closed loop interpolation constraints. The second part of the chapter addresses an alternative singularity problem which arises from near divisions by zero in the indirect closed loop identification method. Remedies for both problems are discussed in the chapter.

The third chapter by Bombois, Anderson and Gevers discusses the problem of mapping parametric confidence bounds obtained from typical identification procedures into the frequency domain. It is shown that this mapping is not bijective due to correlations which exist between the confidence bounds at different frequencies. Implications of this result in the areas of robust control are also discussed.

Part II is concerned with the interaction between model estimation and the intended application.

Chapter Four by Landau shows how controller reduction can be carried out using system identification on closed loop data. The asymptotic properties of the resultant reduced order controllers is also discussed in the context of robust control.

Chapter 5 by Costa, Kokotovic and Mareels considers the problem of using online estimation to make a nonlinear observer adaptive. It considers a specific class of nonlinear observers which are known to provide globally convergent state estimates when the system parameters are known. The chapter presents preliminary results aimed at retaining these convergence properties when parameter estimation is added to the observer.

The sixth chapter by Evans, Krishnamurthy and Nair, is concerned with the tracking of maneuvering targets. This problem can be formulated as that of estimating the state of a partially observed jump Markov linear system. The chapter considers enhancements of existing tracking algorithms to handle flexible multimode measurement sensors and variable bandwidth data links between the sensor and the tracker.

Part III of the book considers specific practical problems and shows how estimation and control principles impact on various applications.

Chapter 7 by Evans, Mareels, Sciacca, Cooper, Middleton , Betz and Kennedy addresses the design of servo tracking loops for large antenna structures which exhibit lightly damped flexible modes. The results are illustrated by reference to 22m., 26m. and 64m. radio astronomy and telecommunication antennae in Australia. A particular aspect of the design is the use of adaptive feedforward action to achieve accurate tracking control.

Chapter 8 by Kennedy, Ding and Hasnie, demonstrates the universitality of feedback control concepts by applying them to a specific problem in telecommunications. In particular, passivity theory is used to examine the stability of generalized decision feedback structures incorporating memoryless nonlinearities. The chapter addresses several design issues including the optimal choice of quantizer deadbands.

Chapter 9 by Johnson and Sethares investigates the effect of regression vector filtering in adaptive IIR filtering. The filtering viewpoint is used to unify many existing schemes and as a mechanism to develop "new" algorithms such as a novel Steiglitz-McBride family of algorithms.

Part IV is concerned with design issues that arise in the context of adaptive control. These results build on the theoretical considerations given earlier but address problems of specific interest to adaptive control.

Chapter 10 by Kosut gives an overview of the iterative adaptive control methodology commonly known as the *Windsurfer Approach.* Specific questions arising in this algorithm relating to model validation are addressed. The paper also raises a fundamental challenge namely how to develop a self consistent theory of identification and validation from a single set of assumptions.

Chapter 11 by Bitmead and Dunstan investigates the nature of model quality in the context of the windsurfer approach to adaptive control. Connections between internal model control and model reference control are exposed and exploited.

Chapter 12 by Brinsmead also considers fundamental design issues in control. It considers the cheap control problem for linear time variable systems and shows that non-minimum phase behaviour induces a fundamental and precomputable limit to the achievable performance. If one thinks of an adaptive controller as a special kind of linear time varying system, then these results may give additional insight into adaptive control.

Graham Goodwin
CIDAC, *The* University *of* Newcastle, Australia
September 2000.

Postgraduate Supervision

Stanford University, USA
1963 - 1966
Supervisor: Robert W. Newcomb

Brian David Outram Anderson
"NETWORK ANALYSIS AND SYNTHESIS"
Stanford University 1966

The University of Newcastle, Australia
1967 - 1981
***Supervisor*: Brian D.O. Anderson**

1968

Murray Wood
"*Detection and estimation of signals in noise*", M.E. 1968.

1970

Konrad Hitz
"*Relations between continuous-time and discrete-time quadratic minimization*", Ph.D. 1970.

Sumeth Vongpanitlerd
"*Analysis of a double-pumped low-noise parametric down-converter*", M.E. 1968.
"*Network analysis and synthesis - A modern system theory approach*", Ph.D. 1970.

1972

Peter Moylan
"*Generalizations of singular optimal control theory*", M.E. 1969.
"*Optimal control of nonlinear systems*", Ph.D. 1972.

1973

Surapong Chirarattananon
"*Optimal and suboptimal linear fixed-lag smoothing*", Ph.D. 1973.

1975

David Clements
"*Nonlinear fixed-lag smoothing of random telegraph type signals*", M.E. 1973.
"*Extensions of linear-quadratic control theory*", Ph.D. 1975.

Nguyen Dinh Diem
"*Spectral factorisation via a Riccati difference equation*", M.E. 1973.
"*Recursive algorithms for solutions of linear-quadratic control, filtering and game problems*", Ph.D. 1975.

1977

Raymond Scott
"*Stable linear multivariable systems*", Ph.D. 1977.

1978

Hoang Bang Doan
"*Approximation procedure for fixed-lag smoother design*", M.E. 1976.

1979

Robert Bitmead
"*Matrix transfer function description of linear systems*", M.E. 1976.
"*Convergence properties of discrete-time stochastic adaptive estimation algorithms*", Ph.D. 1979.

Nguyen Tan Hung
"*Analysis and synthesis of nonlinear reciprocal networks*", M.E. 1977.
"*New multivariable control systems frequency-domain design techniques*", Ph.D. 1979.

1981

M. Viswanathan
"*Determination of effectiveness of water consumption restrictions: methodology and application*", M.E. 1981.

1982

Richard Johnstone
"*Convergence properties of deterministic, discrete-time adaptive algorithms*", Ph.D. 1982.

John Potter
"*Hypothesis testing and estimation with partial prior information*", M.E. 1978.
"*Linear filtering models using transmission lines*", Ph.D. 1982.

Australian National University
1981 – 2000
***Supervisor*: Brian D.O. Anderson**

1984

Soura Dasgupta
"*Adaptive Identification and Control*", Ph.D. 1984.

Geoffrey Latham
"*Frequency weighted optimal Hankel-norm approximation of scalar linear systems*", M.Sc. 1984.

1987

Michael Green
"*Factorizastion techniques in multivariable phase problems*", Ph.D. 1987.

Iven Mareels
"*Dynamics of adaptive control*", Ph.D. 1987.

1988

Rodney Kennedy
"*Operational aspects of decision feedback equalizers*", Ph.D. 1988.

Philip Parker
"*Frequency domain descriptions of linear systems*", Ph.D. 1988.

1989

Yi Liu
"*Frequency-weighted controller and model order reduction methods in linear system design*", Ph.D. 1989.

1990

Michael Frater
"*Estimation of the statistics of rare events in telecommunications systems*", Ph.D. 1990.

1992

Meng-Joo Er
"*Periodic controllers for linear time-invariant systems*", Ph.D. 1992.

Benjamin James
"*Approaches to multiharmonic frequency tracking and estimation*", Ph.D. 1992.

Graham Pulford
"*Development in non-linear equalisation*", Ph.D. 1992.

1994
Wee Sit Lee
"*Iterative identification and control design for robust performance*", Ph.D. 1994.

1995
Mehmet Karan
"*Frequency tracking and hidden markov models*", Ph.D. 1995.

Anton Madievski
"*Sampled-data and reduced order controller implementation*", Ph.D. 1995.

Jeremy Matson
"*Redundancy and multiple objectives in linear robust control*", Ph.D. 1995.

1998
Natasha Linard
"*Identification of nonlinear systems*", Ph.D. 1998.

1999
Louis Shue
"*On performance analysis of state estimators for hidden Markov models*", Ph.D. 1999.

Contents

Part II. Interactions Between Model Identification and Intended Model Use

Part III. Illustrative Applications of Identification and Control

Part I

General Aspects of Model Identification

1 System Identification – General Aspects and Structure

Manfred Deistler

Institut für Ökonometrie, Operations Research und Systemtheorie
Technische Universität Wien, Argentinierstraße 8, A-1040 Wien
Tel: ++43 +1 58801-11910, Fax: ++43 +1 58801-11999
E-mail: Manfred.Deistler@tuwien.ac.at

1.1 Introduction

System identification is concerned with finding a good model from, in general, noisy data, i.e. with data driven modeling. Often the task of identification is so complex that it cannot be performed in a naive way with the naked eye. In addition many identification problems share common features. For these reasons methods and theories have been developed which make system identification a subject of its own. This is the case despite the fact that problems of system identification are treated in different and partly rather separated communities, such as in system and control theory, signal processing, statistics and econometrics; the latter is explained by the fact that system identification is a central issue in many branches of science and has a wide range of areas of applications from control of chemical processes to the analysis of earth quake data or forecasting of sales for firms. Accordingly, we will use the term system identification for data driven modeling in general, i.e. not necessarily relating to system and control theory.

In a sense, system identification is concerned with *developing rules and methods*, in particular algorithms, for attaching a good model to data and with the *evaluation* of such rules and methods.

In identification the following has to be specified:

- A *model class*, that is the class of all a priori feasible candidate systems to be fitted to the data. The model class incorporates the a priori information about the phenomenon under consideration.
- A *class of feasible data.*
- An *identification procedure* which is a set of rules (in the fully automatized case a function) attaching to every feasible data string a system from the model class.

Of course, identification has many different aspects and there is a wide range of theories and methods available. In this contribution the emphasis is on general features and structures for a rather large class of problems and procedures. Given the limitations of space the treatment of the topic is necessarily sketchy; no proofs are presented. The contribution is addressed to readers familiar with system identification.

Here we only deal with discrete time, equally spaced *time series data* $y_t, t = 1, \ldots, T, y_t \in \mathbb{R}^s$ and with, in general, dynamic systems. We mainly treat the so called *main stream approach* (see [9]) for identification of linear systems. The main stream approach still is the most important and best developed case and in addition shows "prototypical features" which are important for other cases too. This main stream case is characterized as follows:

- The model class consists of linear, time-invariant, finite dimensional, causal and stable systems only. The classification of the variables into inputs and outputs is given a priori.
- Stochastic models for noise are used. In particular noise is assumed to be stationary and ergodic with a rational spectral density.
- The observed inputs are assumed to be free of noise and to be uncorrelated with the noise process.
- A *semi-nonparametric approach* to identification is adopted in the sense that a suitable parametric model subclass of the model class is determined by a data driven model selection procedure and that then the finite dimensional parameters in the selected subclass are estimated.
- In evaluation of identification procedures, emphasis is placed on *asymptotic properties* of estimators such as consistency and asymptotic normality.

Let us make a few general remarks on the ingredients of main stream theory as listed above:

First note that stochastic models for noise are not universally suited for modeling of uncertainty or lack of knowledge, since the underlying assumptions may not be innocent, in particular if there is no a priori reasoning for their justification. This has been pointed out already in the famous Keynes-Tinbergen debate; see e.g. [22]. Clearly, stochastic models provide valuable insights into properties of estimation or test procedures. However, the evaluation of such procedures in a stochastic framework should be considered as an evaluation in a test bed rather than as an evaluation in a "real life" situation. In particular there is a danger of "overbreeding" of methods with respect to such stochastic test beds.

The assumption of orthogonality of noise and observed inputs is equivalent (in our context) to assume that the transfer function from observed inputs to outputs corresponds to the best linear least squares approximation of the output in the (Hilbert) space spanned by the inputs.

The semi-nonparametric approach differs from the nonparametric approach where e.g. nonparametric spectral estimators are used.

Finally, the emphasis on asymptotic properties of estimators links the field to mathematical statistics, in particular to asymptotic theory. In a certain sense asymptotic theory provides a "second best" evaluation because small sample properties can hardly be obtained analytically in our context. Of course, finite sample properties may be investigated by simulations. It should also be noted that the standard statistical evaluation criteria, such as the variance of the asymptotic distribution, may not be appropriate quality measures for a number of cases. In addition to statistical, also numerical properties of identification procedures may be of interest.

For the semi-nonparametric approach, and in particular for the main stream case, the identification problem may be decomposed into three modules as follows:

- *Structure theory.* Here an idealized identification problem is treated in so far as we commence from the stochastic processes generating the data or some of their population moments rather than from data directly. In more general terms, the relation between "external behavior" (as described e.g. by population moments of the observed processes) and *"internal parameters"* (describing system and noise properties) is analyzed with respect to properties relevant for identification.
- *Estimation of real valued parameters.* By real valued parameters we mean parameters that may vary e.g. in an open set of an Euclidean space, as opposed to integer valued parameters (such as maximum lag lengths for ARMA systems) which are used for the characterization of a model subclass. In this module the subclass, parametrized by a subset of an Euclidean space, is assumed to be known and the real valued parameters are estimated by "parametric" procedures such as by Gaussian maximum likelihood type estimators.
- *Model selection.* Here, typically the integer valued parameters or other quantities characterizing the subclasses are obtained from data, e.g. by using information criteria or test procedures.

This contribution is organized as follows: In the next sections the three modules described above are treated for the main stream case. In section (1.5) an example for a linear non-mainstream case is described. In section (1.6) a few remarks concerning non-linear systems are made. Finally we conclude with some general comments.

In system identification elements of system theory, the theory of stationary processes and the statistical analysis of time series are dovetailed. Early work on system identification may be found in [29] and Mann and Wald (1943). For general references, in particular for the main stream case, we refer to [14] or [24] and the references there. For the sake of brevity we in general do not give references even to important original literature if it is cited in [14]. For this reason important and seminal papers by Akaike, T. W. Anderson, Hannan, Kalman and others will not be found in the list of references for this contribution. For a general statistical analysis, emphasizing nonlinear systems, the reader is referred to [23] and the references there.

1.2 Structure Theory

Here we deal with structure theory for the main stream case, where the external behavior is described by the population second moments of the observations or by the ("true") transfer functions and the innovation variance. The internal parameters are ARMAX or state space parameters and the innovation variance. This analysis is not statistical in the narrow sense, but is important for the subsequent statistical analysis. The main problems considered here are observational equivalence and identifiability, realization and parametrization.

For brevity of presentation we will only consider the case where we have no observed inputs. The state space forms are assumed to be in innovations representation:

$$x_{t+1} = Ax_t + B\varepsilon_t \tag{1.1}$$

$$y_t = Cx_t + \varepsilon_t \tag{1.2}$$

where y_t are the s-dimensional outputs, x_t the n-dimensional states, the ε_t form the white noise innovations process with $\Sigma = \mathbb{E}\varepsilon_t\varepsilon_t'$ and $A \in \mathbb{R}^{n\times n}$, $B \in \mathbb{R}^{n\times s}$, $C \in \mathbb{R}^{s\times n}$ are the parameter matrices. For this and the next two sections the stability condition

$$|\lambda_{max}(A)| < 1 \tag{1.3}$$

and the miniphase condition

$$|\lambda_{max}(A - BC)| \leq 1 \tag{1.4}$$

are assumed to hold, where λ_{max} denotes an eigenvalue of maximal modulus. In addition we assume

$$\Sigma > 0. \tag{1.5}$$

The transfer function is of the form

$$k(z) = \sum_{j=1}^{\infty} K_j z^j + I; \quad K_j = CA^{j-1}B, \quad z \in \mathbb{C}. \tag{1.6}$$

In this contribution we also refer to ARMA systems, which are of the form (using an obvious notation)

$$a(z)y_t = b(z)\varepsilon_t \tag{1.7}$$

where z is the backward shift, i.e. $z(y_t|t \in \mathbb{Z}) = (y_{t-1}|t \in \mathbb{Z})$ and where the polynomial matrices $a(z)$ and $b(z)$ are assumed to satisfy the stability and miniphase assumption respectively.

As is well known a stationary process $(y_t|t \in \mathbb{Z})$ has a (λ-a.e.) nonsingular rational spectral density if and only if it is the (steady state) solution of a state space system satisfying our assumptions or if and only if it is the solution of an ARMA system satisfying our assumptions. By (1.3) and (1.4) the transfer function (1.6) corresponds to the Wold decomposition and by additionally assuming (1.5), $(k(z), \Sigma)$ are uniquely determined from

$$f(\lambda) = (2\pi)^{-1}k(e^{-i\lambda})\Sigma k^*(e^{-i\lambda}) \tag{1.8}$$

where * denotes the conjugate transpose. A completely analogous statement holds for ARMA systems with $k(0) = I$. In both cases the transfer functions have no poles for $|z| \leq 1$ and no zeros for $|z| < 1$.

However, unless the contrary is stated explicitely, for brevity of presentation, the state space case is considered here; thus the relation between $(k(z), \Sigma)$ and $((A, B, C), \Sigma)$ is analyzed. Let U_A denote the set of all rational $s \times s$ transfer functions (1.6) satisfying our assumptions (i.e. $k(0) = I$, $k(z)$ has no poles for $|z| \leq 1$ and no zeros for $|z| < 1$) and let $M(n) \subset U_A$ denote the subset of all transfer functions of order n. Furthermore let T_A denote the set of all state space systems (A, B, C), for fixed s but variable n, satisfying (1.3) and (1.4), let $S(n) \subset T_A$ denote the subset of all (A, B, C) with n fixed and $S_m(n) \subset S(n)$ the subset of all minimal (A, B, C). Then the mapping

$$\pi : T_A \to U_A \tag{1.9}$$

such that

$$\pi(A, B, C) = C(Iz^{-1} - A)^{-1}B + I \tag{1.10}$$

is surjective, but not injective. For a number of reasons which are discussed below, U_A is broken into bits U_α, $\alpha \in I$ say, such that each bit allows for a convenient parametrization. Desirable properties of parametrizations are:

- For each U_α there is a corresponding $T_\alpha \subset T_A$ such that π restricted to T_α, $\pi_{T_\alpha} : T_\alpha \to U_\alpha$ is bijective. Then there exists a *parametrization*

  $$\psi_\alpha : U_\alpha \to T_\alpha$$

  such that $\psi_\alpha(\pi(A, B, C)) = (A, B, C) \quad \forall (A, B, C) \in T_\alpha$.

- U_α is finite dimensional in the sense that there exists a $n \in \mathbb{N}$ such that $U_\alpha \subset \cup_{i=1}^{n} M(i)$. In general, for given T_α, not all entries in $(A, B, C) \in T_\alpha$ will be free. By $\tilde{T}_\alpha \subset \mathbb{R}^{d_\alpha}$ we denote the set of the corresponding free parameters which is assumed to contain a nonvoid open subset and where we assume in addition that $\rho_\alpha : T_\alpha \to \tilde{T}_\alpha$ is a homeomorphism (which additionally satisfies differentiability requirements). Here, for the sake of brevity of presentation, we will identify T_α and $\tilde{T}_\alpha$ unless the contrary is stated explicitely.
- "Well posedness": The parametrization $\psi_\alpha : U_\alpha \to T_\alpha$ is a homeomorphism. Thereby U_A is endowed with the so called pointwise topology T_{pt} which corresponds to the relative topology in the product space $(\mathbb{R}^{s\times s})^{\mathbb{N}}$ for the $(K_j | j \in \mathbb{N})$. Clearly, this continuity is an important property; by a result obtained by [15], there exists no continuous parametrization for U_A, in other words, U_A has to be broken into bits to allow for a continuous parametrization of the bits.
- U_α is T_{pt}-open in $\bar{U}_\alpha$. Here $\bar{U}_\alpha$ denotes the closure of U_α. Note that if estimators are obtained by optimization, some boundary points cannot be avoided. Now if k_0, the true transfer function, is contained in U_α, and the estimators $\hat{k}_T$ for sample size T satisfy $\hat{k}_T \in \bar{U}_\alpha$ and $\hat{k}_T \to k_0$, then the openness of U_α in $\bar{U}_\alpha$ implies $\hat{k}_T \in U_\alpha$ from a T_0 onwards thus these $\hat{k}_T$ can be parametrized in the corresponding T_α.
- $\cup_{\alpha \in I} U_\alpha$ is a cover for U_A, i.e. every $k \in U_A$ can be described by a suitable choice of α.

In addition some topological and geometrical properties of the boundaries of the sets U_α and T_α are important; see [14] or [12]. Note that $\bar{M}(n) = \cup_{i=1}^{n} M(i)$.

Most, but not all parametrizations commence from further breaking down $M(n)$. Other approaches are AR parametrizations according to the prescribed maximum degree of $a(z)$, ARMA parametrizations according to column degrees of $(a(z), b(z))$ or a parametrization of ARMA systems using the coefficients of c and p where $k(z) = c^{-1}(z)p(z)$ and c is the least common multiple of the denominator polynomials of the entries k_{ij} of k and the degrees of c and p are prescribed. Another case is when the model class is highly structured by a priori information coming e.g. from physical or economic theory. As is well known, $M(n)$ is a real analytic manifold of dimension $2ns$ and cannot be described with one coordinate system for $s > 1$. We now discuss the following parametrizations commencing from $M(n)$:

1. *Canonical forms.* Canonical forms are mappings $c : U_A \to T_A$, selecting from every equivalence class $\pi^{-1}(k)$, $k \in U_A$ a unique representative. As is well known, the classes of all observationally equivalent minimal state space systems (A, B, C) are characterized by linear nonsingular state transformations $\tilde{A} = TAT^{-1}$, $\tilde{B} = TB$, $\tilde{C} = CT^{-1}$, $\det(T) \neq 0$. The classes of observationally equivalent relatively left prime ARMA systems

are characterized by premultiplication with unimodular polynomial matrices. The most common canonical forms are echelon forms and balanced forms; for the latter see e.g. [5]. Both forms lead to a partition of $M(n)$ into sets of different dimension which are parametrized separately (only for s=1 and echelon forms, $M(n)$ is parametrized directly). For echelon forms, the integers in α defining U_α are the Kronecker indices which are defined by a special selection of basis rows from the Hankel matrix of the transfer function. The (A, B, C) then are defined via these basis rows. A main advantage of echelon forms is that certain entries in (A, B, C) are fixed to be zero or one and all other entries are free parameters (for all Kronecker indices strictly positive). There also exist echelon ARMA forms and there is a simple one-to-one relation between the free parameters in both forms. One piece $U_\alpha \subset M(n)$ is open and dense in $M(n)$ and this U_α is called a generic neighborhood. From a numerical point of view, echelon forms seem to be sometimes problematic. Balanced forms are obtained from an SVD of the Hankel matrix of the transfer function by imposing additional restrictions. Again, there is a generic neighborhood, U_n say, of $M(n)$ (which differs from U_α above), where all singular values have multiplicities equal to one, which is open and dense in $M(n)$. As opposed to echelon forms, here the free parameters are rather complicated transformations of the entries in (A, B, C), however as a tradeoff the parameter spaces are simpler; for example, the minimality requirement translates into strict positivity of the singular values of the Hankel matrix (these singular values or their differences are part of the vector of free parameters). This simplicity of parameter spaces may be an advantage for specification search; see e.g. [5]. A disadvantage of balanced compared to echelon forms is that (for the first) in general more pieces are needed to cover $M(n)$; even for $s = 1$, for balanced forms $M(n)$ cannot be described by a single parameter space.
There are no balanced ARMA representations. Note that, despite the fact that there is a one-to-one relation between state space and ARMA equivalence classes, the respective equivalence classes are different; for instance, for $s = 1$, for the usual ARMA parameter space the $k \in M(n)$ are described by singletons in $\mathbb{R}^{2n}$, whereas the state space equivalence classes in $S(n) \subset \mathbb{R}^{2n+n^2}$ are manifolds of dimension n^2 (with some boundary points) .
In both cases the parametrizations have the desirable properties listed above.

2. *The description of the manifold $M(n)$ by local coordinates.* The most common choice here is a choice of local coordinates obtained from selections of basis rows of the Hankel matrix of the transfer function corresponding to the so called dynamical indices. This approach is similar to echelon forms, for instance w.r.t. properties of parameters and parameter spaces. However, the bits U_α which are characterized by the dynamical indices, are all of the same dimension $2ns$, they overlap (thus α is not unique

in general) and they are all open and dense in $M(n)$. A particular U_α and its parametrization even coincide with the generic neighborhood in echelon forms and its parametrization. This "overlapping description" of $M(n)$ is available for state space and ARMA systems. Again, the desirable properties for parametrizations listed above are satisfied.

3. *"Full parametrization"* for state space systems. Here, $S(n) \subset \mathbb{R}^{n^2+2ns}$ or $S_m(n)$ are used as parameter spaces. Note that $S_m(n)$ is open and dense in $S(n)$. We have $\pi(S_m(n)) = M(n)$ and $\pi(S(n)) = \cup_{i=1}^{n} \pi(S(i))$ is the closure of $M(n)$ in U_A. Of course in this situation we do not have identifiability; for $k \in M(n)$ the classes of observational equivalence within $S_m(n)$ are manifolds of dimension n^2. For criteria functions which are constant along equivalence classes then of course the optimum is not unique. Numerically optimizing e.g. over $S_m(n)$ rather than e.g. over a T_α corresponding to an overlapping description has the drawback that there are n^2 more (and thus essentially unnecessary) coordinates. On the other hand there exist algorithms which give a unique parameter estimator, however for different data, different representatives of the same equivalence class may occur.
4. *Data driven local coordinates* have been proposed by Wolodkin, Rangan and Poolla (1997) and [21]. Here we commence from a given initial estimator and its equivalence class in $S_m(n)$. Orthogonal coordinates for the $2ns$ dimensional ortho-complement to the tangent space to the equivalence class at the initial estimator are given. In this way, a "thick" subset, U_l say, of $\bar{M}(n)$ is described. The advantage of the corresponding parameter space is that it is of dimension $2ns$ rather than $2ns + n^2$ (and thus has no unnecessary coordinates) and that it is well conditioned. A further investigation of this approach is needed.

1.3 Estimation for a given subclass

Here we assume that a subclass U_α, which is finite dimensional (i.e. $U_\alpha \subset \bar{M}(n)$ for some n), is given or has been selected. We concentrate on the identifiable case, where a parametrization

$$\psi_\alpha : U_\alpha \to T_\alpha$$

exists which in addition has the desirable properties listed in section (1.2). Let $\tau \in T_\alpha \subset \mathbb{R}^{d_\alpha}$ denote the free parameters for U_α and let $\sigma \in \mathbb{R}^{\frac{n(n+1)}{2}}$ denote the free parameters which are the on and above diagonal elements of the (unrestricted) innovation variances $\Sigma > 0$. We assume that the overall parameter space is of the form $\Theta = T_\alpha \times \underline{\Sigma}$, where $\underline{\Sigma} = \{\sigma | \Sigma = \Sigma', \Sigma > 0\}$, i.e. we assume that there are no cross restrictions between τ and σ. Let $\theta = (\tau, \sigma)$.

For given U_α, an identification procedure is a mapping attaching to the data $y_1, \ldots, y_T$ the parameter estimators $\hat{\tau}_T$ and $\hat{\sigma}_T$ for the true parameters τ_0 and σ_0. We will emphasize estimation of τ_0, since estimation of σ_0 is rather straightforward. A rather general picture of identification in this context is as follows:

Many identification procedures – at least asymptotically – commence from the sample second moments of the data

$$\hat{\gamma}(s) = T^{-1} \sum_{t=1}^{T-s} y_{t+s} y_t' \quad \text{for } s \geq 0. \tag{1.11}$$

Now, direct realization from these sample moments gives a transfer function, $\hat{\hat{k}}_T$ say, of a moving average system, "typically" of order sT. This order will be too high in general, thus information compression is performed by a *model reduction step*, where $\hat{\hat{k}}_T$ is approximated by a transfer function $\hat{k}_T \in U_\alpha$ (or $\bar{U}_\alpha$ if we do not avoid boundary points). The kind of this approximation is essential for the statistical quality of the identification procedure. In a second step, the *realization step*, the estimator $\hat{k}_T$ is realized as a state space or ARMA system, or by the corresponding free parameters: $\hat{\tau}_T = \psi_\alpha(\hat{k}_T)$. Of course, if $\hat{k}_T \in \bar{U}_\alpha - U_\alpha$, then problems as discussed in chapter 2 of [14] will arise.

A nice feature of this approach is that in certain respects estimation here is "coordinate free".

We may distinguish between estimators obtained by optimizing a criterion function over e.g. Θ, so called *M-estimators* (where, as has been stated already, boundary points may not be excluded), and *direct procedures* such as subspace methods where the parameter estimator is an explicit function of the data; however, this distinction is not so clear as it may seem at a first glance.

The most common M-estimators in our context are of the Gaussian maximum likelihood type. Here the likelihood is written down as if the observations were Gaussian, but Gaussianity is not needed for the asymptotic properties considered. (Thus, in a strict sense, our estimation problem is not parametric but semiparametric, because the law of the observations is not determined by θ. However, we do not further bear this point in mind.)

Now $-2T^{-1}$ times the logarithm of the (Gaussian) likelihood is given up to a constant by

$$\hat{L}_T(\theta) = T^{-1} \log \det \Gamma_T(\theta) + T^{-1} y'(T) \Gamma_T(\theta)^{-1} y(T). \tag{1.12}$$

Here $y(T) = (y_1', \ldots, y_T')'$ denotes the stacked vector of the observations, θ is the parameter vector and finally $\Gamma_T(\theta) = \mathbb{E} y(T;\theta) y'(T;\theta)$ where $y(T;\theta)$ is a stacked vector of observations corresponding to a system and a noise process with parameters θ. The maximum likelihood estimation (MLE) then (not taking into account boundary problems) is defined as

$$\hat{\theta}_T = \text{argmin}_{\theta \in \Theta} \hat{L}_T(\theta). \tag{1.13}$$

It should be noted that in a certain sense the MLE, as defined in (1.13), is a theoretical concept since in general there is no explicit formula for the MLE; the actual calculation of (an approximation to) the MLE can be performed by numerical optimization procedures, each of which, strictly speaking, defines an estimator. However, in many cases these estimators have the same asymptotic properties as the MLE.

As can be seen immediately from (1.12), the likelihood function $\hat{L}_T$ depends on τ via k only; thus we may define a coordinate-free likelihood function as $\hat{L}_T(k, \sigma)$.

Also note that in optimization boundary points may not be excluded. In addition the existence of a minimum of $\hat{L}_T$ cannot be guaranteed in general; see [11].

In many cases approximations to the likelihood function, where the resulting M-estimators share the same asymptotic properties with the MLE, are used. An important example is the Whittle likelihood

$$\hat{L}_{W,T}(k, \sigma) = \log \det \Sigma + (2\pi)^{-1} \int_{-\pi}^{\pi} \text{tr} \left[\left(k(e^{-i\lambda}) \Sigma k^*(e^{-i\lambda}) \right)^{-1} I(\lambda) \right] d\lambda \tag{1.14}$$

where $I(\lambda)$ is the periodogram, i.e. the Fourier Transform of the second sample moments. From (1.14) we see how minimizing $\hat{L}_{W,T}$ over $U_\alpha \times \underline{\Sigma}$ smoothes the periodogram or in other words defines the model reduction step mentioned above.

Also prediction error minimization may be interpreted in this way.

For actual calculations (of M-estimators in general) multiple relative optima may cause problems. For this reason good initial estimators are important.

"Coordinate free consistency", i.e. $\hat{k}_T \to k_0$ a.s. and $\hat{\Sigma}_T \to \Sigma_0$ a.s. is obtained if $k_0 \in U_\alpha$ and

$$\lim T^{-1} \sum_{t=1}^{T-s} \varepsilon_{t+s} \varepsilon_t' = \delta_{0,s} \Sigma_0 \quad \text{a.s. for } s \geq 0;$$

see chapter 4 of [14].

The basic idea of the consistency proof can be seen already in [28] where consistency has been shown for the case of independent and identically distributed observations. Of course, the case considered here poses extra problems, because the observations are dependent and the natural parameter spaces are not compact (they are neither closed, nor – for $s > 1$ – bounded). The first correct proofs for this case seem to be due to Hannan (for $s = 1$) and Dunsmuir and Hannan (for $s \geq 1$), respectively. Main ingredients for the consistency proof are to show that there is an asymptotic form of the likelihood function, i.e.

$$\lim_{T\to\infty} \hat{L}_T(k,\sigma) = L(k,\sigma) =$$

$$\log\det\Sigma \; + \\ (2\pi)^{-1}\int_{-\pi}^{\pi} \operatorname{tr}\left[\left(k(e^{-i\lambda})\Sigma k^*(e^{-i\lambda})\right)^{-1}\left(k_0(e^{-i\lambda})\Sigma_0 k_0^*(e^{-i\lambda})\right)\right] d\lambda \text{ a.s.} \tag{1.15}$$

and that this asymptotic likelihood function, for $k_0 \in U_\alpha$ has a unique minimum at (k_0, Σ_0), i.e. $L(k,\sigma) \geq L(k_0,\sigma_0)$ and strict inequality holds for $(k_0,\Sigma_0) \neq (k,\Sigma)$. Of course pointwise convergence in (1.15) alone does not imply convergence of the minimizers, i.e. $(\hat{k}_T, \hat{\Sigma}_T) \to (k_0, \Sigma_0)$. An essential further step to show this is to show that the estimators, from a certain T_0 onwards, are in a compact subset (which is a level set of the parameter space).

The consistency result described above also holds if k_0 is on the boundary of U_α, i.e. if $k_0 \in \bar{U}_\alpha$.

For $k_0 \notin \bar{U}_\alpha$, there are *generalized consistency* results available: First note that in standard situations, e.g. for $\bar{U}_\alpha = \bar{M}(n)$ such sets are non-convex and $L(k,\sigma)$ may have multiple minima for $k_0 \notin \bar{U}_\alpha$, in other words there may be more than one best approximant to (k_0,Σ_0) within $\bar{U}_\alpha \times \underline{\Sigma}$ (this has been shown e.g. by Kabaila). Call D the set of such best approximants; then as has been shown e.g. by Caines and Pötscher $(\hat{k}_T, \hat{\Sigma}_T) \to D$ a.s. (in the sense that for every subsequence, there is a sub-subsequence converging to a point in D).

Now consider the case of parametrizations $\psi_\alpha : U_\alpha \to T_\alpha$ having the desirable properties listed above, such as the parametrizations discussed in (1) and (2) in section (1.2). Actual calculation of the MLE's is performed in coordinates and in addition in many cases the parameters τ are of direct interest. If $k_0 \in U_\alpha$ and $\hat{k}_T \to k_0$ then $\hat{k}_T \in U_\alpha$ from a certain T_0 onwards and thus $\hat{\tau}_T = \psi_\alpha(\hat{k}_T)$ exists and, by the continuity of ψ_α, we have consistency for τ_0, i.e. $\hat{\tau}_T \to \tau_0$ a.s.. For $k_0 \in \bar{U}_\alpha - U_\alpha$ let us discuss the case where U_α is characterized by dynamical indices as described in (2) in section (1.2), as an example. Here we have to distinguish between the case $k_0 \in \pi(\bar{T}_\alpha) - U_\alpha$ and the case $k_0 \in \bar{U}_\alpha - \pi(\bar{T}_\alpha)$. In the first case k_0 corresponds to a lower dimensional system, i.e. $k_0 \in U_\beta$ for some $\beta < \alpha$ (i.e. inequality holds componentwise and is strict for one component) and k_0 corresponds to an equivalence class containing more than one element on the boundary of T_α. If prior bounds on the elements of $\bar{T}_\alpha$ are imposed then $\hat{k}_T \to k_0$, $\hat{k}_T \in T_\alpha$ implies that the corresponding $\hat{\tau}_T$ converge to the equivalence class, but not necessarily to a fixed point within this class. The case $k_0 \in \bar{U}_\alpha - \pi(\bar{T}_\alpha)$ can only occur for $s > 1$. In this case k_0 corresponds to the point of infinity in the one point compactification of $\mathbb{R}^{d_\alpha}$ and correspondingly the estimators $\hat{\tau}_T$ converge to infinity as $\hat{k}_T \to k_0$, $\hat{k}_T \in T_\alpha$.

For a *central limit theorem* coordinates have to be used and in addition those boundary points corresponding to equality in (1.4) in $T_\alpha \times \underline{\Sigma}$ have to be excluded, since for those boundary points the limiting distribution would not be Gaüssian; as examples consider the sets T_α discussed in (1) and (2) in section (1.2), where in addition strict inequality holds in (1.4). Also additional requirements have to be imposed on (ε_t), namely

$$\mathbb{E}(\varepsilon_t | \mathcal{F}_{t-1}) = 0 \tag{1.16}$$

where $\mathcal{F}_{t-1}$ is the σ-algebra generated by ε_s, $s \le t$, and

$$\mathbb{E}(\varepsilon_t \varepsilon_t' | \mathcal{F}_{t-1}) = \Sigma_0. \tag{1.17}$$

Then the vector $T^{\frac{1}{2}}(\hat{\tau}_T - \tau_0)$ has a Gaussian limiting distribution with mean zero and with covariance matrix given by the inverse of the Fisher information matrix. If in addition

$$\mathbb{E}\{\varepsilon_t^4\} < \infty \tag{1.18}$$

and

$$\mathbb{E}\{\varepsilon_{i,t}\varepsilon_{j,t}\varepsilon_{k,t} | \mathcal{F}_{t-1}\} = \mathbb{E}\{\varepsilon_{i,t}\varepsilon_{j,t}\varepsilon_{k,t}\}, \quad 1 \le i,j,k \le s \tag{1.19}$$

are satisfied then $T^{\frac{1}{2}}(\hat{\sigma}_T - \sigma_0)$ has a Gaussian limiting distribution too.

From these results it is seen that the asymptotic properties of the MLE's obtained from the Gaussian likelihood function are valid for a certain class of non-Gaussian observations too. As is known, for Gaussian observations, the MLE's obtained from the Gaussian likelihood function are asymptotically efficient. On the other hand, if the actual distribution of the ε_t were known and non-Gaussian, the corresponding non-Gaussian likelihood could give estimators having a smaller variance of its limiting distribution. Gaussian likelihood estimation thus may be interpreted as minimizing the variance of the asymptotic distribution for the worst case (i.e. the Gaussian case) within a certain class of distributions for the observations. Asymptotic normality for MLE's in case of independent and identically distributed observations has been obtained in [8]; for the case considered here Hannan, Dunsmuir and Hannan, and Ljung and Caines are early references. The proof commences from a Taylor series expansion of the derivative of the likelihood function, as described e.g. in chapter 4 of [14].

Direct identification procedures, i.e. procedures not containing a numerical optimization step, may be used either to obtain initial estimators, e.g. for a likelihood-type procedure, or as a separate procedure. An important novel class of direct procedures are ***subspace identification methods***, see e.g. [1], [17] and [26]. In general terms, subspace identification procedures use algorithms analogous to realization algorithms; since they commence from data rather than from e.g. population second moments, a model reduction step has to be incorporated in addition; usually this model reduction step is performed via a singular value decomposition.

We will shortly describe Larimore's algorithm as an example. First note that, as is well known, a minimal state space representation of state dimension n may be obtained from a factorization of the (block) Hankel matrix H of the transfer function into observability and controllability matrix. This Hankel matrix relates the past of the process (y_t), $Y_t^- = (y_{t-1}', y_{t-2}', \ldots)'$ via the innovations $\varepsilon_{t-1}, \varepsilon_{t-2}, \ldots$ to the prediction of the future $Y_t^+ = (y_t', y_{t+1}', \ldots)'$ of the process; using an obvious notation we may write this as

$$Y_T^+ = HG^{-1}Y_t^- + V_t^+ \tag{1.20}$$

where G^{-1} corresponds to the inverse of the Wold decomposition given by (1.6) and $HG^{-1}Y_t^-$ is the corresponding infinite vector of predictors. Now $HG^{-1} = OK$ where the observability matrix $O \in \mathbb{R}^{\infty \times n}$ and $K \in \mathbb{R}^{n \times \infty}$ both have rank n. Then a minimal state is defined as $x_t = KY_t^-$.

Accordingly, in the first step of Larimore's algorithm, the state is estimated as follows: Let $Y_{t,f}^+ = (y_t', \ldots, y_{t+f-1}')'$ and $Y_{t,p}^- = (y_{t-1}', \ldots, y_{t-p}')'$ denote the finite future and past respectively, for suitably chosen $f, p > n$. Then estimate $\beta \in \mathbb{R}^{fs \times ps}$, the northwest corner of HG^{-1}, from the "truncated form" of (1.20),

$$Y_{t,f}^+ = \beta Y_{t,p}^- + \nu_{t,f} \tag{1.21}$$

by ordinary least squares which gives an estimator $\hat{\beta}$ say. Now, typically $\hat{\beta}$ has rank equal to $\min(fs, ps)$, whereas β has rank n. In order to obtain an estimator of β of rank n, we consider the singular value decomposition

$$\hat{W}_f \hat{\beta} \hat{W}_p = \hat{U}\hat{\Lambda}\hat{V}' = \hat{U}_n \hat{\Lambda}_n \hat{V}_n' + R \tag{1.22}$$

where $\hat{W}_f$ and $\hat{W}_p$ are suitably chosen (square) weighting matrices, $\hat{\Lambda}_n$ is the matrix consisting of the largest n singular values of $\hat{W}_f \hat{\beta} \hat{W}_p$ (i.e. the largest elements at the diagonal of $\hat{\Lambda}$) and $\hat{U}_n \in \mathbb{R}^{fs \times n}$ and $\hat{V}_n \in \mathbb{R}^{ps \times n}$ are the matrices consisting of the corresponding left and right singular vectors, respectively. The matrix R corresponds to the neglected smaller singular values. In this way, we define the rank n approximation to $\hat{\beta}$ as $\hat{O}_f \hat{K}_p$ where

$\hat{O}_f = \hat{W}_f^{-1}\hat{U}_n\hat{\Lambda}_n^{\frac{1}{2}} \in \mathbb{R}^{fs\times n}$ and $\hat{K}_p = \hat{\Lambda}_n^{\frac{1}{2}}\hat{V}_n'\hat{W}_p^{-1} \in \mathbb{R}^{n\times ps}$ and the state estimator if given as $\hat{x}_t = \hat{K}_p Y_{t,p}^-$.

Note that, as has been stated above in general, this procedure is analogous to a particular realization algorithm, additionally consisting of a singular value decomposition as a model reduction step. However, e.g. $\hat{O}_f$ does not have the structure of an observability matrix; thus, in a second step, replacing the state x_t by its estimator $\hat{x}_t$, the parameter matrix C and ε_t in (1.2) are estimated by ordinary least squares, and then A and B in (1.1) are estimated from $\hat{x}_t$ and $\hat{\varepsilon}_t$.

The advantage of subspace methods compared to methods based on optimizing a criterion function such as the likelihood function lies in the substantial reduction of the computational effort due to the use of the structure of the problem as described by a realization algorithm.

Subspace algorithms generically give unique estimators $\hat{A}_T, \hat{B}_T, \hat{C}_T$ for $k_0 \in M(n)$; however, they do not use canonical forms, or, in other words, no a priori prescribed representatives from the equivalence classes in the sense that for different data different representatives from the same equivalence class may be selected; they thus have a "full parametrization" as described in section (1.2).

Statistical properties of Larimore's procedure are derived in Deistler, Peternell and Scherrer (1995), [4] and Bauer, Deistler and Scherrer (1999). Consistency can be shown by showing the continuity of the mapping relating the $(\hat{A}_T, \hat{B}_T, \hat{C}_T)$ to the $\hat{\gamma}(s)$, where the maximum lag p increases at a certain rate with T. Asymptotic normality is shown by linearizing this mapping around the true values. Simulations show that the mean squared error of the parameter estimators of Larimore's procedure (for the case of no observed input) is similar to that of the MLE.

1.4 Model selection

In many applications the original model class is not finite dimensional; here we treat the set U_A as the most important example for this. As has been explained in section (1.2), U_A has to be broken into bits in order to allow for a convenient parametrization. The determination of such bits is called model selection or dynamic specification. The development and evaluation of data-based procedures for model selection is one of the most important contribution to the subject during the last thirty years.

Such procedures may be classified as non-automatic and automatic. In the non-automatic case, as for instance in the procedure developed by Box and Jenkins for the case $s = 1$, some decisions have to be made by the researcher. The advantage of automatic procedures is that they do not require a large amount of experience.

Here we mainly consider the problem of estimating the order n by information criteria. Note that $\bar{M}(n_1) \subset \bar{M}(n_2)$ holds for $n_1 < n_2$ and that

$M(n_1)$ has smaller dimension than $M(n_2)$. The estimators of n are obtained by minimizing a criterion of the form

$$A(n) = \log \det \hat{\Sigma}_T(n) + 2ns \cdot c(T) \cdot T^{-1}; \quad 0 \leq n \leq N \tag{1.23}$$

where $\hat{\Sigma}_T(n)$ is the MLE for Σ_0 over $\bar{M}(n) \times \underline{\Sigma}$, N is a prescribed upper bound for the order and $c(T)$ is a prescribed function. The first term on the right hand side of (1.23), namely $\log \det \hat{\Sigma}_T(n)$ is a measure for the goodness of fit of the best system in $\bar{M}(n)$ to the data. For given T, $\log \det \hat{\Sigma}_T(n)$ will be (not strictly) decreasing for increasing n. The idea is now to counterbalance this by the second term on the right hand side of (1.23) in order to avoid overfitting. This second term contains the dimension $2ns$ of $M(n)$ as a measure of complexity. Note that such a procedure may be interpreted as a sequence of likelihood ratio tests, where the significance levels are prescribed by the second term on the right hand side of (1.23).

It should be noted that complexity is in the eye of the beholder and that we have used a special notion of complexity. Consider as an example the following autoregressive model for $s = 1$:

$$y_t + a_1 y_{t-1} + a_2 y_{t-2} = \varepsilon_t$$

Then the overall parameter space $T = \{(a_1, a_2) \in \mathbb{R}^2 | 1 + a_1 z + a_2 z^2 \neq 0$ for $|z| \leq 1\}$ for the corresponding transfer functions (which is even finite dimensional in this case) according to our philosophy is decomposed into $T_0 = \{(0,0)\}$, $T_1 = \{(a_1, 0) || a_1| < 1, a_1 \neq 0\}$ and $T_2 = T - (T_0 \cup T_1)$. Here T_0 has complexity zero, T_1 complexity 1 and T_2 complexity 2. However, in special applications it may be justified by a priori reasoning to use other decompositions of T, e.g. why not choose T_1 as $\{(0, a_2) \in T\}$ or T_0 as $\{(0.1, 0.1)\}$?

By far the most important choices for $c(T)$ are $c(T) = 2$, in which case $A(n)$ is called *AIC criterion*, and $c(T) = c \cdot \log(T)$, $c \geq 1$ which gives the *BIC criterion*.

Criteria like AIC or BIC have been derived by Akaike, Rissanen and Schwarz by partly different arguments which all are quite fundamental and provide extensions of the classical statistical paradigm. Such criteria partly have been derived for and applied to other cases such as choosing the degrees for polynomial regressions. A description of the derivations of information criteria is beyond the scope of this contribution, we present just a few ideas: The basic problem is the choice among the different dimensions of the model subclasses which are nested (or their closures are, as in our case). In such cases the maximum likelihood principle fails, because it typically leads to choosing the highest possible dimension.

AIC has been derived by Akaike by considering the quality of out-of-sample forecasts with the models obtained; his approach may be interpreted as minimizing an estimate of the Kullback-Leibler divergence.

The derivation of BIC by Schwarz is perhaps the closest to traditional statistical thinking. He commences in a Bayesian framework from priors which are not absolutely continuous for the whole model class; they give positive a priori probabilities to the subclasses (which are of different dimensions) under consideration, whereas for each given subclass the priors on the finite dimensional parameter spaces may be e.g. chosen to be flat (and thus lead to maximum likelihood estimation for a given subclass). BIC then is derived from the analysis of the asymptotic behavior of the Bayes estimator under such priors. It is quite instructive to look at the simple autoregressive example given above from this point of view; in particular note the relation between prior information and the decomposition of the model class into subclasses and realize the difference between e.g. maximum likelihood estimation over T and maximum likelihood estimation for a subclass, after it has been selected by BIC.

Rissanen has proposed the principle of minimal description length (MDL), the idea of which is to describe a time series with a shortest code length, and which leads to BIC-type criteria.

As far as the *statistical evaluations* of e.g. order estimators obtained from information criteria, $\hat{n}_T = \text{argmin}_{0 \le n \le N} A(n)$, are concerned, Hannan has shown that under suitable additional conditions

$$\lim_{T\to\infty} \frac{c(T)}{T} = 0 \text{ and } \liminf_{T\to\infty} \frac{c(T)}{\log T} > 0$$

imply $\hat{n}_T = n_0$ a.s. if the true order n_0 satisfies $0 \le n_0 \le N$. Thus, in particular BIC gives consistent estimators of the order, which in this case means that $\hat{n}_T = n_0$ from a certain T_0 onwards. On the other hand, AIC does not give consistent estimators of the order. However, as has been shown by Shibata, for the autoregressive case when the data generating system is infinite dimensional, AIC has other optimality properties; for instance spectral estimators based on autoregressive models selected by AIC are asymptotically efficient.

Also the Kronecker indices α can be estimated by a criterion function of the form (1.23) and the estimators are consistent e.g. for BIC.

It is important to note that this uncertainty coming from model selection adds uncertainty to the estimators of the real valued parameters. As is analyzed by [18] even the asymptotics of postmodel selection estimators create severe problems and have to be further analyzed in detail.

As a general remark note that if two different decompositions into bits are compared, there is a tradeoff between the quality in estimating integer valued parameters versus estimating real valued parameters. For bits of lower dimension, corresponding to a finer decomposition, the estimators of the real valued parameters will asymptotically have smaller variance, but on the other hand the model selection of the bits is harder. To the author's knowledge this problem has not been analyzed up to now.

For order estimation in the context of Larimore's procedure, the singular values in (1.22) may be used; see e.g. [4]. Criteria of the form (1.23) cannot be used for estimating the dynamical indices α in the overlapping description of $M(n)$, since every corresponding U_α is open and dense in $M(n)$ and of equal dimension $2ns$. Here the decision e.g. may be based on the conditioning of the respective basis rows of the Hankel matrix.

1.5 Linear non-mainstream cases

There are different important cases for linear system identification in a non-mainstream setting; examples are:

- Identification of linear systems with time varying parameters (tracking of such parameters).
- Identification of systems under feedback and identification for control.
- Identification in an errors-in-variables setting, where also observed inputs are corrupted by noise; identification of symmetric systems, where no a priori distinction is made between observed inputs and outputs.
- Identification of infinite dimensional systems.
- Identification of unstable systems.
- Identification of linear systems with nonstochastic noise.

It would be beyond the scope of this contribution to discuss this wide range of cases. We will only present a rather short description of *cointegration*, a case that has attracted great attention in econometric time series analysis during the last twenty-five years, but is rather unknown e.g. in systems and control.

The motivation for cointegration is that on the one hand many economic (in particular macroeconomic) time series show apparent nonstationarities such as trends in means and variances and that on the other hand in many cases the most important economic information is the long run, static equilibrium relation between time series. In the classical econometric analyses often data transformations such as differencing in order to achieve stationarity were used. However, the disadvantage of differencing is that important information at frequency zero is lost. For modeling of nonstationarity so called *integrated processes* $(y_t|t \in \mathbb{N})$ are considered. Such processes are nonstationary, whereas the first differences $(1-z)y_t = y_t - y_{t-1}$ are stationary. (Here, only integration of order one is considered, z is the one-sided backward shift, i.e. $z(y_t|t \in \mathbb{N}) = (0, y_1, y_2, \dots)$ and stationarity is not required for the initial elements of the difference processes).

An integrated vector process (y_t) (i.e. $s > 1$) is called *cointegrated*, see [13], if there exists an $\alpha \in \mathbb{R}^s$, $\alpha \neq 0$ such that (αy_t) is stationary. The idea is that α describes a (static) equilibrium relation between the nonstationary components of (y_t), because "stationary" is interpreted as "relatively small".

The number of linearly independent vectors α such that (αy_t) is stationary is called the *cointegrating rank*.

In the ARMA context (co)integrated processes may be represented as

$$(1-z)y_t = u_t = a^{-1}(z)b(z)\varepsilon_t \tag{1.24}$$

where (ε_t) is white noise and (u_t) is a stationary ARMA process with a spectral density having no zeros at frequency zero. Writing

$$a^{-1}(z)b(z) = k(z) = k(1) + \tilde{k}(z) \tag{1.25}$$

where $\tilde{k}(1) = 0$, we obtain

$$y_t = (1-z)^{-1}k(1)\varepsilon_t + (1-z)^{-1}\tilde{k}(z)\varepsilon_t \tag{1.26}$$

where $(1-z)^{-1}\tilde{k}(z)\varepsilon_t$ is stationary. In the cointegrated case $k(1)$ must be singular and unequal to zero and $(1-z)^{-1}k(1)\varepsilon_t$ can be interpreted as common trends generated by a lower dimensional, integrated factor process. Thus (1.26) may be interpreted as a factor model with stationary noise. Note that the two terms on the right hand side of (1.26) are not uncorrelated. The cointegrating rank then is equal to the corank of $k(1)$ and any matrix A the columns of which form a basis for the left kernel of $k(1)$ describes all static equilibrium relations between components of (y_t) and the second part on the right hand side of (1.26) describes the deviations from equilibrium.

From now on we restrict ourselves to the AR case

$$a(z)y_t = \varepsilon_t \tag{1.27}$$

where $A(0) = I$, $a(z)$ has degree p and where $\det a(z)$ is allowed to have zeros only for $|z| > 1$ or for $z = 1$; in addition we assume $\mathbb{E}\varepsilon_t\varepsilon_t' = \Sigma > 0$. Note that $a(z)$ has a different meaning in (1.27) and (1.24). This autoregressive case is the most important one in econometrics and a rather complete theory for this case is presented in [16].

The system (1.27) may be written as

$$(1-z)y_t = \Gamma_1(1-z)y_{t-1} + \cdots + \Gamma_{p-1}(1-z)y_{t-p+1} + \Pi y_{t-p} + \varepsilon_t \tag{1.28}$$

where $\Pi = -a(1)$. Now a structure theory can be based on Π as follows: If $\Pi = 0$, then (y_t) is integrated, but not cointegrated, if Π has full rank s, then (y_t) is stationary and if Π has rank r, $0 < r < s$, then y_t is cointegrated with cointegrating rank r. In this case $\Pi = BA'$, where $A, B \in \mathbb{R}^{n\times r}$ and the rows of A are the cointegrating vectors, i.e. they form a basis of the nullspace of $a(1)$.

Based on these properties, Johansen developed Gaussian maximum likelihood estimation and test procedures for real valued and integer valued parameters, respectively. This approach commences (for given p) from a Gaussian likelihood in the parameters $\Gamma_1, \ldots, \Gamma_{p-1}, \Sigma, B$ and A (clearly, B and A are not identifiable, unless further restrictions are imposed); these parameters are concentrated out stepwise until a concentrated likelihood function depending on A and r only is obtained. This concentrated likelihood function is the basis for a sequence of likelihood ratio tests to determine r.

We will not describe the asymptotic properties of the procedure described above (which are given in [16]), we will only mention that the results are non-standard as can be already seen from a scalar autoregressive process of order one:

$$y_t = ay_{t-1} + \varepsilon_t.$$

As is well known, in the stationary case ($|a| < 1$) the ordinary least squares estimator $\hat{a}_T$ for a_0 (the true a) is consistent and $\sqrt{T}(\hat{a}_T - a_0)$ converges in distribution to a Gaussian distribution with mean zero and variance $1-a_0^2$. For the integrated case ($a_0 = 1$), the situation is different, as the convergence is faster and the limiting distribution is non-Gaussian; here $T(\hat{a}_T - 1)$ converges in distribution to the distribution of $\frac{1}{2}(W(1)^2 - 1)(\int_0^1 W^2(\nu)d\nu)^{-1}$, where $W(.)$ is the standard Brownian motion.

1.6 Non-linear systems

This is a wide area and we only mention a few things here. Non-linear system identification is a word like non-elephant zoology. This analogy is appropriate in two respects: On the one hand identification of linear systems still is the largest animal and on the other hand there is a great variety of classes of non-linear systems and approaches for their identification. Examples for such classes are bilinear systems, recurrent neural nets, chaotic systems, hidden Markov models or the GARCH type models as developed in finance econometrics.

Identification of linear systems is a non-linear problem already, because the mapping attaching a system to data is non-linear, also for linear systems. In a certain sense principles of parameter estimation and the corresponding asymptotic theory for identification of linear systems can be generalized to parametric classes of non-linear systems as treated in [23]. However, a structure theory, up to now, has only been developed for special cases such as hidden Markov models ([3]) or special recurrent networks (Albertini and Dai Pra (1997)).

1.7 Past, present state and perspectives of system identification

System identification, i.e. data driven modeling, is used for data compression and coding, for signal extraction, spectral estimation, to obtain models for analysis, forecasting, simulation, monitoring or control, to estimate unknown parameters in models obtained from theory and for empirical discrimination between competing theories. There is a large and increasing number of applications in different fields. The dramatic increase in computer capacity and computing speed over the last thirty years has been a main factor for the spread of applications and the development of more appropriate and demanding procedures. Now there is also a number of software packages for system identification available.

System identification may be considered as an enabling technology. In many applications in technology or logistics system identification procedures are incorporated as a part in a technical device or a software program comprising other functions too. Such incorporations may make the overall system more intelligent, may improve quality and efficiency or reduce costs, but are invisible for non-experts. In other cases system identification is used as a tool for scientific investigations, for instance in natural science, economics or medicine.

Cornerstones for the development of methods and theory in the past were (this list is naturally incomplete):

- The introduction of the method of least squares for linear regression by Legendre and Gauss in the early 19th century.
- The development of the linear theory of stationary processes by Kolmogorov, Wiener and others in the thirties and forties of the 20th century.
- The work of the Cowles Commission on (in general multivariate) ARX system identification in the forties of the last century by Wald, Koopmans and others.
- The development of the asymptotic theory for estimators for AR and ARMA models, pioneered by T.W. Anderson, Hannan and Whittle, mainly in the sixties and seventies of the last century.
- The structure theory for state space and ARMA systems, triggered in particular by Kalman at about the same time.
- The development of information criteria for order estimation which (more or less) started with Akaike's work approximately forty years ago.

At present there is a substantial body of methods, fundamental insights and theories available for system identification which make the subject to a certain degree self-contained. This is also documented by a number of textbooks which were published in the late eighties, such as the books by [7], [19] and [25].

In the last decade, in particular in that part of system identification related to control, the development has been rather in the form of further extensions and ramifications of the existing body of methods and theories, partly in form of "small continuous improvements" driven by applications, than in terms of substantial changes in basic ideas and instruments. System identification for control, procedures for non-linear systems or subspace algorithms fall in this category.

In the author's opinion, in this last decade the most innovative developments of methods and theories were in econometrics. This was triggered by developing model classes and methods genuine for economic data and economic theories. An outstanding role thereby has to be attributed to finance econometrics, where genuine features of financial time series, such as volatility clustering, thick tails, leverage effects or non-equidistant (tick by tick) data were dealt with. The boom in finance econometrics was and still is caused by the drastic increase of the importance of financial markets and the development of new products for these markets.

An important second point is that this "demand pull" for econometric methods was met by using and further developing "high tech" instruments and theories from modern statistics, such as functional central limit theorems for the non-normal limiting laws for estimators in cointegration.

The developments of modern statistics and econometrics are only partially adopted on the engineering side of the subject and vice versa. This may have to do with features genuine for a particular field of applications, as described above, but is also caused – in the author's opinion at least – by an unnecessarily high and harmful degree of fragmentation of the subject. An extreme example for this is the neural net community, late comers to the field of system identification who created even their own terminology irrespective of the existing one. On the other hand there has been a significant increase of applications during the last decade in a number of fields. It is important to note that in many applications, in the initial phase, still an experienced expert is needed for designing an appropriate identification procedure.

The fact that system identification is a mature subject now, does not mean that there are no important open problems; such problems, in the author's opinion, are:

- The further development and evaluation of methods for different classes of non-linear systems.
- Identification of models for spatio-temporal processes such as of partial differential equations.
- To increase the degree of automatization in system identification.
- The further development of hybrid procedures which consist of interacting rule-based and algorithmic parts.
- The use of symbolic computation, for instance for systems which are highly structured by a priori information.

System identification is an interesting and important field – to a certain extent. It definitely lacks the excitement of the fundamental discoveries of physics in the first half of the twentieth century or of certain branches of biology now or the excitement coming from spectacular technological innovations. To a good part this is due to the fact that system identification is instrumental – it is more like building microscopes than discovering new bacteria.

System identification is a field with vaguely defined and shifting boundaries. This is a chance and a risk at the same time and makes forecasting of the development of the field and its parts rather difficult. A particularly important chance is the penetration of new "markets" (i.e. areas of application opening up at present). There are competitors on these markets, e.g. certain branches of computer science such as artificial intelligence. In the author's opinion main ingredients for future success are: First a higher degree of vertical integration in terms of deeper penetration of areas of applications and a development of integrated solutions. In my opinion a good part of the excitement in the area is coming from understanding and learning from applications and the feedback they provide for theory and methods. Second, enhanced communication between the different subareas like statistics, econometrics and system and control theory and with the competitors mentioned above.

Acknowledgement

I want to thank T. Ribarits, A. Gleiss and W Scherrer for valuable discussions and comments.

References

1. H. Akaike. Canonical Correlation Analysis of Time Series and the Use of an Information Criterion. In R. K. Mehra and D. G. Lainiotis (eds.): System identification: Advances and Case Studies, Academic Press, New York, 27-96. 1976
2. F. Albertini and P. Dai Pra. Recurrent Neural Networks: Identification and other System Theoretic Properties. *Neural Network Systems Techniques and Applications*, 3, 1-41, 1997.
3. B.D.O Anderson. The Realization Problem for Hidden Markov Models. *Preprint*, 1996.
4. D. Bauer. *Some Asymptotic Theory for the Estimation of Linear Systems using Maximum Likelihood Methods or Subspace Algorithms.* PhD thesis, TU Wien, 1998.
5. D. Bauer and M. Deistler. Balanced Canonical Forms for System Identification. *IEEE Trans. Autom. Control*, 44, 1118-1131, 1999.
6. D. Bauer, M. Deistler and W. Scherrer. Consistency and Asymptotic Normality of some Subspace Algorithms for Systems without Exogenous Inputs. *Automatica*, 35, 1243-1254, 1999.

7. P.E. Caines. *Linear Stochastic Systems.* John Wiley & Sons, New York, 1988.
8. H. Cramer. *Mathematical Methods of Statistics.* Princeton University Press, Princeton, 1946.
9. M. Deistler. Linear System Identification – a Survey. In: Willems J. (ed.): From Data to Model. Springer Verlag, Berlin, 1-25. 1989
10. M. Deistler, K. Peternell and W. Scherrer. Consistency and Relative Efficiency of Subspace Methods. *Automatica*, 31, 1865-1875, 1995.
11. M. Deistler and B.M. Pötscher. The Behaviour of the Likelihood Function for ARMA Models. *Adv. Appl. Prob.*, 16, 843-865, 1984.
12. M. Deistler and L. Wang. The Common Structure of Parametrizations for Linear Systems. *Linear Algebra and its Applications.*, 122/123/124, 921-941, 1989.
13. R.E. Engle and C.W.J. Granger. Cointegration and Error Correction: Representation Estimation and Testing. *Econometrica.*, 55, 251-270, 1987.
14. E.J. Hannan and M. Deistler. *The Statistical Theory of Linear Systems.* John Wiley & Sons, New York, 1998.
15. M. Hazewinkel and R.E. Kalman. Invariants, Canonical Forms and Moduli for Linear, Constant, Finite Dimensional Dynamical Systems. In Marchesini G. and Mitter S. K. (eds.): Lecture Notes in Economics and Mathematical Systems, 131, Springer Verlag, Berlin, 48-60. 1976
16. S. Johansen. Estimation and Hypothesis Testing of Cointegration in Gaussion Vector Autoregressive Models. *Econometrica.*, 53, 1551-1586, 1991.
17. W.E. Larimore. System Identification, Reduced Order Filters and Modelling via Canonical Variate Analysis. In: Rao H. S. and Dorato P. (ed.): Proc. American Control Conf., IEEE Service Center, Piscataway, NJ, 445-451. 1983
18. H. Leeb and B.M. Pötscher. The Finite Sample Distribution of Post-Model-Selection Estimation, and Uniform Versus Non-Uniform Approximations. Mimeo, Institute for Statistics and Decision Support Systems, University Vienna, 2000.
19. L. Ljung. *System Identification: Theory for the User.* Prentice Hall, Englewood Cliffs, New Jersey, 1987.
20. H. B. Mann and A. Wald On the Statistical Treatment of Linear Difference Equations. *Econometrica.*, 11, 173-220, 1943.
21. T. McKelvey and A. Helmersson. A Dynamic Minimal Parametrization of Multivariable Linear Systems and its Application to System Identification. Mimeo Linköping University. 1999
22. M.S. Morgan. *The History of Econometric Ideas.* Cambridge University Press, Cambridge, 1990.
23. B.M. Pötscher and I. Prucha. *Dynamic Nonlinear Econometric Models: Asymptotic Theory.* Springer Verlag, Berlin, 1997.
24. G.C. Reinsel. *Elements of Multivariate Time Series Analysis.* Springer, New York, 1993.
25. T. Söderström and P. Stoica. *System Identification.* Prentice Hall, New York, 1989.
26. P. Van Overschee and B. De Moor. N4sid; Subspace Algorithms for the Identification of Combined Deterministic-Stochastic Systems. *Automatica.*, 30, 75-93, 1994.
27. G. Wolodkin, S. Rangan and K. Poolla. An LFT Approach to Parameter Estimation. *Proc. 11 th IFAC Symposium on System Identification.*, Kitakyushu, 87-92, 1997.

28. A. Wald. Note on the Consistency of the Maximum Likelihood Estimation. *Annals of Mathematical Statistics.*, 20, 595-600, 1949.
29. G.U. Yule. On a Method for Investigating Periodicities in Disturbed Series with Special Reference to Wolfer's Sunspot Numbers. *Philos. Trans. Roy. Soc. London*, ser. A, 226, 267-298, 1927.

2 Singularity Issues in Closed Loop Identification.

Graham C. Goodwin and James S. Welsh

Department of Electrical and Computer Engineering
University of Newcastle, Callaghan, NSW Australia, 2308.

Abstract. This chapter addresses issues relating to the properties of estimation using closed loop data. This has been a topic of interest for at least four decades. Here we address specific problems which occur in these methods due to singularities. It is shown that unless special precautions are included in the algorithms, these singularities will lead to problems with the recovered open loop transfer function estimates. These problems are investigated in this chapter and remedies suggested under finite sample conditions.

2.1 Introduction

It is frequently desirable to carry out estimation under closed loop conditions. This arrangement is often needed for safety reasons and offers minimal perturbation to the normal operating conditions of the plant. For this reason there has been continuing research on this problem for at least four decades. Early work appears in [12], [1], [5], [6], [4], [15], [20], [23], [25], [28] and [14]. As an illustration of these results, it was shown in early collaboration between the first author and Brian Anderson - see [24] that the open loop transfer functions of the forward and reverse paths of a multivariable closed loop system are uniquely recoverable from the joint-input-output stochastic closed loop models provided the driving noise in the forward and reverse paths are uncorrelated.

Our focus here is on closed loop estimation in the frequency domain [17], [27],[26], [22], [8] of a linear time invariant system operating in closed loop. Our specific interest is on issues that arise from singularities in these methods. We restrict attention to the special case where the controller is known to have a given linear transfer function, $C = \frac{P}{L}$.

The first problem occurs when a parametric model is fitted initially to the (closed loop) complimentary sensitivity. Let the estimates be $\hat{T} = \frac{\hat{B}}{\hat{A}}$.

A parametric estimate of the transfer function can the be obtained by noting that, if T is the true complimentary sensitivity, then it is related to the plant transfer function, G, by

[0] This chapter is dedicated to Brian Anderson on the occasion of his 60th birthday. The first author wishes to place on record the pivotal influence that Brian had in his career over the past quarter of a century.

$$T = \frac{GC}{1 + GC} \tag{2.1}$$

Inverting this relationship suggests the following indirect parametric estimate for G:

$$\hat{G} = \frac{\frac{\hat{B}}{\hat{A}}}{C\left(1 - \frac{\hat{B}}{\hat{A}}\right)} \tag{2.2}$$

However, as recently pointed out by Brian Anderson and co-workers in [3], there is a singularity issue associated with the method, namely, the right half plane zeros of C have to exactly cancel with zeros of $\hat{T}$ and right half plane poles of C have to exactly cancel with zeros of $1 - \hat{T}$. If this does not happen, then near pole-zero cancellations will be induced in $\hat{G}$.

Here we assume C is a known linear controller. In this case, the required linear interpolation constraints on $\hat{T}$ and $1 - \hat{T}$ can easily be included in the estimation algorithms as linear equality constraints on the coefficients of $\hat{B}$ and $\hat{A}$ respectively. Details of how this may be achieved are given in section 2.2 of this chapter.

An alternative scheme which seems, at first sight, to overcome the above difficulty is to calculate $\hat{G}_i$ as a frequency domain estimate based on frequency domain estimates of $\hat{T}$. This circumvents the pole-zero cancellation problem. Surprisingly, however, it leads to a second singularity problem. Specifically, if the pointwise frequency domain estimates of T are near unity then the estimate for G will be very large at that frequency. Indeed, it turns out that this problem is generic and arises due to the effects of noise on the estimate of the complimentary sensitivity in the frequency domain. With gaussian noise, this event has a non-negligible probability of occurring. This, in turn, leads to the recovered open loop transfer function having infinite variance. This clearly undermines any simplified approach to obtaining theoretical support for this algorithm.

Fortunately, this second problem also has a simple solution, namely to incorporate an exclusion zone around the point where the complimentary sensitivity estimate is unity. Estimates which fall within this zone are rejected. This leads to frequency domain estimates having finite variance. Indeed, it turns out that, under these conditions, and provided the signal to noise ratio exceeds some threshold, there is no substantive difference between the accuracy achievable from open and closed loop frequency domain estimation. Details are given in the sequel.

2.2 Frequency Domain Estimation of Complimentary Sensitivity

Consider the closed loop system shown in figure 2.1.

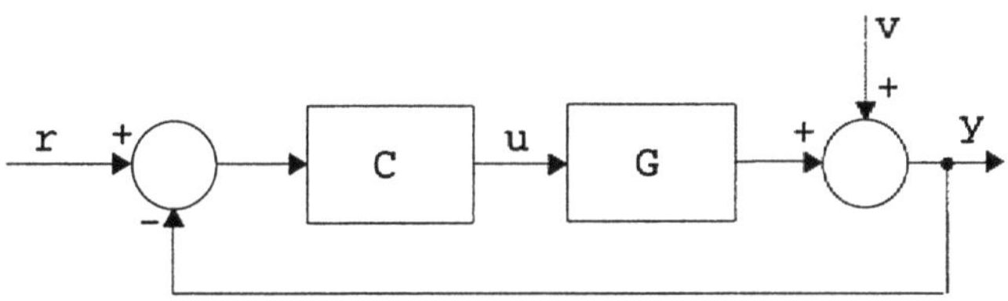

Fig. 2.1. Closed Loop System

The least mean square estimate of the frequency response is given by the ratio of the input and output Discrete Fourier Transform (DFT). This estimate has also been called the Empirical Transfer Function Estimate (ETFE). It is known [17] that this estimate has poor asymptotic properties when the input is a stochastic process. In fact, the variance of the ETFE for a stochastic input has been shown to be infinite [7]. However, the method is known to give excellent results when the input is periodic [17].

As mentioned above, the least mean squares estimate of the closed loop transfer function, $\hat{T}$, from r to y is given by the ratio of the DFT of y to that of r, ie

$$\hat{T}_i = \frac{Y_i}{R_i} \tag{2.3}$$

where Y_i, R_i denote the DFT at frequency ω_i of the output and reference respectively and $\hat{T}_i$ denotes the estimate of the closed loop transfer function at frequency ω_i.

We will next examine the accuracy of the frequency domain estimate, $\hat{T}_i$, given in (2.3).

2.2.1 Open Loop Nonparametric Estimation

We begin by summarising known results on the nonparametric estimation of linear frequency responses from open loop data. We make the following assumptions:

i. The reference signal r is periodic with period M, having energy at frequencies $\omega_1, \ldots \omega_p$ and corresponding amplitude A_i; $i = 1, \ldots p$.
ii. The observation period N is an integer multiple of M.

iii. The noise v (See Figure 2.1) is a stationary stochastic process having gaussian probability function and spectral density $\mathcal{S}_v(\omega)$ at frequency ω.

As a preliminary to our first result, we present

Lemma 1. PART (I).
The finite sample properties of the Fourier Transform of the noise are described by:

$$\mathbf{E}\{V_i V_i^*\} = N\Delta\mathcal{S}_1(\omega_i) \tag{2.4}$$

$$\mathbf{E}\{V_i V_i\} = \Delta\mathcal{S}_2(\omega_i) \tag{2.5}$$

where $\mathcal{S}_1(\omega_i)$ is the noise spectral density, $\mathcal{S}_v(\omega_i)$, convolved with a $sinc^2$ function centered on ω_i, i.e.

$$\mathcal{S}_1(\omega_i) = \frac{1}{2\pi}\int_{-\pi}^{\pi} \mathcal{S}_v(\omega) \otimes \Gamma_1(\omega - \omega_i)d\omega \tag{2.6}$$

$$\Gamma_1(\omega) = \Delta \sum_{k=-\infty}^{\infty} \gamma_1(k)e^{-j\omega k} \tag{2.7}$$

$$\gamma_1(k) = \begin{cases} \frac{|N-k|}{N} & \text{for } k \le N, \\ 0 & \text{otherwise} \end{cases} \tag{2.8}$$

and $\mathcal{S}_2(\omega_i)$ is given by

$$\mathcal{S}_2(\omega_i) = \Delta \sum_{k=1}^{N-1} \mathcal{R}_v(k) f(k,i) \tag{2.9}$$

with

$$\mathcal{R}_v(k) = \mathbf{E}\{v(i)v(i-k)\} \tag{2.10}$$

$$f(k,i) = -2\left(\cot\frac{2\pi i}{N} + j\right)\sin\frac{2\pi k i}{N} \tag{2.11}$$

PART (II).
The finite sample properties of the Discrete Fourier Transform of the periodic excitation signal applied to the reference input (See Figure 2.1) are described by:

$$R_j R_j^* = g_i^2 + h_i^2 = \left(\frac{N\Delta}{2}\right)^2 A_i^2 \qquad \text{for } j = i,\ N-i \tag{2.12}$$

$$R_j R_j = g_i^2 - h_i^2 + 2jg_i h_i \qquad \text{for } j = i,\ N-i \tag{2.13}$$

where g_i and h_i are simple functions of the deterministic input, namely

$$g_i = Re\left\{R_i\right\} \tag{2.14}$$
$$h_i = Im\left\{R_i\right\} \tag{2.15}$$

Proof. *See [29].*

■

Remark 1. We note that, provided $\mathcal{S}_v(\omega)$ is smooth, then

$$\lim_{N\to\infty} \mathcal{S}_1(\omega) \to \mathcal{S}_v(\omega) \tag{2.16}$$

due to the sampling properties of the sinc^2 function.

■

We now temporarily focus on the **open loop estimation problem**. Then the open loop analogue of equation (2.3) is

$$\hat{G}_i = \frac{Y_i}{R_i} \tag{2.17}$$

The following result summarises properties of the **open loop estimator** (2.17).

Lemma 2. *Let G_i denote the true open loop frequency response of the system at frequency ω_i and define $\tilde{G}_i = \hat{G}_i - G_i$. Then*

$$\tilde{G}_i = \frac{V_i}{R_i}; \qquad i = 1, 2, ..., P \tag{2.18}$$

where V_i, R_i denote the DFT (at ω_i) of the noise and of the reference input, respectively.

We then have:

i.

$$\mathbf{E}\left\{\tilde{G}_i\right\} = 0 \tag{2.19}$$

ii.

$$\mathbf{E}\left\{\begin{pmatrix} Re\left(\tilde{G}_i\right) \\ Im\left(\tilde{G}_i\right)\end{pmatrix}\left(Re\left(\tilde{G}_i\right)\ Im\left(\tilde{G}_i\right)\right)\right\} = \begin{bmatrix}\gamma_{1i} & \gamma_{3i} \\ \gamma_{3i} & \gamma_{2i}\end{bmatrix} = D \tag{2.20}$$

where

$$\gamma_{1i} = \frac{2\mathbf{E}\{V_i V_i^*\}}{(N\Delta)^2 A_i^2} + \frac{1}{2} Re \left[\frac{\mathbf{E}\{V_i V_i\}}{(g_i^2 - h_i^2) + 2jg_ih_i}\right] \tag{2.21}$$

$$\gamma_{2i} = \frac{2\mathbf{E}\{V_i V_i^*\}}{(N\Delta)^2 A_i^2} - \frac{1}{2} Re \left[\frac{\mathbf{E}\{V_i V_i\}}{(g_i^2 - h_i^2) + 2jg_ih_i}\right] \tag{2.22}$$

$$\gamma_{3i} = \frac{1}{2} Im \left[\frac{\mathbf{E}\{V_i V_i\}}{(g_i^2 - h_i^2) + 2jg_ih_i}\right] \tag{2.23}$$

and the g_i, h_i, $\mathbf{E}\{V_i V_i^*\}$ *and* $\mathbf{E}\{V_i V_i\}$ *are as given in Lemma 1.*

Proof. See [29].

■

Remark 2. We note from Lemma 1 that for $i \neq 0$,

$$\gamma_{1i} = \sigma_i^2 + k_{1i} \tag{2.24}$$

$$\gamma_{2i} = \sigma_i^2 - k_{1i} \tag{2.25}$$

$$\gamma_{3i} = k_{2i} \tag{2.26}$$

where

$$\sigma_i^2 = \frac{2S_1(\omega_i)}{N\Delta A_i^2} \tag{2.27}$$

and k_{1i}, k_{2i} are of order $\frac{1}{N^2}$.
Hence, for $i \neq 0$, we have that

$$\lim_{N\to\infty} N \begin{bmatrix} \gamma_{1i} & \gamma_{3i} \\ \gamma_{3i} & \gamma_{2i} \end{bmatrix} = \begin{bmatrix} \frac{2S_1(\omega_i)}{\Delta A_i^2} & 0 \\ 0 & \frac{2S_1(\omega_i)}{\Delta A_i^2} \end{bmatrix} \tag{2.28}$$

■

Remark 3. From Lemma 2, we see that asymptotically

$$\frac{\mathbf{E}\left\{\tilde{G}_i \tilde{G}_i^*\right\}}{G_i G_i^*} = \frac{4S_1(\omega_i)}{|G_i|^2 N\Delta A_i^2} \tag{2.29}$$

This has a nice physical interpretation, namely asymptotically the normalised mean square error in the frequency response estimate is two times the ratio of the output noise spectral density (windowed by a $sinc^2$ function) to the output energy due to the test signal at frequency ω_i.

■

Remark 4. Actually the result in equation (2.29) gives a powerful reason for using frequency domain non-parametric estimators as an intermediate step prior to obtaining a parametric model. The reason is that, with a periodic input and reasonable observation period, the non-parametric estimates of the frequency response will be substantially noise free. This means that the subsequent step of fitting a parametric model to the frequency domain data can be carried out without paying too much attention to noise effects. Of course, it is possible to exactly account for the noise, see [18], but one can be reasonably casual (eg use simple least squares [16]) without incurring significant errors. Of course, the trade-off is that the frequency domain estimates need to be restricted to a relatively small number of frequencies to keep the overall signal power manageable. However, in the context for identification for control system design, there is usually strong reasons to model the frequency response at particular points. This idea has been implicitly used in contemporary work on controller autotuning [11] and [2].

■

2.2.2 Application to Non Parametric Estimate of Complementary Sensitivity

We now consider the closed loop system of Figure 2.1. We apply open loop methods to the estimation of the complementary sensitivity function, T_i. In particular, we utilise equation (2.3) where Y_i is now the DFT at ω_i of the output as measured **under closed loop conditions.**

Clearly the results of section 2.2.1 are directly applicable to the quantification of the errors in $\hat{T}_i$. The only changes are that

i. The noise needs to be transformed to the closed loop output where it has spectral density

$$\mathcal{S}_w(\omega_i) = |S_o(\omega_i)|^2 \mathcal{S}_v(\omega_i) \tag{2.30}$$

where

$$S_o(\omega) = \frac{1}{1 + G(\omega) C(\omega)} \tag{2.31}$$

ii. The output signal energy becomes

$$output\ \ energy = \frac{|T_i|^2 N \Delta A_i^2}{2} \tag{2.32}$$

An immediate consequence of Lemma 2 is then:

Corollary 1. *Let T_i denote the true complementary sensitivity at frequency ω_i and define $\tilde{T}_i = \hat{T}_i - T_i$. Then*

$$\tilde{T}_i = \frac{W_i}{R_i} \tag{2.33}$$

where W_i, R_i denote the DFT (at ω_i) of the closed loop output noise and reference input respectively.

We then have

i).

$$\mathbf{E}\left\{\tilde{T}_i\right\} = 0 \tag{2.34}$$

ii).

$$\mathbf{E}\left\{\begin{pmatrix} Re\left(\tilde{T}_i\right) \\ Im\left(\tilde{T}_i\right)\end{pmatrix}\begin{pmatrix} Re\left(\tilde{T}_i\right) & Im\left(\tilde{T}_i\right)\end{pmatrix}\right\} = \left|S_o(\omega_i)\right|^2 D = \acute{D} = \begin{bmatrix} \acute{\gamma}_{1i} & \acute{\gamma}_{3i} \\ \acute{\gamma}_{3i} & \acute{\gamma}_{2i}\end{bmatrix} \tag{2.35}$$

where D is as in equation (2.20)

iii).

$$\lim_{N\to\infty} \acute{D} \to \begin{bmatrix} \frac{2|S_o(\omega_i)|^2 S_1(\omega_i)}{N\Delta A_i^2} & 0 \\ 0 & \frac{2|S_o(\omega_i)|^2 S_1(\omega_i)}{N\Delta A_i^2}\end{bmatrix} = \begin{bmatrix} \sigma^2 & 0 \\ 0 & \sigma^2\end{bmatrix} \tag{2.36}$$

Proof. Immediate from Lemma 2.

■

Remark 5. It follows from Corollary 1 (parts (i) and (ii)), that if the noise is gaussian, then $\tilde{T}_i$ has a gaussian distribution with mean and covariance as in equations (2.34) and (2.35). We denote this distribution as $p\left(\tilde{T}_i\right)$.

■

2.3 Plant Identification via Parametric Modelling of the Complementary Sensitivity

Given the frequency domain estimates $\hat{T}_i$, there exists many techniques for estimating a parametric model. A recent survey is contained in [18]. We will not delve into details. Instead, we turn our focus to how one can use this parametric model to obtain a model for the plant.

Let the parametric model for $\hat{T}$ be $\frac{\hat{B}}{\hat{A}}$. Then it is readily seen that a parametric estimate $\hat{G}_p$ for the plant transfer function is given by equation (2.2).

However, as correctly pointed out in [3] it is important that interpolation constraints be satisfied by $\frac{\hat{B}}{\hat{A}}$ and $1 - \frac{\hat{B}}{\hat{A}}$ otherwise spurious near pole-zero cancellations in the RHP will be included in $\hat{G}$. The required interpolation constraints (see [9]) are:

i. $\hat{B}(z_i) = 0$ where z_i is any RHP zero of C.
ii. $\hat{A}(p_i) - \hat{B}(p_i) = 0$ where p_i is any RHP pole of C.

In the case when C is known, there is a simple way to ensure that these interpolation constraints are satisfied. Specifically, (i) is equivalent to

$$\hat{B}(z_i) = \hat{b}_m z_i^m + \hat{b}_{m-1} z_i^{m-1} + \dots \hat{b}_0 = 0 \tag{2.37}$$

Similarly (ii) is equivalent to

$$\begin{aligned}\hat{A}(p_i) - \hat{B}(p_i) &= \hat{a}_n p_i^n + \hat{a}_{n-1} p_i^{n-1} + \dots \left(\hat{a}_m - \hat{b}_m\right) p_i^m + \dots \left(\hat{a}_0 - \hat{b}_0\right) \\ &= 0\end{aligned} \tag{2.38}$$

Where z_i and p_i are known, then equations (2.37) and (2.38) are simply linear constraints on the coefficients. Least squares estimation and other related estimation procedures are easily modified to incorporate this kind of constraint. For example, the following result is easily established [10].

Consider a linear least squares problem in which Y is regressed on θ via the regressor X. The unconstrained least squares estimate is

$$\hat{\theta} = \left(X^T X\right)^{-1} X^T Y \tag{2.39}$$

Say that the estimate is required to satisfy s linear side constraints expressed in the following form:

$$L\theta = \Gamma \tag{2.40}$$

where L is an $s \times p$ matrix which is assumed to have rank s. Then the least squares estimate subject to these constraints is

$$\hat{\theta}_c = \hat{\theta} - \left(X^T X\right)^{-1} L^T \left(L \left(X^T X\right)^{-1} L^T\right)^{-1} \left(L\hat{\theta} - \Gamma\right) \tag{2.41}$$

2.4 Simulation Example

To illustrate the above ideas we choose the following first order plant as an example,

$$G = \frac{1}{s+3} \tag{2.42}$$

A controller is designed with a RHP pole and zero that will stabilise this plant. (The controller is somewhat artificial but is intended to illustrate the principles):

$$C = \frac{-4(s-10)}{10(s-1)} \tag{2.43}$$

The closed loop system is illustrated in figure 2.2.

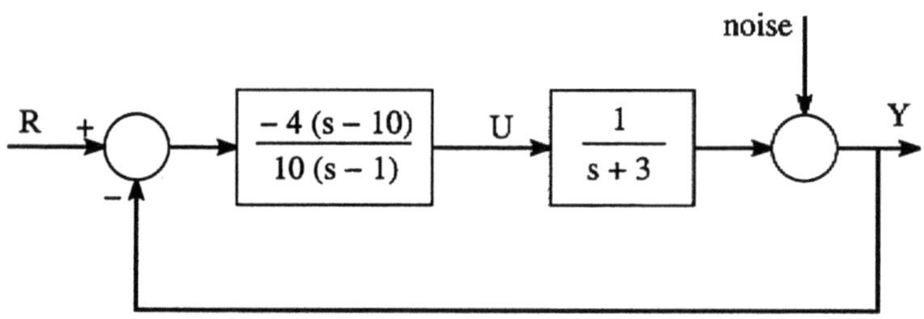

Fig. 2.2. Closed Loop System

The system has the nominal complimentary sensitivity:

$$T_0 = \frac{-4s+40}{10s^2+16s+10} \tag{2.44}$$

The system was simulated in closed loop with a periodic input and added output noise. The complementary sensitivity was then estimated from the frequency response data using equation (2.3). A parametric model was then fitted to this data via ordinary least squares [16].

The resulting unconstrained least-squares estimate for T is

$$\hat{T} = \frac{0.0773s+3.11}{s^2+1.31s+0.791} \tag{2.45}$$

Given the controller (2.43), $\hat{G}$ may be estimated via equation (2.2). This yields

$$\hat{G} = -0.193\frac{(s-1)(s+40.23)}{(s-1.03)(s+2.26)(s-10)} \tag{2.46}$$

Note that, spurious RHP poles and zeros result as predicted.

The system may be constrained to reflect the knowledge of C. We note that C has a RHP zero at 10 and a RHP pole at 1. The interpolation constraints when expressed in the form of equations (2.37), (2.38) and (2.40) become,

$$L\theta = \Gamma \tag{2.47}$$

In our special case, these constraints take the form:

$$\begin{bmatrix} 10 & 1 & 0 & 0 \\ 1 & 1 & -1 & -1 \end{bmatrix} \begin{bmatrix} \hat{b}_1 \\ \hat{b}_0 \\ \hat{a}_1 \\ \hat{a}_0 \end{bmatrix} = \begin{bmatrix} 0 \\ 1 \end{bmatrix} \tag{2.48}$$

Using equation (2.41), the constrained estimate for T is,

$$\hat{T}_c = \frac{-0.398(s-10)}{s^2+1.578s+1.002} \tag{2.49}$$

It is readily checked that the interpolation constraints are exactly satisfied. The resultant estimate for the plant is

$$\hat{G}_c = \frac{0.995}{s+2.98} \tag{2.50}$$

Figure 2.3 shows the measured closed loop data in the frequency domain and the constrained and unconstrained frequency domain fits.

2.5 Modelling of Plant via Nonparametric Methods

An alternative to the parametric method outlined above is to use a frequency domain equivalent of equation (2.2) to obtain a nonparametric estimate of the plant. The appropriate relationship is:

$$\hat{G}_i = \frac{\hat{T}_i}{C_i\left(1-\hat{T}_i\right)} \tag{2.51}$$

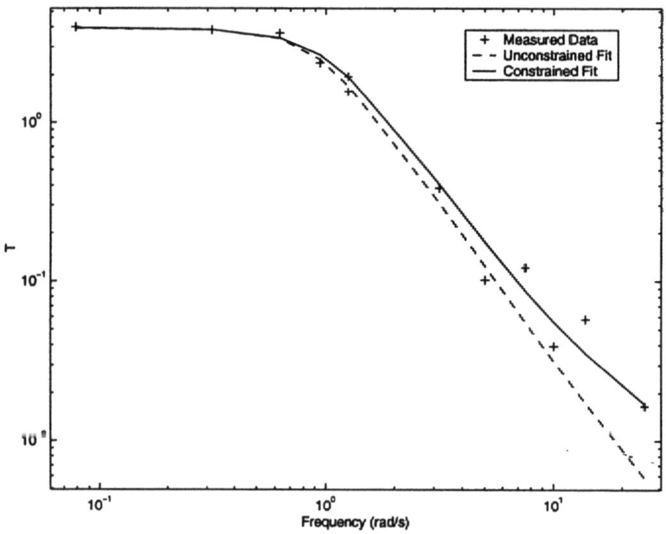

Fig. 2.3. Closed Loop Data in the Frequency Domain

where $\hat{T}_i$ is as in equation (2.3).

The above frequency domain formula sidesteps the singularity due to the interpolation constraints. Indeed, given $\hat{G}_i$, over a range of frequencies, then we can obtain a parametric model for the plant using standard parametric methods. However, equation (2.51) suffers from a second singularity issue. Namely, noise can cause $\hat{T}_i$ to be near unity at some frequencies. A consequence of this possibility is that the variance of $\hat{G}_i$ will be infinite. This is clearly an unacceptable result.

Our suggested remedy to the above difficulty is to assume that the frequency of interest has been chosen so that the true value for the complementary sensitivity, T_i, is bounded away from 1. This allows us to discard any value of $\hat{T}_i$ that falls within an exclusion zone around 1. We then utilise this idea to develop an expression for variance which takes account of the fact that neither T_i nor $\hat{T}_i$ will lie in a region surrounding the point 1 in the complex plane. A similar idea of using exclusion zones to deal with a potential singularity has recently been used in [19]. However this was in a different context namely errors in variables estimation.

Reference [13] has addressed the issue of bias without an exclusion zone. The inherent singularity is dealt with by using the Cauchy Central Value for the bias integral. This interpretation leads to a finite value for the bias. However, this result needs to be carefully interpreted. In particular, there will be large (theoretically unbounded) variability in the results due to the unbounded variance resulting from the singularity. An exclusion zone remedies this difficulty without significantly changing the bias result.

Another technical difficulty is that, with finite samples, the real and imaginary parts of the estimate, $\hat{T}_i$ are correlated (see Corollary 1). Asymptotically this correlation becomes negligible. However, a rigorous analysis of finite sample properties depends on explicitly accounting for this correlation. We will show how this can be achieved in the sequel.

We will use the above results to obtain asymptotic expressions for the bias and variance. This result confirms that bias is not an issue at high signal to noise ratio. The result also rigorously confirms that the asymptotic variance is proportional to the ratio of output noise spectral density to the input energy. Note, however, that the idea of exclusion to avoid the frequency domain singularity is essential in arriving at these conclusions.

2.6 Modified Indirect Estimation of G

We propose to essentially use the indirect estimate of G as in equation (2.51). However, this formula clearly has a singularity at $\hat{T}_i = 1$. We therefore introduce the following further assumptions that allow us to remove this singularity from the estimate:

i. At frequency ω_i, the true value of the complementary sensitivity function, namely T_i, satisfies $|T_i - 1| \geq \epsilon$.
ii. The true complementary sensitivity satisfies $|T| < L$.

The procedure we use for the indirect estimation of G is:

i. Apply a periodic excitation signal to the reference input (r) as shown in Figure 2.1.
ii. Simultaneously sample the reference (r) and output (y) signals.
iii. Obtain the complementary sensitivity function estimate, $\hat{T}$, at the frequencies of the periodic input signal applied to the reference by taking the ratio of the Discrete Fourier Transforms (DFT's) of the output signal (y) to the reference signal (r).
iv. Estimate G by solving equation (2.51), assuming knowledge of the controller C.

Furthermore, we modify the indirect estimator by discarding any experiment in which $\left|\hat{T}_i - 1\right| < \epsilon$. This is easily accomplished as the experiment yields the estimate $\hat{T}_i$.

In the sequel, we will focus on a single frequency, ω_i. Thus, for clarity, we remove the explicit dependence of $\hat{T}$ etc on i.

The removal of the singularity by excluding a zone around $\hat{T} = 1$ of radius ϵ is shown in Figure 2.4.

Using the exclusion zone, defined above, we will, in the sequel, show that the resulting estimate of $\hat{G}$, is essentially unbiased for a good signal to noise ratio and that the variance in the estimate is identical to that achievable with open loop data.

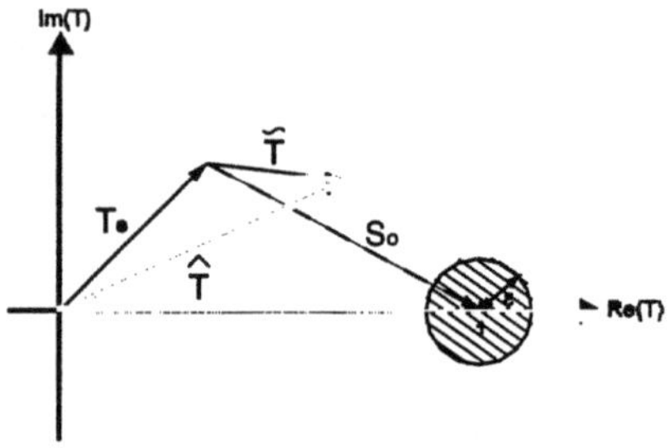

Fig. 2.4. Exclusion Zone in T Space

2.7 Heuristic Analysis of Finite Sample Properties

It is easily seen that the difference between G and the true value G is given by

$$\tilde{G} = \frac{\tilde{T}}{C\left(1-T\right)\left(1-T-\tilde{T}\right)} \tag{2.52}$$

Now provided, $T \neq 1$ and $\hat{T} \neq 1$, then there is no singularity in the denominator. This is ensured by the method described in section 2.6. Also, with this exclusion zone and high signal to noise ratio situations, we can essentially ignore the $\tilde{T}$ in the denominator (since it is subtracted from $1-T$). Consequently we can write

$$\tilde{G} \approx \frac{\tilde{T}}{C\left(1-T\right)^2} \tag{2.53}$$

Hence, under reasonable signal to noise conditions, we might expect that

$$\mathbf{E}\left\{\tilde{G}\right\} \approx 0 \tag{2.54}$$

and

$$\mathbf{E}\left\{\left|\tilde{G}\right|^2\right\} \approx \frac{1}{\left|C\right|^2\left|1-T\right|^4}\mathbf{E}\left\{\left|\tilde{T}\right|^2\right\} \tag{2.55}$$

Hence using equation (2.36)

$$\frac{\mathbf{E}\left\{\left|\tilde{G}\right|^2\right\}}{|G|^2} \approx \frac{|S|^2}{|G|^2 |C|^2 |1-T|^4} \frac{4S_1(\omega_i)}{N\Delta A_i^2} \tag{2.56}$$

$$= \frac{4S_1(\omega_i)}{|T|^2 N\Delta A_i^2} \tag{2.57}$$

Remark 6. An interesting observation based on the above result is that $\frac{|T|^2 \Delta A_i^2}{2}$ is actually the output signal energy due to the test signal. Hence, if one fixes the output energy due to the test signal, then we see that there is no essential difference in terms of estimation accuracy between the open and closed loop cases.

■

In the next section, we will present rigorous results which confirm and embellish the above heuristic conclusions.

2.8 Rigorous Analysis of Bias and Variance

Here we present rigorous results relating to bias and variance. Note that the use of an exclusion zone is central to obtaining these results.

2.8.1 Probability Density Function for $\tilde{T}$ Without Exclusion Zone

In our modified indirect estimator we are removing a zone from the space of all possible estimates by discarding any result that falls within the zone. We thus need to re-normalise the probability density function.

In this section we will first find the probability density function of the whole space, then calculate, in section 2.8.2, the re-normalisation constant to account for the excluded zone.

We first observe that since the noise is assumed zero mean gaussian and since $\hat{T}$ is a linear function of the data, then $\hat{T}$ and $\tilde{T}$ will also have a gaussian distribution. Indeed, from Corollary 1, we have that $\tilde{T}$ has a gaussian distribution with zero mean and covariance (2.35). We denote this density function $p\left(\tilde{T}\right)$.

We will next express the probability density function in terms of polar co-ordinates. This requires a transformation from rectangular co-ordinates. We will use the following preliminary result,

Lemma 3. *Say that a 2 variate gaussian random variable z has zero mean and radially symmetric covariance*

$$\mathbf{E}\left\{\begin{pmatrix} Re\,(z) \\ Im\,(z) \end{pmatrix} \begin{pmatrix} Re\,(z) & Im\,(z) \end{pmatrix}\right\} = \begin{bmatrix} \sigma^2 & 0 \\ 0 & \sigma^2 \end{bmatrix} \tag{2.58}$$

Let $z = Re^{j\lambda}$, then the probability density function for (R, λ) is a Rayleigh distribution:

$$\acute{p}\,(R, \lambda : \sigma) = \frac{R}{2\pi\sigma^2} e^{\frac{-R^2}{2\sigma^2}} \tag{2.59}$$

Proof. *This follows from standard results on transformation of probability densities [21].*

■

We now wish to apply the above result to $\tilde{T}$. However, a difficulty with the distribution for $\tilde{T}$ is that it is not radially symmetrical with finite sample size. To deal with this problem, we will use bounding techniques on the distribution. Let $p_T\,(R, \lambda)$ be the transformed (true) probability density function for $\tilde{T}$ without an exclusion zone. Then

$$p_T\,(R, \lambda) = Jp\left(\tilde{T}\right) \quad ; \quad \tilde{T} = Re^{j\lambda} \tag{2.60}$$

where J is the Jacobian [21] relating $\tilde{T}$ to $Re^{j\lambda}$.
We then have

Lemma 4. *Let $\tilde{T} = Re^{j\lambda} = Re[\tilde{T}] + jIm[\tilde{T}]$. Then the probability density function for $\tilde{T}$ can be bounded as follows:*

$$k_{min}\acute{p}\,(R, \lambda : \sigma_{min}) \;\leq\; p_T\,(R, \lambda) \;\leq\; k_{max}\acute{p}\,(R, \lambda : \sigma_{max}) \tag{2.61}$$

where

$$k_{min} = \frac{\sigma_{min}^2}{\sqrt{det(\acute{D})}} \tag{2.62}$$

$$k_{max} = \frac{\sigma_{max}^2}{\sqrt{det(\acute{D})}} \tag{2.63}$$

$$\sigma_{min}^2 = \min\left(\acute{\gamma}_{1i}, \acute{\gamma}_{2i}\right) - |\acute{\gamma}_{3i}| \tag{2.64}$$

$$\sigma_{max}^2 = \max\left(\acute{\gamma}_{1i}, \acute{\gamma}_{2i}\right) + |\acute{\gamma}_{3i}| \tag{2.65}$$

where $\acute{D}$ is as in equation (2.35) *and $\acute{p}\,(R, \lambda : \sigma)$ is the Rayleigh Distribution, given in equation* (2.59). *Where σ^2 takes on the value of either σ_{min}^2 or σ_{max}^2 depending on the side of the inequality.*

Proof. First we establish an overbound on the probability density function as follows,

$$p\left(\tilde{T}\right) = p\left(Re\left[\tilde{T}\right], Im\left[\tilde{T}\right]\right) \tag{2.66}$$

$$= \frac{1}{2\pi\sqrt{det\left(\acute{D}\right)}} e^{\left(-\frac{1}{2}\left(Re[\tilde{T}], Im[\tilde{T}]\right)\acute{D}^{-1}\left(Re[\tilde{T}], Im[\tilde{T}]\right)^{T}\right)} \tag{2.67}$$

$$\leq \frac{1}{2\pi\sqrt{det\left(\acute{D}\right)}} e^{\left(\frac{-\|\tilde{T}\|^2}{2\lambda_{max}(\acute{D})}\right)} \tag{2.68}$$

$$\leq \frac{1}{2\pi\sqrt{det\left(\acute{D}\right)}} e^{\left(\frac{-\|\tilde{T}\|^2}{2\sigma_{max}^2}\right)} \tag{2.69}$$

$$= \frac{\sigma_{max}^2}{\sqrt{det\left(\acute{D}\right)}} \frac{1}{2\pi\sigma_{max}^2} e^{\left(\frac{-\|\tilde{T}\|^2}{2\sigma_{max}^2}\right)} \tag{2.70}$$

where

$$\sigma_{max}^2 = \max\left(\acute{\gamma}_{1i}, \acute{\gamma}_{2i}\right) + |\acute{\gamma}_{3i}| \geq \lambda_{max}(\acute{D}) \tag{2.71}$$

$$det(\acute{D}) = \acute{\gamma}_{1i}\acute{\gamma}_{2i} - \acute{\gamma}_{3i}^2 \tag{2.72}$$

Similarly we can determine a lower bound on the probability density function as

$$p\left(\tilde{T}\right) \geq \frac{\sigma_{min}^2}{\sqrt{det\left(\acute{D}\right)}} \frac{1}{2\pi\sigma_{min}^2} e^{\left(\frac{-\|\tilde{T}\|^2}{2\sigma_{min}^2}\right)} \tag{2.73}$$

where

$$\sigma_{min}^2 = \min\left(\acute{\gamma}_{1i}, \acute{\gamma}_{2i}\right) - |\acute{\gamma}_{3i}| \tag{2.74}$$

It is then trivial to express bounding distributions in terms of polar coordinates using the result from Lemma 3.

■

Remark 7. In the limit as $N \to \infty$, we see for part(iii) of Corollary 1, that $\sigma_{max}^2 \to \sigma^2$ and $\sigma_{min}^2 \to \sigma^2$. Also, $\frac{\sigma_{max}^2}{\sqrt{det(\acute{D})}} \to 1$ and $\frac{\sigma_{min}^2}{\sqrt{det(\acute{D})}} \to 1$.

■

2.8.2 Probability Density Function for $\tilde{T}$ Accounting for the Exclusion Zone

In the modified indirect estimator we remove the singularity in the estimate of $\hat{G}$ by excluding a zone around $\hat{T}$. Hence, in this zone, we are forcing the probability to zero.

In the sequel, it will be convenient to re-centre and scale the error in the complementary sensitivity function. We thus define $\tilde{T}' = \frac{\tilde{T}}{S_o}$. In the symmetrical case, the real and imaginary parts of the $\tilde{T}'$ have variance $\sigma_{sc}^2 = \frac{\sigma^2}{|S_o|^2}$. Also, the exclusion zone, when transformed to the $\tilde{T}'$ space, remains a circle centered at $(1, 0)$ but having radius $\epsilon_1 = \frac{\epsilon}{|S_o|}$.

We will develop below an expression for re-normalisation constants c_{max} and c_{min} for the bounding probability density functions when the excluded zone is removed. The excluded zone is shown as the shaded area 'A' in Figure 2.5. We have seen that the density function $\acute{p}(\tilde{R}, \lambda : \sigma)$ is bounded from above and below by a scaled Rayleigh Distribution. The Rayleigh density functions, with the excluded zone removed will be expressed as:

$$\bar{p}(R, \lambda : \sigma) = \frac{\acute{p}(R, \lambda : \sigma)}{c_h} \tag{2.75}$$

where σ^2 is replaced by either σ_{max}^2 or σ_{min}^2 then c_h takes the value c_{max} or c_{min} respectively.

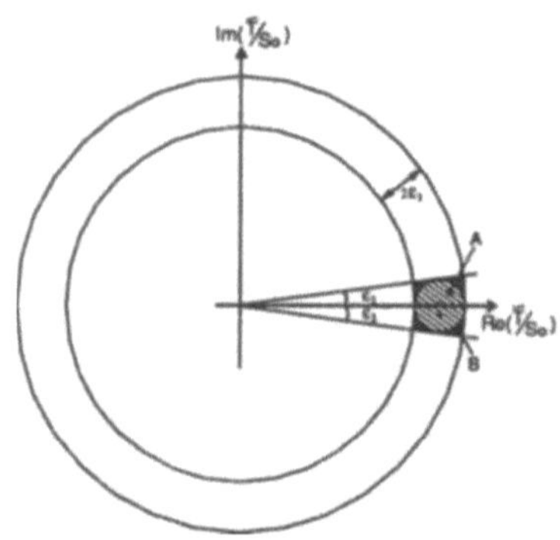

Fig. 2.5. Exclusion Zone in $\tilde{T}$ Space

Lemma 5. *The probability of lying outside the exclusion zone for the probability density function given in* (2.59) *is*

$$c_h = 1 - e^{\frac{-1}{2\sigma_{sc}^2}} \sum_{k=0}^{\infty} \frac{1}{(k!)^2 \, 2^k \sigma_{sc}^{2k}} \gamma\left(k+1, \frac{\epsilon_1^2}{2\sigma_{sc}^2}\right) \tag{2.76}$$

where

$$\gamma\left(k+1, \frac{\epsilon_1^2}{2\sigma_{sc}^2}\right) \text{ is an incomplete Gamma function} \tag{2.77}$$

Proof. *See [29].*

■

Remark 8. For the case of an elliptical shaped distribution, we denote the normalised distribution as $\bar{p}_T(R,\lambda)$ where

$$\bar{p}_T(R,\lambda) = \frac{1}{c_1} p_T(R,\lambda) \tag{2.78}$$

with $p_T(R,\lambda)$ as previously defined. We will not explicitly evaluate c_1 but note

$$c_{min} \leq c_1 \leq c_{max} \tag{2.79}$$

where c_{max} and c_{min} are as in equation (2.76).

■

2.8.3 Quantification of Finite Sample Bias

In this section we develop an expression for the bias in the estimator as described in Section 2.6. We know from Corollary 1 that asymptotically the distribution of $\tilde{T}$ becomes symmetrical. Here we first determine an expression for the bias corresponding to a symmetrical distribution. Then we bound the error between the true bias under finite sampling and that of the bias assuming a symmetrical distribution.

In the symmetrical case, it is shown that the estimate is (effectively) unbiased when the signal to noise ratio is of a sufficient level. This implies that an excitation signal is a necessity in the case of nonparametric estimation, supporting other work [17] and [26]. It is also shown that a rapid transition occurs between an unbiased and a biased result with respect to the signal to noise ratio. In general, this indicates that the bias is quite well behaved until a threshold is reached in the signal to noise ratio. We see also that the bias decreases like e^{-N} where N is the number of data observations recorded during the experiment.

Theorem 1. *For the asymptotic case, where the distribution for $\tilde{T}$ has symmetrical covariance as in part (iii) of Corollary 1, then the bias in the indirect estimate of the plant transfer function is:*

$$\mathbf{E}\left\{\tilde{G}_h\right\} = \frac{-1}{CS_o} + \frac{1}{c_h CS_o}\left(1 - e^{\frac{-(1-\epsilon_1)^2}{2\sigma_{sc}^2}}\right) + K_{b1} + K_{b2} \tag{2.80}$$

where $\tilde{G}_h$ *is the bias for the asymptotic radially symmetric distribution for* $\tilde{T}$ *and where, as before,* $\sigma_{sc}^2 = \frac{\sigma^2}{|S_o|^2}$, c_h *is as in equation* (2.76). K_{b1} *and* K_{b2} *are vanishingly small constants related to the size of the exclusion zone.*

Proof. *See [29].* ■

We next consider the actual case when the distribution for $\tilde{T}$ is not radially symmetric. This significantly complicates the result. Thus, we will relate the bias in this case to the asymptotic result developed above for the radially symmetric case. Specifically we have:

Theorem 2. *In a closed loop indirect identification experiment, the bias in the case of an asymmetric distribution,* $\mathbf{E}\{\tilde{G}\}$, *is related to the bias in the asymptotic radially symmetric case,* $\mathbf{E}\{\tilde{G}_h\}$, *as follows:*

$$\begin{aligned}
&\left|\mathbf{E}\{\tilde{G}\} - \mathbf{E}\{\tilde{G}_h\}\right| \le \\
&K_e \int_0^{1-\epsilon_1} \left(R + R^2 + R^3 + \cdots\right)\left(k_{max}\bar{p}_{max}(R) - k_{min}\bar{p}_{min}(R)\right) dR \\
&+ K_e \int_{1+\epsilon_1}^{\infty} \left(1 + \frac{1}{R} + \frac{1}{R^2} + \cdots\right)\left(k_{max}\bar{p}_{max}(R) - k_{min}\bar{p}_{min}(R)\right) dR \\
&\qquad + K_{e1} + K_{e2}
\end{aligned} \tag{2.81}$$

where, k_{max}, k_{min} *are as in equations* (2.63), (2.62),

$$K_e = \frac{1}{|C|\,|S_o|\,\sqrt{det(D)}} \tag{2.82}$$

$$\bar{p}_{min}(R) = \frac{1}{c_{min}} e^{\frac{-R^2}{2\sigma_{scmin}^2}} \tag{2.83}$$

$$\bar{p}_{max}(R) = \frac{1}{c_{max}} e^{\frac{-R^2}{2\sigma_{scmax}^2}} \tag{2.84}$$

$$\sigma_{scmax}^2 = \frac{\sigma_{max}^2}{|S_o|^2} \tag{2.85}$$

$$\sigma_{scmin}^2 = \frac{\sigma_{min}^2}{|S_o|^2} \tag{2.86}$$

K_{e1} and K_{e2} are vanishingly small constants related to the size of the exclusion zone.

Proof. See [29].

■

Remark 9. The integrals in equation (2.81) can be evaluated by using the Error Function (erf(x)).

■

Remark 10. Note that the true bias, $\mathbf{E}\{\tilde{G}\}$, is displaced from $\mathbf{E}\{\tilde{G}_h\}$ both in magnitude and phase but rapidly approaches the bias, $\mathbf{E}\{\tilde{G}_h\}$, obtained with asymptotic radial symmetry. Hence, for all practical purposes it suffices to use the simpler expression $\mathbf{E}\{\tilde{G}_h\}$.

■

Remark 11. It is also evident from equation (2.80) that the bias for the asymptotic symmetrical case essentially lies on a straight line between G_o, the true model, and $\frac{-1}{C}$. The amount the estimate is displaced from this line is due to the computable terms K_b1 and K_b2 which are dependent on the size of the chosen exclusion zone around the singularity, and hence can be made arbitrarily small. The bias result is implicit in earlier publications [13] but has been defined via the Cauchy Central Value. We have rigorously accounted for the singularity in the expression by using an exclusion zone and have accounted for the asymmetry in the distribution for $\hat{T}$.

We can also rewrite the bias for the asymptotic radially symmetric case in the following informative fashion:

$$\mathbf{E}\left\{\hat{G}_h\right\} = \mathbf{E}\left\{G_o + \tilde{G}_h\right\} \tag{2.87}$$

$$= G_o - \frac{1}{CS_o} + \frac{1}{c_1 CS_o}\left(1 - e^{-KN}\right) + K_{b1} + K_{b2} \tag{2.88}$$

$$= \frac{G_o\left(1 - e^{-KN}\right)}{c_1} + \frac{\left(1 - e^{-KN}\right) - c_1}{c_1 C} + K_{b1} + K_{b2} \tag{2.89}$$

$$= \beta G_o + (1 - \beta)\left(\frac{-1}{C}\right) + K_{b1} + K_{b2} \tag{2.90}$$

where

$$\beta = \frac{\left(1 - e^{-KN}\right)}{c_h} \tag{2.91}$$

$$K = \frac{-\left(1 - \epsilon_1\right)^2}{2\sigma_{sc}^2} \tag{2.92}$$

■

Remark 12. We see for equation (2.80) that a rapid bifurcation occurs between an unbiased and a biased estimate depending on the Signal to Noise Ratio (SNR) as captured by σ_{sc}^2. Shown in Figure 2.6 is a plot of Bias Vs $\frac{1}{\mathrm{SNR}}$ which displays this rapid transition.

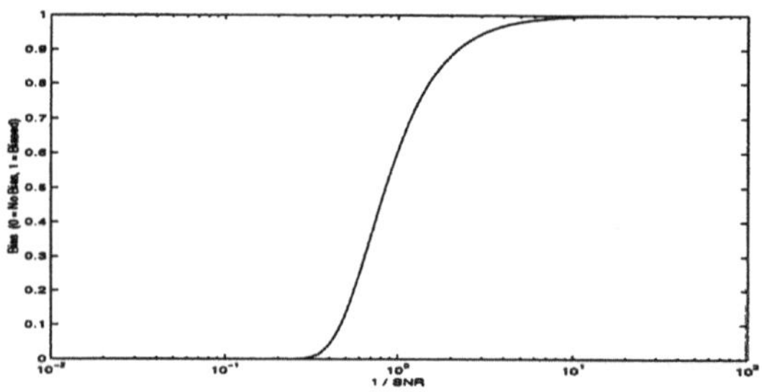

Fig. 2.6. Bias Vs $\frac{1}{\mathrm{SNR}}$

■

Corollary 2. *The estimate is asymptotically unbiased, specifically:*

$$\lim_{N\to\infty} \left|\mathbf{E}\left\{\tilde{G}\right\}\right| = 0 \tag{2.93}$$

Note that this corresponds to the heuristic result in section 2.7.

Proof. See [29].

■

Remark 13. The rate of which the bias (equation (2.80)) converges to zero is of an exponential order.

$$\lim_{N\to\infty} \left|\mathbf{E}\left\{\tilde{G}\right\}\right| = \mathcal{O}\left(e^{-N}\right) \tag{2.94}$$

■

2.8.4 Quantification of Finite Sample Variance

In this section we develop an expression for the variance of the estimator described in Section 2.6.

Here, the skewed distribution caused by the finite samples is overbounded using the maximum value of the probability density function to obtain our result.

Theorem 3. *In a closed loop indirect identification experiment the variance of the estimation error in the estimate of the plant can be bounded as follows:*

$$\mathbf{E}\left\{\tilde{G}\tilde{G}^*\right\} \leq K_v \int_0^{1-\epsilon_1} R^2\left(1+R^2+R^4+\cdots\right) k_{max}\bar{p}_{max}(R)dR$$
$$+ K_v \int_{1+\epsilon_1}^{\infty} \left(1+\frac{1}{R^2}+\frac{1}{R^4}+\cdots\right) k_{max}\bar{p}_{max}(R)dR$$
$$+ K_{v1} + K_{v2} \qquad (2.95)$$

where k_{max} is as in equation (2.63),

$$K_v = \frac{k_{max}}{|C|^2\,|S_o|^2} \qquad (2.96)$$

$$\bar{p}_{max}(R) = \begin{cases} \frac{R}{c_{max}\sigma^2_{scmax}} e^{\frac{-R^2}{2\sigma^2_{scmax}}} & for(R,\lambda) \in (\mathbb{C}-A) \\ 0 & for(R,\lambda) \in A. \end{cases} \qquad (2.97)$$

$$(2.98)$$

and K_{v1} and K_{v2} are vanishingly small constants related to the size of the exclusion zone.

Proof. See [29].

■

Remark 14. The integrals in (2.95), which involve the power series, can be evaluated using the following reduction formulae. These reduction formulae are easily derived using integration by parts.

$$\int x^m e^{-x^2}\,dx = \frac{-x^{m-1}}{2}e^{-x^2} + \frac{m-1}{2}\int x^{m-2}e^{-x^2}\,dx \qquad (2.99)$$

$$\int x^{-m} e^{-x^2}\,dx = \frac{x^{1-m}}{1-m}e^{-x^2} + \frac{2}{1-m}\int x^{2-m}e^{-x^2}\,dx \qquad (2.100)$$

■

Corollary 3. *As the number of observations (N) tends to infinity the normalised variance of the estimation error converges to a value of two times the ratio of the noise spectral density to the output energy due to the test signal at frequency ω_i. Specifically the following result holds:*

$$\lim_{N\to\infty}\left|\frac{\mathbf{E}\left\{N\tilde{G}\tilde{G}^*\right\}}{GG^*}-\frac{4S_1(\omega_i)}{|T_o|^2\,\Delta A_i^2}\right|=0 \tag{2.101}$$

Note that this confirms the heuristic result presented in section 2.7.

Proof. See [29].

■

Remark 15. The rate of which the variance (equation (2.95)) converges to the ratio of the noise spectral density and the energy present in the output due to the excitation signal applied at the reference is of the order of $\frac{1}{N}$.

$$\lim_{N\to\infty}\left|\mathbf{E}\left\{\tilde{G}\tilde{G}^*\right\}\right|=\mathcal{O}\left(\frac{1}{N}\right) \tag{2.102}$$

■

Remark 16. The result in Corollary 3 is the same as the low order approximations for the variance given in section 2.7 - see also [11] and [27].

■

2.9 Simulation

To test the theory presented here, we reexamine the system of section 2.4 with $G=\frac{1}{s+3}$ and $C=\frac{-4(s-10)}{10(s-1)}$.

A test signal consisting of Schroeder phased sinusoids was applied to the reference input with their amplitude made inversely proportional to their frequency. Additive noise was normally distributed with zero mean and a spectral density = 0.1. An experiment length of 80 seconds was used with a sampling period of 5 msecs. A Monte Carlo study was carried out utilising 2500 tests with different noise seeds. The averaged estimated gains are shown in Figure 2.7 together with the theoretical bias obtained using the results of section 2.6.

From Figure 2.7 it can be seen that as the amplitude of the test signal decreases, then the estimate becomes biased towards $\frac{-1}{C}$. The bias was estimated using the result for the asymptotic symmetrical distribution. The plot shows that this bias estimate predicts the behaviour of the bias to a high degree of accuracy in our estimate, $\hat{G}$, under the test conditions.

2.10 Conclusion

This chapter has discussed two singularity problems which occur in estimation using closed loop data. Remedies for these singularity issues have been presented for the special case when the controller is a known linear transfer function.

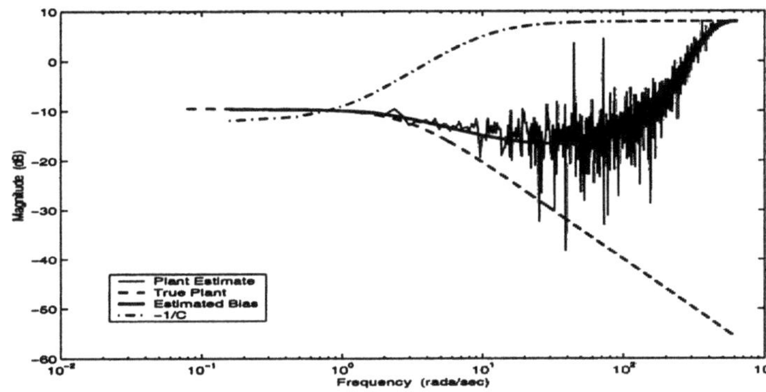

Fig. 2.7. Estimated Bias

References

1. H. Akaike. On the use of a linear model for the identification of feedback systems. *Ann. Inst. Stat. Math.*, 20:425–439, 1968.
2. K.J. Åström and T. Hagglund. *PID Controllers: Theory, Design and Tuning. 2nd Edition.* Instrument Society of America, America, 1995.
3. B.D.O. Anderson B. Codrons and M. Gevers. Closed-loop identification with an unstable or nonminimum phase controller. *Proceedings IFAC System Identification*, June 2000.
4. V.N. Fetisov B.G. Vorchik and S.E. Schteinberg. Identification of closed loop stochastic systems. *Automation Rem. Control*, 34(7), July 1973.
5. T. Bohlin. On the problem of ambiguities in maximum likelihood identification. *Automatica*, 7:199–210, 1971.
6. G.E.P. Box and J.F. Macgregor. The analysis of closed loop dynamic stochastic systems. *Tech Report 309, Dept. Stat. University of Wisconsin*, 1972.
7. P.M.T. Broersen. A comparison of transfer function estimators. *IEEE Trans. Instrumentation and Measurement*, 44:657–661, 1995.
8. U. Forssell. *Closed-loop Identification: Methods, Theory and Applications.* Phd. thesis, Linkoping University, 1999.
9. G.C. Goodwin, S.F. Graebe, and M. Salgado. *Principles of Control System Design.* Prentice Hall, Englewood Cliffs, to appear 2000.
10. G.C. Goodwin and R.L. Payne. *Dynamic System Identification: Experiment Design and Data Analysis.* Academic Press, New York, 1977.
11. G.C. Goodwin and J.S. Welsh. Analysis of a novel method of autotuning a multivariable plant based on quantisation. *Proceedings American Control Conference*, pages 3347–3351, June 1999.
12. C.W.J. Granger. Economic process involving feedback. *Inform. Control*, 6:28–48, 1963.
13. W. Heath. Statistical properties of non-parametric transfer function estimates for open and closed loop systems. *CIDAC Technical Report 9902, University of Newcastle*, 1999.
14. L. Ljung I. Gustavsson and T. Söderström. Identification of process in closed loop: Identifiability and accuracy aspects. *Automatica*, 13:59–76, 1977.

15. I.Gustavsson L. Ljung and T. Söderström. Identification of linear multivariablesystem operating under linear feedback control. *IEEE Trans. Aut. Control*, AC-19(6), Dec. 1974.
16. E.C. Levy. Complex curve fitting. *IRE Trans. Aut. Control*, AC-4:37–44, 1959.
17. L. Ljung. *System Identification: Theory for the User.* Prentice Hall, Upper Saddle River, New Jersey, 2nd edition, 1999.
18. T. McKelvey. Frequency domain identification. *Proceedings IFAC System Identification*, June 2000.
19. I. Kollar P. Guillaume and R. Pintelon. Statistical analysis of nonparametric transfer function estimates. *IEEE Trans. Instrumentation and Measurement*, 45(2):594–600, 1996.
20. M.S. Phadke and S.M. Wu. Identification of multi input multi output transfer function and noise model of a blast furnace from closed-loop data. *IEEE Trans. Aut. Control*, AC-19(6):944, 1974.
21. M.B. Priestley. *Spectral Analysis and Time Series.* Academic Press, New York, 1981.
22. T. Söderström and P. Stoica. *System Identification.* Prentice Hall, Hertfordshire (UK), 1989.
23. I. Gustavsson T. Söderström and L. Ljung. Identifiability conditions for linear systems opearting in closed loop. *Int. J. Control*, 21(2):243–255, 1975.
24. G.C. Goodwin T.S. Ng and B.D.O. Anderson. Identifiability of mimo linear dynamic systems operating in closed loop. *Automatica*, 13:477–485, 1977.
25. V.G. Vorchik. Plant identification in a closed loop system. *Aut. Telemekhanika*, 4:32–48, April 1975.
26. P. Wellstead. Reference signals for closed loop identification. *Int. J. Control*, 26(6):945–962, 1977.
27. P. Wellstead. Non-parametric methods of system identification. *Automatica*, 17(1):55–69, 1981.
28. P.E. Wellstead and J.M. Edmunds. Least-squares identification of closed loop systems. *Int. J. Control*, 21(4):688–699, 1975.
29. J.S. Welsh and G.C. Goodwin. Finite sample properties of indirect nonparametric closed loop identification. *CIDAC Technical Report 0030, University of Newcastle*, 2000.

3 Mapping Parametric Confidence Ellipsoids to Nyquist Plane for Linearly Parametrized Transfer Functions*

Xavier Bombois[1], Brian D.O. Anderson[2], and Michel Gevers[1]

[1] Centre for Systems Engineering and Applied Mechanics (CESAME) Université Catholique de Louvain, B-1348 Louvain-la-Neuve, Belgium
[2] Department of Systems Engineering, Research School of Information Sciences and Engineering, The Australian National University, Canberra ACT 0200, Australia

Abstract. This chapter considers linearly parametrized plants whose parameters are normally distributed and addresses the problem of analyzing the image in the Nyquist plane of a set of these plants defined by a confidence ellipsoid in the parameter space. The image in the Nyquist plane of such set of plants is made up of ellipses at each frequency. However, the connection between different frequencies makes the mapping nontrivial. We show that the probability level linked to this image in the Nyquist plane is larger than that of the confidence region in the parameter space. This is due to the fact that the mapping between the parametric and frequency domain spaces is not bijective.

3.1 Introduction

In many recent works [3,8,12,2], robustness analysis and robust control design have been achieved on frequency domain uncertainty regions containing the true system at a prescribed probability level. This frequency domain uncertainty region represents the frequency responses of parametrized transfer functions [10,9,7]. The parameters of these transfer functions have a Gaussian probability density function that is the result of a prediction error identification experiment. It is therefore important to understand the properties of the mapping from parameter space to Nyquist plane. In this paper, we analyze the particular case of linearly parametrized model structures, which is the case treated e.g. in [8,12,2,7]. For that particular case, we deduce the link between the frequency domain and parametric representations, their differences and the consequences of these differences on the probability level.

* The authors acknowledge the Belgian Programme on Inter-university Poles of Attraction, initiated by the Belgian State, Prime Minister's Office for Science, Technology and Culture. The scientific responsibility rests with its authors. The research reported here has been performed jointly by the three authors. However, the writing has been done by X. Bombois and M. Gevers as a surprise gift to B.D.O. Anderson. Let him share the credit for the ideas, and not be blamed for the text.

For model structures that are linear in the parameter vector θ, the image in the Nyquist plane of a parametric confidence region defined by an ellipsoid U_θ in the parameter space is an uncertainty region made up of ellipses at each frequency in the Nyquist plane. However, the mapping between the parametric and frequency domains is not bijective. We prove that the corresponding uncertainty region in the frequency domain is the image of more parameters θ than those in U_θ and that the probability level linked to the frequency domain confidence region is larger than the one linked to U_θ. We also consider the inverse mapping. We show that, if we consider the ellipses in the Nyquist plane frequency by frequency, they are the image of sets of parameters θ that are different at each frequency.

Chapter outline. In Section 2, we define the set $\mathcal{D}$ that contains the linearly parametrized systems whose parameter vector is constrained to lie in an ellipsoid. In Section 3, we present two theorems that describe the image of an ellipsoid by a nonbijective mapping, as well as the inverse image defined by such mapping. In Section 4, we present the frequency domain set $\mathcal{L}$, image of the set $\mathcal{D}$ in the Nyquist plane. In Section 5, we analyze the inverse image of the set $\mathcal{L}$. In Section 6, we define the probability level linked to $\mathcal{L}$ and give the value of this probability level. In Section 7, we give some comments about the case of model structures that are not linearly parametrized and we finish by an illustration and some conclusions.

3.2 Problem statement

As stated in the introduction, we consider linearly parametrized transfer functions. The case of nonlinearly parametrized transfer functions will be briefly discussed in Section 7. Let us thus consider the following system description:

$$G(z,\theta) = \bar{G}(z) + \Lambda(z)\theta \tag{3.1}$$

with $\theta \in \mathbf{R}^{k\times 1}$ the parameter vector, $\bar{G}(z)$ a known transfer function and $\Lambda(z)$ a known row vector of transfer functions. Let us further assume that θ has a Gaussian probability density function with zero mean[1] and covariance $P_\theta \in \mathbf{R}^{k\times k}$ i.e.

$$\theta \sim \mathcal{N}(0, P_\theta) \tag{3.2}$$

We have therefore:

[1] In the case where $\theta \sim \mathcal{N}(\hat{\theta}, P_\theta)$, one can always write $G(z,\theta) = \bar{G} + \Lambda\hat{\theta} + \Lambda\tilde{\theta} = \bar{G}_{bis} + \Lambda\tilde{\theta}$ with $\tilde{\theta} = \theta - \hat{\theta} \sim \mathcal{N}(0, P_\theta)$

$$\theta^T P_\theta^{-1} \theta \sim \chi^2(k) \tag{3.3}$$

where $\chi^2(k)$ is the chi-square probability density function with k degrees of freedom.

Let us now write the frequency response $g(e^{j\omega}, \theta)$ of $G(z, \theta)$ at the frequency ω in the following form:

$$\begin{aligned} g(e^{j\omega}, \theta) &\triangleq \begin{pmatrix} Re(G(e^{j\omega}, \theta)) \\ Im(G(e^{j\omega}, \theta)) \end{pmatrix} \\ &= \overbrace{\begin{pmatrix} Re(\bar{G}(e^{j\omega})) \\ Im(\bar{G}(e^{j\omega})) \end{pmatrix}}^{\bar{g}(e^{j\omega})} + \overbrace{\begin{pmatrix} Re(\Lambda(e^{j\omega})) \\ Im(\Lambda(e^{j\omega})) \end{pmatrix}}^{T(e^{j\omega})} \theta \end{aligned} \tag{3.4}$$

The frequency response vector $g(e^{j\omega}, \theta)$ has thus a Gaussian probability density function with mean $\bar{g}(e^{j\omega})$ and covariance $P_g(\omega) = cov((g(e^{j\omega}, \theta) - \bar{g}(e^{j\omega}))(g(e^{j\omega}, \theta) - \bar{g}(e^{j\omega}))^T) = T(e^{j\omega}) P_\theta T(e^{j\omega})^T \in \mathbf{R}^{2\times 2}$. We have thus

$$\begin{aligned} g(e^{j\omega}, \theta) &\sim \mathcal{N}(\bar{g}(e^{j\omega}), P_g(\omega)) \\ g(e^{j\omega}, \theta)^T P_g(\omega)^{-1} g(e^{j\omega}, \theta) &\sim \chi^2(2) \end{aligned} \tag{3.5}$$

The results presented in (3.5) are very common and can e.g. be found in [7]. However, these results do not give a response to some important questions. If we design a confidence ellipsoid in the parameter space using (3.3), is the image of such confidence ellipsoid in the Nyquist plane a confidence region with the same probability level? How can we relate this image with the known probability density function of the frequency response (3.5)? If we design a confidence ellipse at each frequency using (3.5) and define a set by connecting all these ellipses, what is the inverse image of that set in parameter space? In order to answer these questions, we will consider throughout this paper the following confidence ellipsoid in parameter space and the corresponding region in transfer function space. We will choose a probability level of 0.95 for these confidence regions.

Definition 1. Let us consider the parametrized model structure given in (3.1) and the probability density function of the parameter vector θ given in (3.2). The ellipsoid U_θ of size χ:

$$U_\theta = \{\theta \mid \theta^T P_\theta^{-1} \theta < \chi\}, \tag{3.6}$$

with χ such that $Pr(\chi^2(k) < \chi) = 0.95$, is a confidence ellipsoid of probability 0.95 in the parameter space. We define the set $\mathcal{D}$ of transfer functions that correspond to the parameters $\theta \in U_\theta$:

$$\mathcal{D} = \{G(z, \theta) \mid \theta \in U_\theta\} \tag{3.7}$$

The probability level $\alpha(\mathcal{D})$ linked to $\mathcal{D}$ is thus given by $\alpha(\mathcal{D}) \triangleq Pr(G(z,\theta) \in \mathcal{D}) = 0.95$. ■

In the next sections, we describe the image in the Nyquist plane of the uncertainty region $\mathcal{D}$ and we analyze the properties of such image, as well as its inverse image, with respect to the probability level.

3.3 Linear algebra preliminaries

We first present two theorems that describe properties of a mapping T between a real vector y and another real vector x of lower dimension. This mapping has the following expression

$$x = Ty \tag{3.8}$$

where $y \in \mathbf{R}^{k\times 1}$, $x \in \mathbf{R}^{n\times 1}$ $(n < k)$ are real vectors, and $T \in \mathbf{R}^{n\times k}$ is a real matrix of rank n.

Let us first recall a well-known lemma that will be useful to prove the first theorem.

Lemma 1. *Let us consider the partitioned symmetric positive definite matrix* $P \in \mathbf{R}^{k\times k}$*:*

$$P = \begin{pmatrix} P_{11} & P_{12} \\ P_{12}^T & P_{22} \end{pmatrix}$$

with $P_{11} \in \mathbf{R}^{n\times n}$*,* $P_{12} \in \mathbf{R}^{n\times(k-n)}$ *and* $P_{22} \in \mathbf{R}^{(k-n)\times(k-n)}$*. Let us also consider two real vectors* $x \in \mathbf{R}^{n\times 1}$ *and* $\bar{x} \in \mathbf{R}^{(k-n)\times 1}$ *and an ellipsoid* $U_{x\bar{x}}$ *defined as:*

$$U_{x\bar{x}} = \left\{ \begin{pmatrix} x \\ \bar{x} \end{pmatrix} \mid \begin{pmatrix} x \\ \bar{x} \end{pmatrix}^T P^{-1} \begin{pmatrix} x \\ \bar{x} \end{pmatrix} < 1 \right\}.$$

Then the set U_x

$$U_x \triangleq \{x \mid \begin{pmatrix} x \\ \bar{x} \end{pmatrix} \in U_{x\bar{x}}\} \tag{3.9}$$

is also an ellipsoid given by

$$U_x = \{x \mid x^T P_{11}^{-1} x < 1\} \tag{3.10}$$

Proof. The inverse of the block matrix P can be written (see e.g. [13, page 22])

$$P^{-1} = \begin{pmatrix} K_{11} & K_{12} \\ K_{12}^T & K_{22} \end{pmatrix}$$

where $K_{11} = P_{11}^{-1} + P_{11}^{-1}P_{12}\Delta^{-1}P_{12}^TP_{11}^{-1}$, $K_{12} = -P_{11}^{-1}P_{12}\Delta^{-1}$, $K_{22} = \Delta^{-1}$ and $\Delta = P_{22} - P_{12}^TP_{11}^{-1}P_{12}$.

Using these notations and introducing the vector $z = K_{22}^{-1}K_{12}^Tx + \bar{x}$, we have the following equivalences:

$$\begin{pmatrix} x \\ \bar{x} \end{pmatrix}^T P^{-1} \begin{pmatrix} x \\ \bar{x} \end{pmatrix} < 1 \Longleftrightarrow x^T(K_{11} - K_{12}K_{22}^{-1}K_{12}^T)x + z^TK_{22}z < 1$$
$$\Longleftrightarrow x^TP_{11}^{-1}x + z^TK_{22}z < 1 \tag{3.11}$$

Using this last expression, we can now write that

i. if $(x^T \ \bar{x}^T)^T \in U_{x\bar{x}}$, then $x^TP_{11}^{-1}x < 1$. Indeed

$$\begin{pmatrix} x \\ \bar{x} \end{pmatrix}^T P^{-1} \begin{pmatrix} x \\ \bar{x} \end{pmatrix} < 1 \Longrightarrow x^TP_{11}^{-1}x < (1 - z^TK_{22}z) < 1$$

ii. if $x^TP_{11}^{-1}x < 1$ then there exists $\bar{x}$ such that $(x^T \ \bar{x}^T)^T \in U_{x\bar{x}}$. Indeed, take as $\bar{x}$, the vector $\bar{x}$ such that $z = 0$ (i.e. $\bar{x} = -K_{22}^{-1}K_{12}^Tx$). Then, $\begin{pmatrix} x \\ -K_{22}^{-1}K_{12}^Tx \end{pmatrix} \in U_{x\bar{x}}$.

This completes the proof. ∎

Note that U_x is not the intersection of $U_{x\bar{x}}$ with the subspace $\bar{x} = 0$; it is a larger set. Let us now present our two theorems about the mapping T defined in (3.8).

Theorem 1. *Let us consider the mapping T defined in (3.8) and the ellipsoid U_y of size χ in the y-space:*

$$U_y = \{y \mid y^TP_y^{-1}y < \chi\}, \tag{3.12}$$

with $P_y \in \mathbf{R}^{k\times k}$ a positive definite matrix. The image U_x of U_y by the mapping T i.e. $U_x \triangleq \{x \mid x = Ty \text{ with } y \in U_y\}$ is an ellipsoid in the x-space given by

$$U_x = \{x \mid x^TP_x^{-1}x < \chi\}, \tag{3.13}$$

with $P_x = TP_yT^T \in \mathbf{R}^{n\times n}$.

Proof. Let us first complete the mapping T by generating a nonsingular mapping $\widetilde{T}$:

$$\begin{pmatrix} x \\ \bar{x} \end{pmatrix} = \overbrace{\begin{pmatrix} T \\ \overline{T} \end{pmatrix}}^{\widetilde{T}} y \tag{3.14}$$

such that $\widetilde{T} \in \mathbf{R}^{k\times k}$ has rank k. Using $\widetilde{T}$, we have that

$$y^T P_y^{-1} y < \alpha \Longleftrightarrow \begin{pmatrix} x \\ \bar{x} \end{pmatrix}^T \overbrace{\widetilde{T}^{-T} P_y^{-1} \widetilde{T}^{-1}}^{P^{-1}} \begin{pmatrix} x \\ \bar{x} \end{pmatrix} < \chi \tag{3.15}$$

Proving Theorem 1 is thus equivalent to proving that (3.13) is the domain where x is constrained to lie when (3.15) holds. This follows immediately from Lemma 1, noting that if $P = \widetilde{T} P_y \widetilde{T}^T$, then $P_x = P_{11} = T P_y T^T$. ∎

Theorem 2. *Let us consider the mapping T and the ellipsoids U_y and U_x defined in (3.8), (3.12) and (3.13), respectively. Define the inverse image C_y of U_x using the mapping T as*

$$C_y \triangleq \{y \mid x = Ty \in U_x\}, \tag{3.16}$$

Then C_y is a volume given by

$$C_y = \{y \mid y^T R_C y < \chi\}, \tag{3.17}$$

with $R_C = T^T P_x^{-1} T$, a singular matrix $\in \mathbf{R}^{k\times k}$. Moreover, the volume C_y has the following properties:

- *The matrix R_C defining C_y has rank n i.e. it has $k-n$ zero eigenvalues. The volume C_y has therefore $k-n$ infinite main axes. The directions y_i ($i = 1...k-n$) of these infinite main axes are the eigenvectors corresponding to the null eigenvalues of R_C. Moreover, these eigenvectors y_i belong to the null space of T i.e. $Ty_i = 0$.*
- *The ellipsoid U_y is included in C_y.*

Proof. We first prove that the inverse image of U_x by the mapping (3.8) is given by (3.17). This follows directly from:

$$x^T P_x^{-1} x < \chi \Longleftrightarrow y^T T^T P_x^{-1} T y < \chi \tag{3.18}$$

The volume C_y is thus the inverse image of U_x since y has to satisfy the right-hand side of (3.18) in order to have x in U_x.

It follows from $R_C = T^T P_x^{-1} T \in \mathbf{R}^{k \times k}$ with T of rank $n < k$ that R_C has $k-n$ null eigenvalues and that the corresponding eigenvectors are in the null-space of the mapping T.

Theorem 1 and the definition (3.16) of C_y show that U_y is included in C_y. Indeed, we know by Theorem 1 that each y in U_y has an image (i.e. Ty) in U_x. Therefore, each y in U_y lies in C_y defined by (3.16). ■

Comments.

- Since the matrix T has rank $n < k$, the mapping (3.8) is not bijective. This explains the fact that the image of U_y by the mapping (3.8) is exactly U_x and that the inverse image of U_x is a larger volume C_y containing U_y.
- In the particular case where $k = 3$ and $n = 2$, U_x is then an ellipse (Theorem 1) and C_y is a cylinder with infinite axis. The axis of the cylinder is in the direction of the eigenvector corresponding to the single null eigenvalue (Theorem 2).

3.4 Image of $\mathcal{D}$ in the Nyquist plane

Theorem 1 tells us that the image of an ellipsoid by a linear mapping into a smaller dimensional space is also an ellipsoid. This theorem will now be used in order to find the frequency domain region (or dynamic region) that is the image of $\mathcal{D}$ in the Nyquist plane. This frequency domain region is defined via a constraint on the frequency response of the plants in this region at every frequency. The general expression of a frequency domain region can e.g. be written as follows:

$$\mathcal{L} = \{G(z) \mid g(e^{j\omega}) \in U(\omega) \ \forall \omega\}, \tag{3.19}$$

where $g(e^{j\omega}) = \left(Re(G(e^{j\omega})) \ Im(G(e^{j\omega})) \right)^T$ and $U(\omega)$ is the particular domain where the frequency response vector of the plants $G(z) \in \mathcal{L}$ is constrained to lie at the frequency ω.

We are thus looking for the frequency domain region $\mathcal{L}$ that corresponds to the image of the set $\mathcal{D}$ in the Nyquist plane. Let us first define this notion properly.

Definition 2. Consider the set $\mathcal{D}$ of transfer functions defined in (3.7) and the general expression of a frequency domain region $\mathcal{L}$ given in (3.19). The image of $\mathcal{D}$ in the Nyquist plane is the frequency domain region $\mathcal{L}$ defined by (3.19) with $U(\omega)$ defined as follows, at each frequency ω:

$$U(\omega) = \{g(e^{j\omega}) \mid g(e^{j\omega}) = g(e^{j\omega}, \theta) \ for\ some\ \theta \in U_\theta\} \tag{3.20}$$

with $g(e^{j\omega}, \theta)$ defined in (3.4). ■

Important comments. Definition 2 tells us

- that the image $\mathcal{L}$ of $\mathcal{D}$ in the Nyquist plane is a set containing the image of all plants in $\mathcal{D}$;
- that all "points $g(e^{j\omega}) \in U(\omega)$" at a frequency ω are the image of some plant in $\mathcal{D}$.

However, if we randomly select frequency functions $f(e^{j\omega}) \in \mathcal{L}$, for $\omega \in [0\ \pi]$, then most of such functions will not be in $\mathcal{D}$, i.e. for most of such functions $f(e^{j\omega}) \in \mathcal{L}$, there will not exist a θ such that $f(e^{j\omega}) = g(e^{j\omega},\theta)\ \forall\omega$ with $g(e^{j\omega},\theta)$ defined by (3.4).

Using the mapping (3.4) between the space of parametrized transfer functions $G(z,\theta)$ (or parameter space) and the frequency domain space, and the results of Theorem 1, we can construct an explicit expression of the image $\mathcal{L}$ of $\mathcal{D}$ in the Nyquist plane.

Theorem 3. *Consider the set $\mathcal{D}$ of transfer functions $G(z,\theta) = \bar{G}(z) + \Lambda(z)\theta$ presented in Definition 1, and the mapping (3.4) between parameter space and frequency domain space. The image of $\mathcal{D}$ in the Nyquist plane (see Definition 2) is a frequency domain region $\mathcal{L}$ having the following expression.*

$$\mathcal{L} = \{G(z) \mid g(e^{j\omega}) \in U(\omega)\ \forall\omega\} \tag{3.21}$$

$$U(\omega) = \{g \in \mathbf{R}^{2\times 1} \mid (g - \bar{g}(e^{j\omega}))^T P(\omega)^{-1} (g - \bar{g}(e^{j\omega})) < \chi\} \tag{3.22}$$

with $P(\omega) = T(e^{j\omega}) P_\theta T(e^{j\omega})^T$,

$$g(e^{j\omega}) = \begin{pmatrix} Re(G(e^{j\omega})) \\ Im(G(e^{j\omega})) \end{pmatrix} \text{ and } \bar{g}(e^{j\omega}) = \begin{pmatrix} Re(\bar{G}(e^{j\omega})) \\ Im(\bar{G}(e^{j\omega})) \end{pmatrix}.$$

The image $\mathcal{L}$ of $\mathcal{D}$ in the Nyquist plane is thus made up of ellipses $U(\omega)$ at each frequency around the frequency response of the known transfer function $\bar{G}(z)$. The ellipse $U(\omega)$ at a particular frequency can therefore be considered as the image of $\mathcal{D}$ in the Nyquist plane at this frequency.

Proof. In order to establish the proof of Theorem 3, we need to prove that the expression (3.22) of $U(\omega)$ is equivalent with (3.20). The results follow directly from Theorem 1 by considering the mapping (3.4) (i.e. $g(e^{j\omega},\theta) - \bar{g}(e^{j\omega}) = T(e^{j\omega})\theta$) at a particular frequency ω. ∎

Remarks. It is to be noted that the matrix $P(\omega)$ defining $U(\omega)$ is equal to the covariance matrix $P_g(\omega)$ of $g(e^{j\omega},\theta)$ (see (3.5)). It is also to be noted that, at the frequencies $\omega = 0$ and $\omega = \pi$, the ellipse $U(\omega)$ degenerates into a line segment.

3.5 Inverse image of $\mathcal{L}$

In the previous section, we have determined the frequency domain region $\mathcal{L}$, image of the set $\mathcal{D}$ of parametrized transfer functions $G(z, \theta)$. This set $\mathcal{L}$, made up of ellipses $U(\omega)$ at each frequency, is defined by the property (3.20). In particular, $\mathcal{L}$ contains all plants in $\mathcal{D}$. The set $\mathcal{L}$ is nevertheless not equivalent to $\mathcal{D}$. Indeed, we prove that there are more plants in $\mathcal{L}$ than those in $\mathcal{D}$. These additional plants are plants having a structure different from $G(z, \theta)$ (i.e. they cannot be described as $G(z, \theta)$ for any θ (see (3.1))), but also plants having the structure $G(z, \theta)$ but for $\theta \notin U_\theta$.

In this chapter, we will focus on the additional plants in $\mathcal{L}$ having the structure $G(z, \theta)$ given in (3.1) but for $\theta \notin U_\theta$. The fact that such additional plants exist in $\mathcal{L}$ is a consequence of the fact that the mapping (3.4) is not bijective[2] since (3.4) maps a k-dimensional space into the 2-dimensional frequency domain space. In order to establish that additional plants $G(z, \theta)$ lie in $\mathcal{L}$, the inverse image of $\mathcal{L}$ in the space of parametrized transfer functions $G(z, \theta)$ has to be determined. For this purpose, it is useful to first analyze the inverse image $\mathcal{D}(U(\omega))$, via the mapping (3.4), of one ellipse $U(\omega)$ of $\mathcal{L}$ in the space of parametrized transfer functions $G(z, \theta)$.

Proposition 1. *Consider a particular frequency ω and the ellipse $U(\omega)$ defined in (3.22) which is the image of the set $\mathcal{D}$ in the Nyquist plane at the frequency ω. Using the mapping (3.4) from θ to $g(e^{j\omega}, \theta)$, define the inverse image of $U(\omega)$ in the parameter space as*

$$C_\theta(U(\omega)) = \{\theta \mid g(e^{j\omega}, \theta) \in U(\omega)\}. \tag{3.23}$$

Correspondingly, define the inverse image of $U(\omega)$ in the space of parameterized transfer functions $G(z, \theta)$ as

$$\mathcal{D}(U(\omega)) = \{G(z, \theta) \mid g(e^{j\omega}, \theta) \in U(\omega)\}. \tag{3.24}$$

Then the set $C_\theta(U(\omega))$ is a volume in the θ-space with $k-2$ infinite axes defined as:

$$C_\theta(U(\omega)) = \{\theta \in \mathbf{R}^{k\times 1} \mid \theta^T T(e^{j\omega})^T P(\omega)^{-1} T(e^{j\omega})\theta < \chi\}. \tag{3.25}$$

Moreover, $U_\theta \subset C_\theta(U(\omega))$ and $\mathcal{D} \subset \mathcal{D}(U(\omega))$.

Proof. The expression (3.25) of $C_\theta(U(\omega))$ follows directly from Theorem 2 by substituting $U(\omega)$ for U_x, U_θ for U_y and $C_\theta(U(\omega))$ for C_y. It then follows

[2] The mapping $T(e^{j\omega})$ is only bijective if the size k of the vector θ is equal to two.

from the last part of Theorem 2 that U_θ is a subset of $C_\theta(U(\omega))$. Now observe from (3.23) and (3.24) that $\mathcal{D}(U(\omega))$ can equivalently be described as

$$\mathcal{D}(U(\omega)) = \{G(z,\theta) \mid \theta \in C_\theta(U(\omega))\} \tag{3.26}$$

It then follows from $U_\theta \subset C_\theta(U(\omega))$ and the definitions (3.7) and (3.26) that $\mathcal{D} \subset \mathcal{D}(U(\omega))$. ■

Proposition 1 tells us that the ellipse $U(\omega)$ is the image of more plants $G(z,\theta)$ than those in $\mathcal{D}$. These additional plants $G(z,\theta_{out})$ with $\theta_{out} \in C_\theta(U(\omega)) \setminus U_\theta$, have the property that $\exists\, \theta_{in} \in U_\theta$ such that, *at frequency* ω,

$$g(e^{j\omega},\theta_{out}) = g(e^{j\omega},\theta_{in}),$$

since $U(\omega)$ is defined by (3.20).

It is also important to note that the inverse image $\mathcal{D}(U(\omega))$ of $U(\omega)$ in the space of parametrized transfer functions $G(z,\theta)$ is different at each frequency, because the inverse image $C_\theta(U(\omega))$ in parameter space is different at each frequency. In other words, $U(\omega)$ is the image of a set $\mathcal{D}(U(\omega))$ of plants $G(z,\theta)$ that are different at each frequency.

In Proposition 1, we have computed the inverse image $C_\theta(U(\omega))$ in parameter space of one ellipse $U(\omega)$, via the inverse of mapping (3.4). We now determine the inverse image $U_\theta(\mathcal{L})$ in parameter space of the whole set $\mathcal{L}$ defined by (3.21) and (3.22).

Theorem 4. *Consider the frequency domain set $\mathcal{L}$ defined by (3.21) and (3.22). Define the inverse image $U_\theta(\mathcal{L})$ of $\mathcal{L}$ in parameter space, via the mapping (3.4), as:*

$$U_\theta(\mathcal{L}) = \{\theta \mid G(z,\theta) \in \mathcal{L}\}. \tag{3.27}$$

Then

$$U_\theta(\mathcal{L}) = \bigcap_{\omega\in[0\ \pi]} C_\theta(U(\omega)), \tag{3.28}$$

where $C_\theta(U(\omega))$ is defined in (3.23) and (3.25). Moreover,

$$U_\theta \subseteq U_\theta(\mathcal{L}). \tag{3.29}$$

Proof. First observe that, by the definition of $\mathcal{L}$ in (3.21), the set $U_\theta(\mathcal{L})$ defined in (3.27) is equivalent with

$$U_\theta(\mathcal{L}) = \{\theta \mid g(e^{j\omega},\theta) \in U(\omega)\ \forall\omega\}.$$

The result (3.28) then follows immediately from Definition (3.23). The inclusion (3.29) then follows from the main result of Proposition 1, namely $U_\theta \subset C_\theta(U(\omega))\ \forall\omega$. ■

Corollary 1. *Consider the frequency domain set $\mathcal{L}$ defined by (3.21) and (3.22). Define the inverse image $\mathcal{D}(\mathcal{L})$ of $\mathcal{L}$ in the space of parametrized transfer functions $G(z,\theta)$, via the mapping (3.4), as*

$$\mathcal{D}(\mathcal{L}) = \{G(z,\theta) \mid G(z,\theta) \in \mathcal{L}\}. \tag{3.30}$$

Then $\mathcal{D} \subseteq \mathcal{D}_{\mathcal{L}}$.

Proof. By (3.30) and (3.27), it follows that

$$\mathcal{D}(\mathcal{L}) = \{G(z,\theta) \mid \theta \in U_\theta(\mathcal{L})\}. \tag{3.31}$$

The result then follows from the result (3.29) of Theorem 4, and the definition (3.7) of $\mathcal{D}$. ∎

Corollary 2. *With definitions as above, we have:*

$$U_\theta \subseteq U_\theta(\mathcal{L}) \subset C_\theta(U(\omega)) \ \forall\omega \tag{3.32}$$

$$\mathcal{D} \subseteq \mathcal{D}(\mathcal{L}) \subset \mathcal{D}(U(\omega)) \ \forall\omega. \tag{3.33}$$

Proof. The first inclusions follow from Theorem 4 and Corollary 1. The second inclusion in (3.32) follows from (3.28), and the second inclusion of (3.33) from (3.31), (3.26) and (3.32). ∎

Theorem 4 tells that the ellipsoid U_θ which defines $\mathcal{D}$ is a subset of $U_\theta(\mathcal{L}) = \bigcap_{\omega\in[0\ \pi]} C_\theta(U(\omega))$. We shall illustrate by an example in Section 8 that it may be a strictly proper subset of $U_\theta(\mathcal{L})$. As a consequence, $\mathcal{D}$ may be a strictly proper subset of $\mathcal{D}(\mathcal{L})$, and the frequency domain region $\mathcal{L}$ is therefore the image in the Nyquist plane of a set $\mathcal{D}(\mathcal{L})$ containing more plants $G(z,\theta)$ than those in $\mathcal{D}$. It is to be noted that, according to the definition of $\mathcal{L}$ (Definition 2), these additional plants $G(z,\theta_{out})$ with $\theta_{out} \in U_\theta(\mathcal{L}) \setminus U_\theta$, must have the property that, at each frequency ω, there exists θ_{in} in U_θ such that $G(e^{j\omega},\theta_{out}) = G(e^{j\omega},\theta_{in})$.

3.6 Probability level linked to the confidence region $\mathcal{L}$

In the previous sections, we have shown that the image of a set $\mathcal{D}$ in the Nyquist plane is a frequency domain region $\mathcal{L}$ made up of ellipses $U(\omega)$ at each frequency. We have also shown that the sets $U(\omega)$ and the whole region $\mathcal{L}$ are (or may be) the image of more plants $G(z,\theta)$ than those in $\mathcal{D}$. Let us now consider both sets (i.e. $U(\omega)$ and $\mathcal{L}$) as confidence regions. The ellipse $U(\omega)$ is a confidence region for the frequency response vector $g(e^{j\omega}, \theta)$ of the plants $G(z, \theta)$ and the set $\mathcal{L}$ is a confidence region for the plants $G(z, \theta)$. Since the parameter vector θ has a probability density function (see (3.2)), we can relate a probability level to both confidence regions.

Definition 3. Consider the parametrized transfer functions $G(z, \theta)$ given in (3.1), whose parameter vector θ has the probability density function (3.2). Consider also the sets $U(\omega)$ and $\mathcal{L}$ defined in (3.21)-(3.22). The probability level $\alpha(U(\omega))$ linked to $U(\omega)$ is defined as :

$$\alpha(U(\omega)) = Pr(g(e^{j\omega}, \theta) \in U(\omega)),$$

where $g(e^{j\omega}, \theta)$ is defined in (3.4). The probability level $\alpha(\mathcal{L})$ linked to $\mathcal{L}$ is defined as:

$$\alpha(\mathcal{L}) = Pr(G(z, \theta) \in \mathcal{L}).$$

■

These probability levels $\alpha(U(\omega))$ and $\alpha(\mathcal{L})$ will be larger than the probability level $\alpha(\mathcal{D})$ linked to $\mathcal{D}$ (i.e. $\alpha(\mathcal{D}) = 0.95$) since $\mathcal{D} \subseteq \mathcal{D}(\mathcal{L}) \subset \mathcal{D}(U(\omega))\ \forall\omega$ (see Corollary 2). Theorem 5 gives an exact computation of $\alpha(U(\omega))$, as well as upper and lower bounds for $\alpha(\mathcal{L})$.

Theorem 5. *Consider the parametrized transfer functions $G(z, \theta)$ given in (3.1), whose parameter vector θ has the probability density function (3.2). Consider also the sets $U(\omega)$ and $\mathcal{L}$ defined in (3.21)-(3.22). Then the probability level $\alpha(U(\omega))$ linked to $U(\omega)$ (see Definition 3) is given by:*

$$\alpha(U(\omega)) = Pr(G(z, \theta) \in \mathcal{D}(U(\omega))) \tag{3.34}$$

$$= Pr(\chi^2(2) < \chi)\ \forall\omega, \tag{3.35}$$

where $\mathcal{D}(U(\omega))$ is defined in (3.24). The probability level $\alpha(\mathcal{L})$ linked to $\mathcal{L}$ (see Definition 3) is bounded by:

$$\alpha(\mathcal{D}) \leq \alpha(\mathcal{L}) < \alpha(U(\omega)) \tag{3.36}$$

where $\alpha(\mathcal{D})$ is the probability level linked to the set $\mathcal{D}$ presented in Definition 1 and of which the set $\mathcal{L}$ is the image in the Nyquist plane ($\alpha(\mathcal{D}) = 0.95$).

Proof. That $\alpha(U(\omega))$ is equal to $Pr(G(z, \theta) \in \mathcal{D}(U(\omega)))$ follows from Proposition 1. That $\alpha(U(\omega))$ is also equal to (3.35) is a direct consequence of the probability density function of $g(e^{j\omega}, \theta)$ given in (3.5) since the covariance matrix $P_g(\omega)$ of $g(e^{j\omega}, \theta)$ is equal to the matrix $P(\omega)$ defining the ellipse $U(\omega)$.

Since the inverse image of $\mathcal{L}$ in the space of parametrized transfer functions $G(z,\theta)$ is $\mathcal{D}(\mathcal{L})$, we can write the following about the probability level $\alpha(\mathcal{L})$ linked to $\mathcal{L}$:

$$\alpha(\mathcal{L}) = Pr(G(z,\theta) \in \mathcal{D}(\mathcal{L})).$$

The upper bound in (3.36) proceeds then from the fact that $\mathcal{D}(\mathcal{L}) \subset \mathcal{D}(U(\omega))\ \forall\omega$ and the lower bound from the fact that $\mathcal{D} \subseteq \mathcal{D}(\mathcal{L})$ (see Theorem 4). ∎

Important comments. Theorem 5 shows that the probability level $\alpha(\mathcal{L})$ linked to the image of $\mathcal{D}$ in the Nyquist plane is larger than the probability level linked to $\mathcal{D}$ (i.e. $\alpha(\mathcal{D}) = 0.95$). This is a consequence of the fact that $\mathcal{L}$ is the image of more plants than those in $\mathcal{D}$ because of the singularity of the mapping (3.4).

It is also interesting to note that if we consider the ellipses $U(\omega)$ frequency by frequency, these ellipses are the image in the Nyquist plane of a set $\mathcal{D}(U(\omega))$, different at each frequency, and having a probability level $\alpha(U(\omega))$ which follows from the probability density function (3.5) of $g(e^{j\omega},\theta)$. However, since the sets $\mathcal{D}(U(\omega))$ are different at each frequency, when we collect together all ellipses $U(\omega)$ to make up $\mathcal{L}$, the probability level $\alpha(\mathcal{L})$ is smaller than $\alpha(U(\omega))$. This last remark shows that the probability density function of $g(e^{j\omega},\theta)$ given in (3.5) is only relevant for one particular frequency. Theorem 5 shows therefore that, in order to design a confidence region $\mathcal{L}$ with a probability level $\alpha(\mathcal{L})$ larger than 95%, one has to first design a confidence region $\mathcal{D}$ having the desired probability level (i.e. $\alpha(\mathcal{D}) = 0.95$) and then take its image $\mathcal{L}$ in the Nyquist plane.

As a consequence, in the paper [7], the probability density function of the frequency response can be used in order to design a confidence ellipse of 95% *at a particular frequency*. However, these 95%-ellipses can not be connected in order to make up a frequency domain confidence region at 95% for the parametrized transfer functions, as proposed in [1,11]. For this purpose, as said above, one has to first design a confidence region $\mathcal{D}$ having the desired probability level (i.e. $\alpha(\mathcal{D}) = 0.95$) and then take its image $\mathcal{L}$ in the Nyquist plane, as proposed in [2].

Remarks. The plants having another structure than $G(z,\theta)$ and that lie in $\mathcal{L}$ do not modify the probability level $\alpha(\mathcal{L})$ since only the parameter vector θ has a probability density function.

3.7 Case of not linearly parametrized model structures

Until now, we have treated the case of systems $G(z,\theta)$ that can be written as in (3.1) and whose parameters have the probability density function (3.2). We have shown for this type of model structure the link between a set $\mathcal{D}$ of transfer functions $G(z,\theta)$ and its image $\mathcal{L}$ in the Nyquist plane. If the model structure is not linearly parametrized as in (3.1), our conclusions do not hold i.e. the image at a frequency ω is not guaranteed to be an ellipse. In [9,6,3], a first order approximation was used to map the parametric confidence ellipsoid into ellipses in the Nyquist plane. However, using such an approach, no probablity level can be guaranteed for the obtained frequency domain region.

As a consequence, it is very difficult to have a clear idea of the image in the Nyquist plane of a set $\mathcal{D}_{gen}$ of rational transfer functions with parameters appearing in both numerator and denominator. Some partial results have been presented in [4,5]. In [4], the authors have presented a way to compute, at each frequency, the largest and the smallest modulus and phase of the plants in a region $\mathcal{D}_{gen}$. In [5], we have given an LMI procedure that computes at each frequency the smallest overbounding ellipse that contains the frequency response of the plants in such set $\mathcal{D}_{gen}$.

3.8 Simulation example

In order to illustrate the results of this chapter, we present the following example. Let us consider the following system description:

$$\begin{aligned} G(z,\theta) &= \frac{0.08z^{-1}+0.1009z^{-2}+0.0359z^{-3}}{1-1.5578z^{-1}+0.5769z^{-2}} \\ &\qquad + \frac{\theta_1 z^{-1}+\theta_2 z^{-2}+\theta_3 z^{-3}}{1-1.5578z^{-1}+0.5769z^{-2}} \\ &= \bar{G}(z) + \overbrace{\frac{1}{1-1.5578z^{-1}+0.5769z^{-2}}\begin{pmatrix} z^{-1} & z^{-2} & z^{-3}\end{pmatrix}}^{\Lambda(z)} \overbrace{\begin{pmatrix}\theta_1\\ \theta_2\\ \theta_2\end{pmatrix}}^{\theta} \end{aligned}$$

where the parameter vector θ is assumed to have a Gaussian probability density function with zero mean and covariance P_θ given by:

$$P_\theta = 10^{-3} \times \begin{pmatrix} 1.0031 & 0.0263 & -0.0111 \\ 0.0263 & 1.0039 & 0.0268 \\ -0.0111 & 0.0268 & 1.0039 \end{pmatrix}.$$

We consider the 95 % confidence ellipsoid U_θ in the parameter space that defines a corresponding region $\mathcal{D}$ in the space of transfer function:

$$U_\theta = \{\theta \mid \theta^T P_\theta^{-1} \theta < 7.81\},$$

$$\mathcal{D} = \{G(z,\theta) \mid \theta \in U_\theta\}$$

Using Theorem 3, we can design the image $\mathcal{L}$ of $\mathcal{D}$ in the Nyquist plane. This image $\mathcal{L}$ is made up of ellipses at each frequency around the frequency response of $\bar{G}(z)$ and is represented in Figure 3.1. According to Theorem 3, the expression of the ellipse $U(\omega)$ at the frequency ω is given by:

$$U(\omega) = \{g \in \mathbf{R}^{2\times 1} \mid (g - \bar{g}(e^{j\omega}))^T P(\omega)^{-1}(g - \bar{g}(e^{j\omega})) < 7.81\}$$

with $P(\omega) = T(e^{j\omega})P_\theta T(e^{j\omega})^T$ and

$$\bar{g}(e^{j\omega}) = \begin{pmatrix} Re(\bar{G}(e^{j\omega})) \\ Im(\bar{G}(e^{j\omega})) \end{pmatrix}, T(e^{j\omega}) = \begin{pmatrix} Re(\Lambda(e^{j\omega})) \\ Im(\Lambda(e^{j\omega})) \end{pmatrix}.$$

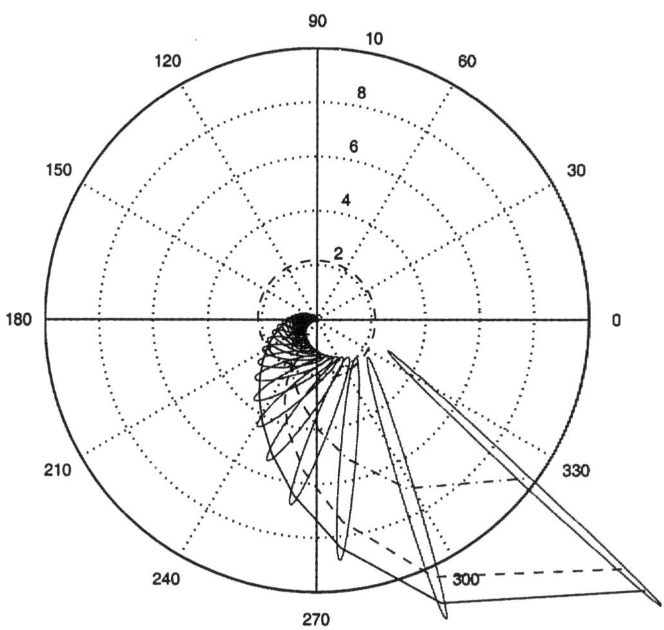

Fig. 3.1. Frequency domain representation of $\mathcal{D}$ in the Nyquist plane with ellipses $U(\omega)$ at some frequencies, frequency response of $\bar{G}(z)$ (dashdot), frequency response of $G(z,\theta_{out})$ (dashed) and frequency response of $G(z,\theta_{bis})$ (solid)

All plants in $\mathcal{D}$ lie in $\mathcal{L}$, and $\mathcal{L}$ has the property (3.20). However, the mappings between $\mathcal{D}$ and $\mathcal{L}$ and between $\mathcal{D}$ and $U(\omega)$ are not bijective as shown in Theorem 4 and Proposition 1, respectively. In order to illustrate the results presented in these theorems, we will show two things:

i. there exist plants $G(z, \theta_{out})$ outside $\mathcal{D}$ whose frequency response vector $g(e^{j\omega}, \theta_{out})$ lies in some ellipses $U(\omega)$ but not in all of them;
ii. there exist plants $G(z, \theta_{bis})$ outside $\mathcal{D}$ that lie in the whole region $\mathcal{L}$.

Since the size of θ is 3, we know that the vectors θ that are projected into $U(\omega)$ at the frequency ω are those lying in the cylinder $C_\theta(U(\omega))$ whose axis direction is given by the normed eigenvector $\theta_{null}(\omega)$ corresponding to the null eigenvalue of the mapping $T(e^{j\omega})$ (see Theorem 2 and Proposition 1). Using this property, we can find a plant $G(z, \theta_{out})$ such that $\theta_{out} \notin U_\theta$, but such that its frequency response $g(e^{j\omega_0}, \theta_{out})$ at ω_0 lies in $U(\omega_0)$ for a particular frequency ω_0, say $\omega_0 = 0.25$. Indeed, let us choose as vector θ_{out} a vector in the same direction as $\theta_{null}(0.25)$ but outside the ellipsoid U_θ:

$$\theta_{out} = \begin{pmatrix} 1.8084 \\ -3.5043 \\ 1.8084 \end{pmatrix}$$

This vector is well outside the ellipsoid U_θ since we have that:

$$\theta_{out}^T P_\theta^{-1} \theta_{out} = 19525 > 7.81$$

but we also have that:

$$g(e^{j0.25}, \theta_{out}) = \bar{g}(e^{j0.25}) + \overbrace{T(e^{j0.25})\theta_{out}}^{=0} = \bar{g}(e^{j0.25}),$$

and therefore $g(e^{j0.25}, \theta_{out})$ lies in $U(0.25)$. However, this plant does not lie in all ellipses as can be seen in Figure 3.1 where it circles around the origin at high frequencies.

There also exist plants $G(z, \theta_{bis})$ whose parameter vectors $\theta_{bis} \notin U_\theta$, but that lie completely in $\mathcal{L}$. According to Theorem 4 and Corollary 1, these are the plants whose parameter vectors θ_{bis} lie in $U_\theta(\mathcal{L}) = \bigcap_{\omega \in [0\ \pi]} C_\theta(U(\omega))$. In order to find one of those particular vectors θ_{bis}, we proceed like we did to find θ_{out}. We choose a particular frequency ω_0 and we choose a vector in the direction $\theta_{null}(\omega_0)$ of the axis of the cylinder $C_\theta(U(\omega_0))$. But, here, we choose this frequency ω_0 in the middle of the frequency range: $\omega_0 = \pi/2$ and we choose the vector **just** outside the ellipsoid U_θ:

$$\theta_{bis} = \begin{pmatrix} 0.0684 \\ 0 \\ 0.0684 \end{pmatrix}, \ \theta_{bis}^T P_\theta^{-1} \theta_{bis} = 9.4501 > 7.81 .$$

In Figure 3.1, we see that the frequency response of the plant $G(z, \theta_{bis})$ lies in $U(\omega)$ for each of the plotted ellipses. Since we only plot the ellipses at a certain number of frequencies, Figure 3.1 alone does not prove that $G(z, \theta_{bis})$ is in $\mathcal{L}$. In Figure 3.2, we have therefore plotted the value of the function

$$(g(e^{j\omega}, \theta_{bis}) - \bar{g}(e^{j\omega}))^T P(\omega)^{-1} (g(e^{j\omega}, \theta_{bis}) - \bar{g}(e^{j\omega}))$$

at each frequency. We see that these values are, at each frequency, smaller than 7.81, the size of the ellipses $U(\omega)$. As a consequence, we can conclude that $G(z, \theta_{bis})$ has its frequency response in $\mathcal{L}$ even though $G(z, \theta_{bis})$ does not lie in $\mathcal{D}$.

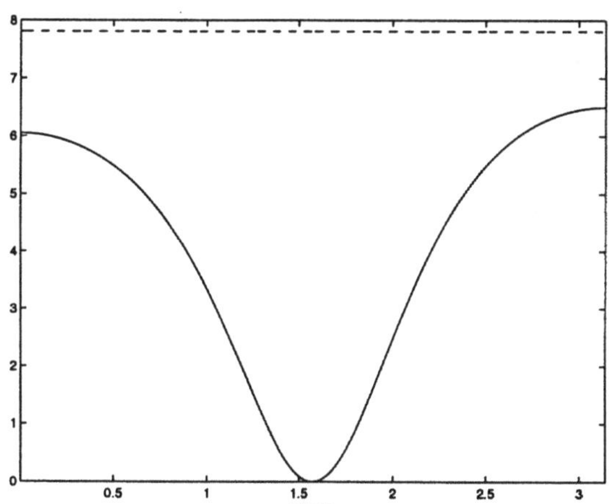

Fig. 3.2. Values of $(g(e^{j\omega}, \theta_{bis}) - \bar{g}(e^{j\omega}))^T P(\omega)^{-1} (g(e^{j\omega}, \theta_{bis}) - \bar{g}(e^{j\omega}))$ as a function of the frequency (solid) and size of the ellipses $U(\omega)$ (dashed)

3.9 Conclusions

In this chapter, we have considered linearly parametrized plants $G(z, \theta)$ whose parameters are normally distributed and we have presented results about the image $\mathcal{L}$ in the Nyquist plane of a confidence region $\mathcal{D}$ in the space of parametrized transfer functions. We have shown that this image is made of ellipses at each frequency. However, since the mapping between these two spaces is not bijective, the image $\mathcal{L}$ in the Nyquist plane contains more plants $G(z,\theta)$ than the initial confidence region $\mathcal{D}$. The image in the Nyquist plane is thus also a confidence region for the parametrized plants $G(z,\theta)$ but with a probability level larger than that of the initial confidence region $\mathcal{D}$.

Acknowledgments

We would like to thank Glenn Vinnicombe, Paresh Date and John Steele of the University of Cambridge who have motivated us to analyze this problem.

References

1. P. Andersen, S. Toffner-Clausen, and T.S. Pedersen. Estimation of frequency domain model uncertainties with application to robust control design. In *Proc. IFAC Symposium on System Identification*, pages 603–608, Copenhagen, 1994.
2. X. Bombois, M. Gevers, and G. Scorletti. Controller validation for stability and performance based on a frequency domain uncertainty region obtained by stochastic embedding. Accepted for presentation at CDC 2000, Sydney, Australia.
3. X. Bombois, M. Gevers, and G. Scorletti. Controller validation for a validated model set. In *CD-ROM Proc. European Control Conference*, paper 869, Karlsruhe, Germany, 1999.
4. G. Chesi, A. Garulli, A. Tesi, and A. Vicino. Exact bounds for the frequency response of an uncertain plant with ellipsoidal perturbations. In *CD-ROM Proc. IFAC Symposium on System Identification*, paper WeMD1-5, Santa Barbara, California, 2000.
5. M. Gevers, X. Bombois, B. Codrons, F. De Bruyne, and G. Scorletti. The role of experimental conditions in model validation for control. In A. Garulli, A. Tesi, and A. Vicino, editors, *Robustness in Identification and Control - Proc. of Siena Workshop, July 1998*, volume 245 of *Lecture Notes in Control and Information Sciences*, pages 72–86. Springer Verlag, 1999.
6. M. Gevers, B. Codrons, and F. De Bruyne. Model validation in closed-loop. In *Proc. American Control Conference,*, pages 326–330, San Diego, California,1999.
7. G.C. Goodwin, M. Gevers, and B. Ninness. Quantifying the error in estimated transfer functions with application to model order selection. *IEEE Trans. Automatic Control*, 37:913–928, 1992.
8. G.C. Goodwin, L. Wang, and D. Miller. Bias-variance trade-off issues in robust controller design using statistical confidence bounds. In *Proc. IFAC World Congress*, Beijing, 1999.
9. L. Ljung. Identification for control - what is there to learn? *Workshop on Learning, Control and Hybrid Systems, Bangalore*, 1998.
10. L. Ljung. Model error modeling and control design. In *CD-ROM Proc. IFAC Symposium on system identification*, paper WeAM1-3, Santa Barbara, California, 2000.
11. S. Toffner-Clausen, P. Andersen, J. Stoustrup, and H.H. Niemann. Estimated frequency domain model uncertainties used in robust controller design - a μ-approach. In *Proc. IEEE Conference on Control Applications*, pages 1585–1590, Glasgow, 1994.

12. L. Wang, G.C. Goodwin, and D.H. Owens. Robust controller design for unstable systems using statistical confidence bounds. In *Proc. European Control Conference*, Karlsruhe, 1999.
13. K. Zhou, J.C. Doyle, and K. Glover. *Robust and Optimal Control*. Prentice Hall, New Jersey, 1995.

Part II

Interactions Between Model Identification and Intended Model Use

4 On the Use of Real Data for Controller Reduction

I. D. Landau

Laboratoire d'Automatique de Grenoble (CNRS-INPG-UJF), ENSIEG, BP 46, 38402 Saint Martin d'Hères, FRANCE

Keywords : Controller reduction, Digital control, Identification in closed loop, Active suspension

Abstract. The paper emphasizes the use of real data for directly estimating the parameters of a reduced order digital controller using closed loop type identification algorithms. The asymptotic properties of the resulting reduced order controllers show that the critical performances of the nominal closed loop system are preserved and furthermore the robustness of the reduced controller is enhanced with respect to modeling errors. The paper also recalls that identification in closed loop may provide better plant models in view of their use for controller reduction.
Simulation and experimental results on an active suspension illustrate the performances of controller reduction techniques using real data.

4.1 Introduction

It is a great pleasure for me to contribute to this book dedicated to Brian Anderson on the occasion of his sixtieth birthday.

Brian's work has long been a source of inspiration for me. I first became familiar with his work through his paper on passivity and its relation to hyperstability [2]. This paper, along with [1] helped me to formulate my approach to these subjects. Later on, his studies of adaptive control were of great value to me as they were to many others. At the beginning of the 1990s, his work on the interplay between identification and control design again attracted my attention. Motivated by practical considerations, my recent work on controller reduction has its roots in Brian's papers on the subject [4], [3].

The objective of this paper is to emphasize the importance of using real data for controller reduction.

Controller design leads often to high order controllers. On one hand this may be the consequence of the complexity of the model used for design, on the other hand robust controller design often lead to complex high order controllers even if the design model is of reasonable size. See for example [11].

Controller reduction is a very important issue in many control applications either because the size of the controller is limited by the hardware and the computation time or simply because they are easier to implement and to understand. There are a number of approaches and methods for obtaining reduced order controllers. See for example [14,3,4].

As pointed out by Brian, in [3,4] the most important aspect is to keep in mind that controller reduction should be done with the aim to preserve as much as possible the closed loop properties. Direct simplification of the controller using standard techniques (pole-zeros cancellation within a certain radius, balanced reduction) without taking in account the closed loop behaviour produces in general unsatisfactory results.

There are basically two approaches for controller reduction:

Indirect approach

— *Obtain a reduced order model which will capture the essential characteristics of the nominal model in the critical frequency regions for design.*
— *Design a new controller using the new low-order model.*

Direct approach

— *Obtain an approximate reduced order controller which will preserve the nominal closed loop properties.*

The indirect approach is subject to a number of criticisms [3,4]. First of all the use of a reduced order model does not necessarily guarantee that the resulting designed controller will be of enough low order(design specification usually become more complex when using model approximation), secondly the errors caused by model approximation will propagate in the subsequent design steps.

The direct approach to controller reduction seems more appropriate because the approximation is carried out in the final step of the controller design and the results can be easily understood. It should be noted that the controller resulting from an indirect reduction procedure can be eventually further reduced by applying in the last step a direct reduction approach.

Till recently, in most of the reduction techniques the plant model was supposed known and all operations involved the use of the plant model and of the nominal designed controller.

In many situations and thanks to the availability of high performance real-time prototyping facilities it is possible to implement a high order controller and to operate it on the real plant. However for the industrial production one needs a severe controller reduction in order to fit the real-time computation constraints.

Therefore one may ask if this new technological set-up allowing to collect real data using high order (nominal) controllers can help for controller reduction. The answer is definitely yes and the present paper will try to illustrate this.

4.2 Notations

Consider the system shown in Fig. 4.1 where the plant model (discrete time) is given by:

$$G(z^{-1}) = \frac{z^{-d}B(z^{-1})}{A(z^{-1})} \tag{4.1}$$

and the nominal controller by:

$$K(z^{-1}) = \frac{R(z^{-1})}{S(z^{-1})} \tag{4.2}$$

where:

$$\begin{aligned} A(z^{-1}) &= 1 + a_1 z^{-1} + \ldots + a_{n_A} z^{-n_A} \\ B(z^{-1}) &= b_1 z^{-1} + \ldots + b_{n_B} z^{-n_B} \\ R(z^{-1}) &= r_0 + r_1 z^{-1} + \ldots + r_{n_R} z^{-n_R} \\ S(z^{-1}) &= 1 + s_1 z^{-1} + \ldots + s_{n_S} z^{-n_S} \end{aligned}$$

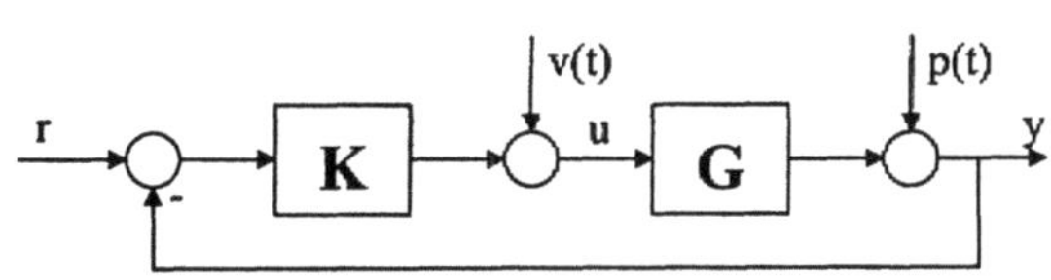

Fig. 4.1. The closed-loop system

One defines the following sensitivity functions:

- $S_{yp}(z^{-1}) = \dfrac{1}{1+KG} = \dfrac{A(z^{-1})S(z^{-1})}{P(z^{-1})}$;
- $S_{up}(z^{-1}) = \dfrac{-K}{1+KG} = \dfrac{-A(z^{-1})R(z^{-1})}{P(z^{-1})}$;
- $S_{yv}(z^{-1}) = \dfrac{G}{1+KG} = \dfrac{z^{-d}B(z^{-1})S(z^{-1})}{P(z^{-1})}$;
- $S_{yr}(z^{-1}) = \dfrac{KG}{1+KG} = \dfrac{z^{-d}B(z^{-1})R(z^{-1})}{P(z^{-1})}$.

where

$$P(z^{-1}) = A(z^{-1})S(z^{-1}) + z^{-d}B(z^{-1})R(z^{-1}) \tag{4.3}$$

The system of Fig.4.1 will be denoted the "true closed loop system". Throughout the paper we will consider feedback systems which will use either an estimation of G (denoted $\hat{G}$) or a reduced order estimation of K (denoted $\hat{K}$) or both. The corresponding sensitivity functions will be denoted as follows:

- S_{xy} - Sensitivity function of the true closed loop system (K, G).
- $\hat{S}_{xy}$ - Sensitivity function of the nominal simulated closed loop system (nominal controller K + estimated plant model $\hat{G}$).
- $\hat{\hat{S}}_{xy}$ - Sensitivity function of the simulated closed loop system using a reduced order controller (reduced controller $\hat{K}$ + estimated plant model $\hat{G}$).

Similar notations are used for $P(z^{-1})$: $\hat{P}(z^{-1})$ when using K and $\hat{G}$, $\hat{\hat{P}}(z^{-1})$ when using $\hat{K}$ and $\hat{G}$.

4.3 The algorithms

Figures 2a and 2b represent the basic diagrams for identification in closed loop using an "output error" type predictor for the closed loop. The objective is to find the best plant model which will minimize the closed loop output error (CLOE algorithms) [7,10]. In Fig. 2a the external excitation is added to the input of the controller while in Fig. 2b the external excitation is added to the output of the controller. While this does not affect the structure of the algorithm, it will affect some of its asymptotic properties.

Figures 3a and 3b represent two basic diagrams for direct estimation of reduced order controllers using real data. The objective is to estimate a reduced order controller which will minimize either the closed loop output error (output matching, CLOM algorithm, Fig. 3a) or the closed loop input input error (input matching, CLIM algorithm, Fig. 3b) between the true closed loop system and the closed loop predictor using the reduced order controller [6,8].

As it was shown in [9] the algorithms for direct estimation of reduced order controllers associated with the schemes shown in Figs. 3a, 3b and the algorithms for plant model identification in closed loop associated with Figs. 2a, 2b feature a "duality" character. This allows to obtain straightforwardly one family of algorithms from the other family by just interchanging the adjustable controller by the adjustable plant model and vice-versa (but taking in account the fact that $R(q^{-1})$ starts with q^0 and that $q^{-d}B(q^{-1})$ starts with q^{-d-1}).

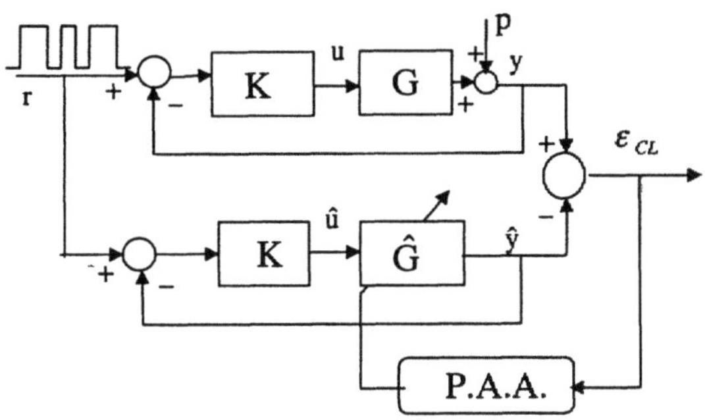

(a)
External excitation added to the controller input

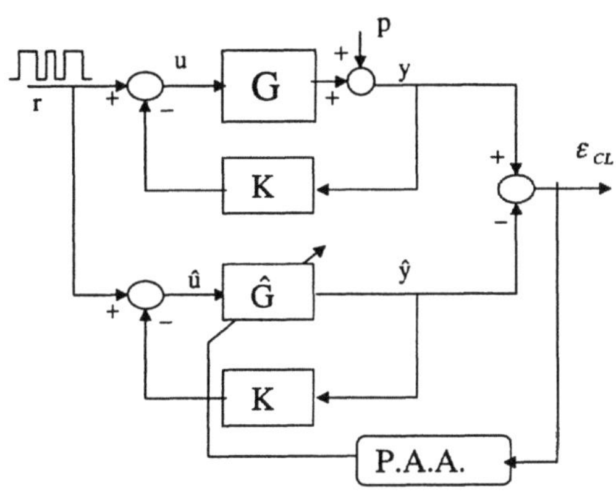

(b)
External excitation added to the controller output

Fig. 4.2. Basic schemes for closed loop output error identification (CLOE algorithms)

Therefore, this duality allows to formulate a general structure for the estimation algorithm (P.A.A in Figs. 2and 3):

$$\widehat{\theta}(t+1) = \widehat{\theta}(t) + F(t)\Phi(t)\varepsilon_{CL}(t+1) \tag{4.4}$$

$$F^{-1}(t+1) = \lambda_1(t)F^{-1}(t) + \lambda_2(t)\Phi(t)\Phi^T(t) \tag{4.5}$$

$$0 < \lambda_1(t) \leq 1\,;\, 0 \leq \lambda_2(t) < 2\,;\, F(0) > 0$$

$$\varepsilon_{CL}(t+1) = \frac{\varepsilon^0_{CL}(t+1)}{1 + \Phi^T(t)F(t)\Phi(t)} \tag{4.6}$$

where

$$\widehat{\theta}^T(t) = [\widehat{a}_1, \ldots, \widehat{a}_{n_A}, \widehat{b}_1, \ldots, \widehat{b}_{n_B}], \tag{4.7}$$

in the case of identification in closed loop, or

$$\widehat{\theta}^T(t) = [\widehat{s}_1, \ldots, \widehat{s}_{n_{\widehat{S}}}, \widehat{r}_0, \ldots, \widehat{r}_{n_{\widehat{R}}}], \tag{4.8}$$

in the case of direct estimation of reduced order controllers.

$$\varepsilon^0_{CL}(t+1) = y(t+1) - y^0(t+1) \tag{4.9}$$

for closed loop output error minimization and

$$\varepsilon^0_{CL}(t+1) = u(t+1) - u^0(t+1) \tag{4.10}$$

for closed loop input error minimization, where $y^0(t+1)$ and $u^0(t+1)$ are the *a priori* outputs of the corresponding closed loop predictors i.e.:

$$\widehat{y}^0(t+1) = \widehat{\theta}^T(t)\phi(t) \qquad \text{(closed loop output predictor)} \tag{4.11}$$

and

$$u^0(t+1) = \widehat{\theta}^T(t)\phi(t) \qquad \text{(closed loop input predictor)}. \tag{4.12}$$

For identification in closed loop the vector $\phi(t)$ in Eq. (4.11) is given by:

$$\phi^T(t) = [-\widehat{y}(t), \ldots, -\widehat{y}(t-n_A+1), \widehat{u}(t-d), \ldots, \widehat{u}(t-d-n_B+1)]. \tag{4.13}$$

For direct estimation in closed loop of reduced order controllers using CLOM algorithm, $\phi(t)$ in Eq. (4.11) is given by:

$$\phi^T(t) = [-\widehat{y}(t), \ldots, -\widehat{y}(t-n_{\widehat{S}}+1), \widehat{x}(t+1), \ldots, \widehat{x}(t-n_{\widehat{R}}+1)]. \tag{4.14}$$

When using CLIM algorithm for controller order reduction, $\phi(t)$ in Eq. (4.12) is given by:

$$\phi^T(t) = [-\widehat{u}(t), \ldots, -\widehat{u}(t-n_{\widehat{S}}+1), \widehat{c}(t+1), \ldots, \widehat{c}(t-n_{\widehat{R}}+1)]. \tag{4.15}$$

In Eqs. (4.8), (4.14) and (4.15) $n_{\widehat{R}}$ and $n_{\widehat{S}}$ are the orders of the numerator and denominator of the reduced order controller.

In general the observation vector $\phi(t)$ used in the parametric adaptation algorithm (4.4), (4.5), (4.6) will have the form

$$\Phi(t) = L(q^{-1}, t)\phi(t) \tag{4.16}$$

where $\phi(t)$ is given by (4.13), (4.14) or (4.15) and $L(q^{-1}, t)$ is either a unit gain, a constant filter or a time varying filter based on the current estimate of

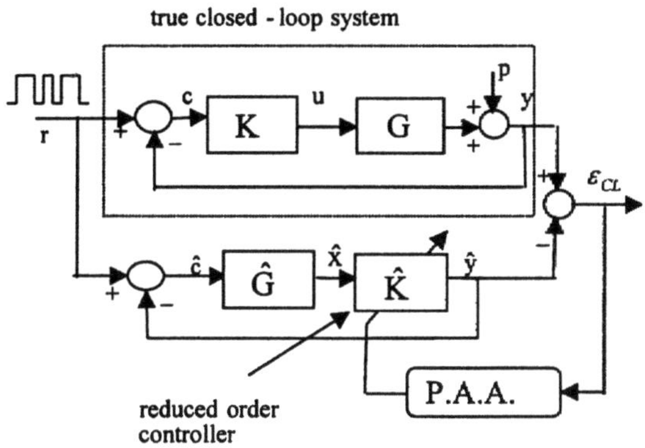

(a)

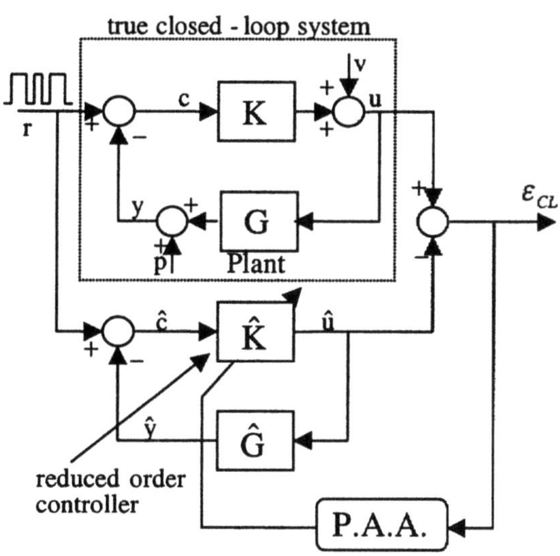

(b)

Closed loop input matching (CLIM)

Fig. 4.3. Basic schemes for identification of reduced order controllers

$\widehat{\theta}(t)$, introduced in order to satisfy a passivity condition for the convergence and robustness of various algorithms. See [10,6,8].

When carrying on identification in closed loop with real data, except in extremely limited situations, the true plant model is not in the model set and one can say that (almost) always we are searching for an identified model of reduced order.

Therefore it is crucial to examine the asymptotic properties of the estimated model in the frequency domain (the asymptotic distribution of the bias [13]).

Similarly, when carrying on estimation of a reduced order controller, the nominal controller is not in the controller set. Therefore it is also crucial to examine the asymptotic properties of the estimated controller (more exactly of the resulting closed loop) in the frequency domain and see the relationship with the classical criteria for the controller reduction.

As it can be seen from Figs. 3a and 3b in the case of controller reduction using real data one has also to examine the property of the algorithm taking into account that there is a difference between the true plant model and the plant model used in the closed loop predictor.

To examine these properties we will use the technique of Ljung [13] which converts the criterion which is minimized in the time domain into a criterion in the frequency domain by using Parseval theorem and some additional hypotheses.

4.4 Plant model identification in closed loop

We will denote by $\widehat{\theta}^*$ the asymptotic value of the estimated parameter vector and by $\widehat{G}(\widehat{\theta}^*)$ the corresponding transfer function of the estimated model (the argument $j\omega$ will be dropped out to simplify the notation).

For the CLOE algorithm associated with the scheme of Fig. 2a one has

$$\begin{aligned}\widehat{G}(\widehat{\theta}^*) &= \arg\min_{\hat{G}(\theta)} \int_{-\pi}^{\pi} |S_{yp}|^2 \left[|G - \widehat{G}|^2 |\widehat{S}_{up}|^2 \phi_r(\omega) + \phi_p(\omega) \right] d\omega \\ &= \arg\min_{\hat{G}(\theta)} \int_{-\pi}^{\pi} \left[|S_{yr} - \widehat{S}_{yr}|^2 \phi_r(\omega) + |S_{yp}|^2 \phi_p(\omega) \right] d\omega \end{aligned} \tag{4.17}$$

where $\phi_r(\omega)$ is the spectral density of the excitation signal and $\phi_p(\omega)$ is the spectral density of the output disturbance and for the CLOE algorithm, associated with the scheme of Fig. 2b, one has:

$$\begin{aligned}\widehat{G}(\widehat{\theta}^*) &= \arg\min_{\hat{G}(\theta)} \int_{-\pi}^{\pi} |S_{yp}|^2 \left[|G - \widehat{G}|^2 |\widehat{S}_{yp}|^2 \phi_r(\omega) + \phi_p(\omega) \right] d\omega \\ &= \arg\min_{\hat{G}(\theta)} \int_{-\pi}^{\pi} \left[|S_{yv} - \widehat{S}_{yv}|^2 \phi_r(\omega) + |S_{yp}|^2 \phi_p(\omega) \right] d\omega \end{aligned} \tag{4.18}$$

These expressions show that:

- The noise does not affect the estimation of the plant model parameters;
- The estimated plant model will minimize the two norm of the difference between either the complementary sensitivity functions S_{yr} and $\widehat{S}_{yr}$[1] (Fig. 2a) or the sensitivity functions S_{yv} and $\widehat{S}_{yv}$ (Fig. 2b) when $r(t)$ is a white noise.
- The frequency distribution of the error (bias) between the true plant model and the estimated plant model is weighted by the true and estimated sensitivity functions. This implies that the error will be small at the critical frequencies where the modulus of the corresponding sensitivity functions is high;
- The frequency distribution of the bias can be further tuned by the choice of $r(t)$.

The practical consequence of these properties can be summarized as follows:

- Given a design model (obtained either by physical modeling or open loop identification) and the nominal controller, one can get by identification in closed loop a new model of the same complexity which in general will produce a smaller error between the true closed loop system and its predictor (the two models can be compared by using validation techniques for closed loop identification [10]). This improved model can then be used in a technique for direct controller reduction.
- Given a design model of a certain complexity and the nominal controller, by identification in closed loop one can get a control oriented reduced complexity model which captures with a good precision the characteristics of the true model in the critical frequencies for design (where the modulus of the sensitivity functions have large value). Again, the level of reduction and the quality of the corresponding identified model can be checked by the validation techniques for identification in closed loop. This new model can then be used for indirect controller reduction. See [5] for the use of this approach for the controller reduction.

4.5 Properties of the estimated reduced order controllers

We will denote by $\widehat{\theta}^*$ the asymptotic value of the estimated parameter vector characterizing the reduced order controller and by $\widehat{K}(\widehat{\theta}^*)$ the corresponding transfer function of the estimated reduced order controller. For the CLOM algorithm (Fig 3a with $v = 0$) one has [6]:

$$\hat{K}(\hat{\theta}^*) = \arg\min_{\hat{K}(\theta)} \int_{-\pi}^{\pi} \left\{ |S_{yp} - \hat{\hat{S}}_{yp}|^2 \phi_r(\omega) + |S_{yp}|^2 \, \phi_p(\omega) \right\} d\omega \tag{4.19}$$

$$= \arg\min_{\hat{K}(\theta)} \int_{-\pi}^{\pi} \left\{ |\hat{\hat{S}}_{yp}|^2 |GK - \hat{G}\hat{K}|^2 |S_{yp}|^2 \phi_r(\omega) + |S_{yp}|^2 \phi_p(\omega) \right\} d\omega$$

[1] Note also that in (4.17) $S_{yr} - \widehat{S}_{yr} = S_{yp} - \widehat{S}_{yp}$.

For the CLIM algorithm (Fig. 3b) one has [8]:

$$\hat{K}(\hat{\theta}^*) = \arg\min_{\hat{K}(\theta)} \int_{-\pi}^{\pi} \left\{ |S_{up} - \hat{\hat{S}}_{up}|^2 \phi_r(\omega) + |S_{yp}|^2 \, \phi_{v'}(\omega) \right\} d\omega \qquad (4.20)$$

with:

$$v'(t) = v(t) - Kp(t) \qquad (4.21)$$

where $v(t)$ and $p(t)$ represent the input and output disturbances(noise) respectively.
These expressions shows that

- The noise does not affect the estimation of the controller parameters;
- The two norms of the difference between either S_{yp} and $\hat{\hat{S}}_{yp}$ (CLOM) or S_{up} and $\hat{\hat{S}}_{up}$ (CLIM) is minimized when $r(t)$ is a white noise
- When using simulated data (one assumes that $\hat{G} = G$) the two norm of the difference between either $\hat{S}_{yp} = S_{yp}$ and $\hat{\hat{S}}_{yp}$ (CLOM) or $\hat{S}_{up} = S_{up}$ and $\hat{S}_{up}$ (CLIM) is minimized when $r(t)$ is a white noise
- The frequency distribution of the bias can be further tuned by the choice of $r(t)$;
- When using the real data, the closed loop system with the reduced order controller approximates the real closed loop system instead of the nominal one. It means that the reduced order closed loop system represents better the behaviour of the real closed loop system than the nominal one.

The key point in (4.19) is that the real data allow to take in account (in the good direction) the modeling error which always exists between the plant model used for controller reduction and the true plant model. Consider the case that $|G| < |\hat{G}|$ at some frequencies , therefore minimizing $|KG - \hat{K}\hat{G}|^2$ leads to estimate a reduced order controller whose gain is less than the gain of the nominal controller at these frequencies ($|\hat{K}| < |K|$). This will reduce the gain of the input sensitivity function of the reduced order closed loop system at these frequencies and thus improves the robustness. A simulation example will illustrate this property.

4.6 A simulation example

In this section a simulation example is presented in order to show the behaviour of the proposed algorithms for direct estimation of reduced order controllers when there is a difference between the true plant model and the plant model used for design and closed loop prediction. The CLOM algorithm is considered.

A fifth-order discrete-time system is considered as the true model of the plant (G) and a fourth-order model as the design model ($\hat{G}$). The magnitudes

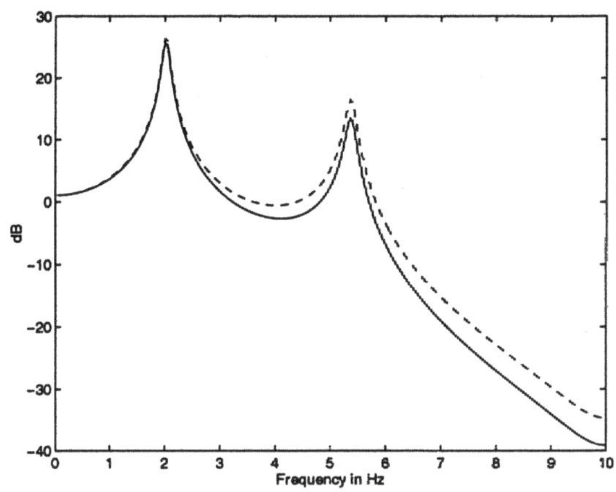

Fig. 4.4. True model (–) and design reduced order model (– –)

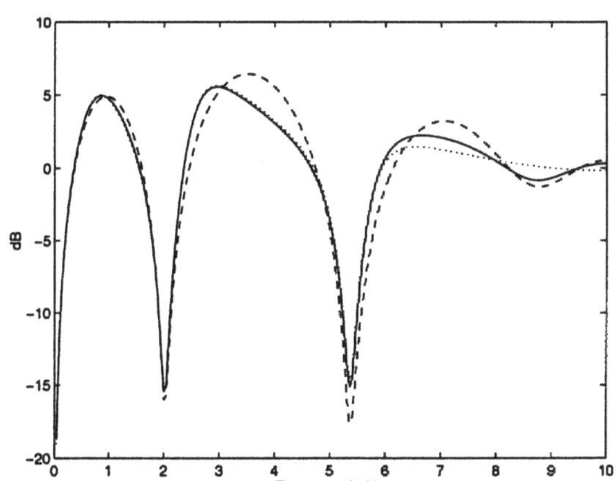

Fig. 4.5. The output sensitivity functions: S_{yp} (–); $\hat{S}_{yp}$ (– –); $\hat{\hat{S}}_{yp}$ ($\cdots$)

of the frequency responses of these two models are compared in Fig. 4.4. As shown in this figure the effect of unmodeled dynamics show up only in the high frequencies. Based on the design model, a controller (K) using the pole placement technique has been computed. The controller contains an integrator and places the dominant closed loop poles in the same frequency of the first oscillatory mode of the plant ($2Hz$) with a damping factor of 0.9. The true plant model with this controller is excited in closed loop with a PRBS of 512 length. This is in fact the simulation of a real data acquisition. Then a reduced order controller ($\hat{K}$) is identified using the CLOM algorithm. Fig. 4.5 shows the real sensitivity function of the closed loop $S_{yp} = (1 + KG)^{-1}$, the designed sensitivity function $\hat{S}_{yp} = (1 + K\hat{G})^{-1}$ and the sensitivity function with the reduced order controller $\hat{\hat{S}}_{yp} = (1 + \hat{K}\hat{G})^{-1}$. It is clear from this

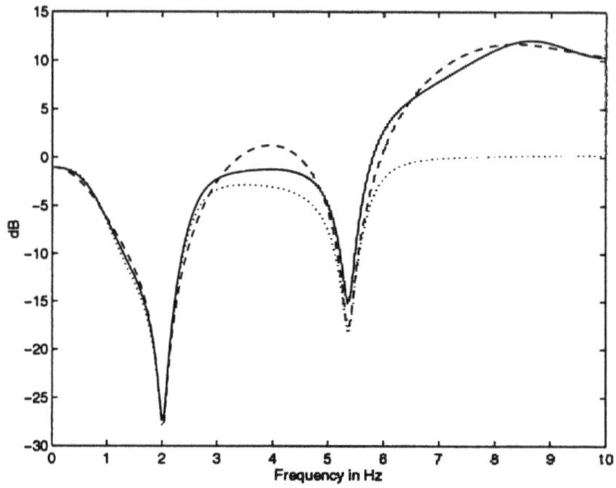

Fig. 4.6. The input sensitivity functions: S_{up} (–); $\hat{S}_{up}$ (– –); $\hat{\hat{S}}_{up}$ (···)

figure that $\hat{\hat{S}}_{yp}$ tries to approximate the real sensitivity function instead of the sensitivity function of the designed loop. On the other hand the comparison of the input sensitivity functions in Fig. 4.6 shows that the magnitude of the sensitivity function for the reduced order controller is reduced in the high frequencies (where there are the modeling errors) which means that the reduced order controller is more robust than the nominal controller with respect to the additive uncertainties.

4.7 Experimental results on an active suspension

The procedure for obtaining better control oriented plant model by identification in closed loop (CLOE algorithm) and the procedure for direct estimation of reduced order controller (CLOM algorithm) using real data will be illustrated for the case of the control of an active suspension.

4.7.1 The active suspension

The active suspension is shown in Fig. 4.7 and the corresponding block diagram in Fig.4.8. The controller through the power amplifier will act upon the piston in order to reduce the residual acceleration.The frequency spectrum of the vibrations which have to be reduced is limited by an upper frequency lower than 200 Hz. The system is controlled by a PC via an I/O board. The sampling frequency is 800 Hz.

The primary acceleration has been generated using a shaker. Its input is a signal given by the computer. The frequency characteristic of the identified primary path model, between the excitation of the shaker and the residual acceleration is presented in figure 4.9. The goal is to compute a controller which minimizes the residual acceleration around the first vibration mode and try to distribute the amplification of the disturbances over the high frequency range. There exist several vibration modes (see figure 4.9), the first one being at $31.47Hz$ with a damping factor 0.093.

The identified plant model, between the piston's position and the residual acceleration (the secondary path), has the following complexity: $n_B = 11$, $n_A = 12$, $d = 2$. The system contains a double derivator. In Fig. 4.10 are presented the frequency characteristics of two plant models: the open loop identified model used for design and the model identified in closed loop using the nominal controller (CLOE algorithm with external excitation added to the controller output). The closed loop identified model provided better closed loop validation results as indicated in Table 1.

Plant model	C.L. Error variance	$\|RN(i)\|_{max}$	$\delta(S_{yv}, \widehat{S}_{yv})$
Open loop	1.5368	0.4802	0.7624
Closed loop	0.1477	0.0346	0.3587

Table 4.1. Closed loop validation results

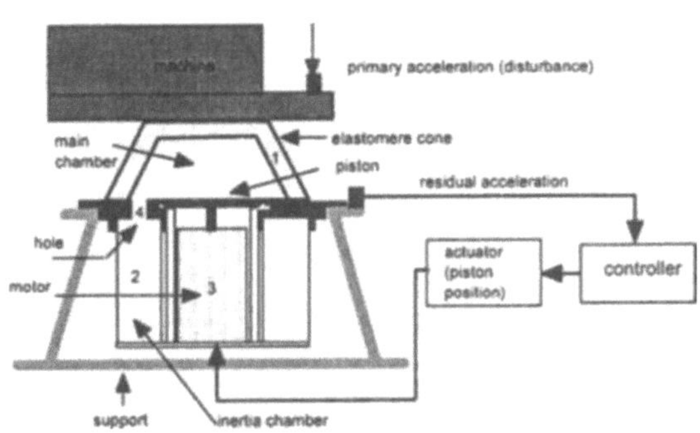

Fig. 4.7. The active suspension system

In Table 1, column 2 gives the variance of the residual closed loop error, column 3 gives the maximum value of the normalized cross-correlations between ε_{CL} and $\widehat{y}$ and column 4 give the ν-gap between the identified S_{yv} sensitivity function of the true system and the computed $\widehat{S}_{yv}$ sensitivity function using the estimated model.

The model has six vibration modes. Five of the six vibration modes have a very low damping (< 0.085).

Note that an important attenuation on S_{yp} is required at the frequency of the first vibration mode.

The nominal controller has been designed using the method of "pole placement with sensitivity function shaping by convex optimization" [12].

A pair of dominant poles has been fixed at the frequency of the first vibration mode and with a damping $\xi = 0.8$. In addition a fixed part $H_R = 1 + q^{-1}$ has been introduced in the controller ($R = H_R R'$) which assures the opening of the loop at $0.5 f_s$. The resulting nominal controller satisfying the specifications on the residual acceleration has been obtained. Its complexity is given by the orders of polynomial R and S: $n_R = 27, n_S = 28$. Note that if standard pole placement is used, by solving the Bezout equation it will result a controller having the orders: $n_R = 12$, $n_S = 13$ (considering the fixed part H_R).

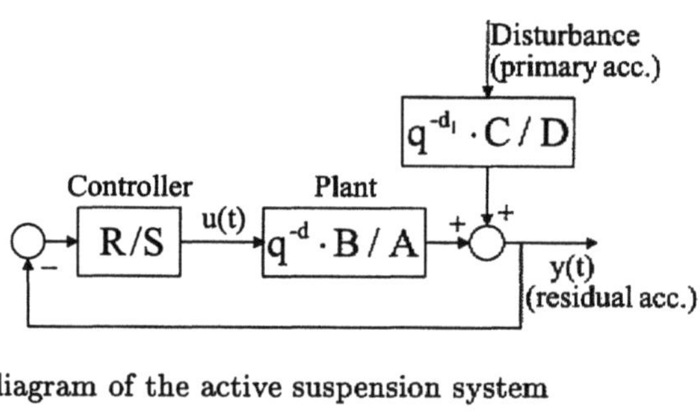

Fig. 4.8. Block diagram of the active suspension system

4.7.2 Identification of reduced order controllers using CLOM algorithm on real data

The external input was a PRBS generated by 10-bit shift register and with a clock frequency of $f_s/2$ (the data length is $L = 4096$ samples). A variable forgetting factor with $\lambda_1 = 0.95$ and $\lambda_0 = 0.9$ has been used ($\lambda_1(t) = \lambda_0 \lambda_1(t-1) + 1 - \lambda_0$).

Four controllers have been considered for a comparison. The nominal controller (K_n) with $n_R = 27$, $n_S = 28$ and three reduced order controllers:K_1 with $n_R = 17$, $n_S = 18$, K_2 with $n_R = 12$, $n_S = 13$ and K_3 with $n_R = 9$, $n_S = 10$ (a fixed part $H_R = 1 + q^{-1}$ has been imposed in the reduced order controllers).

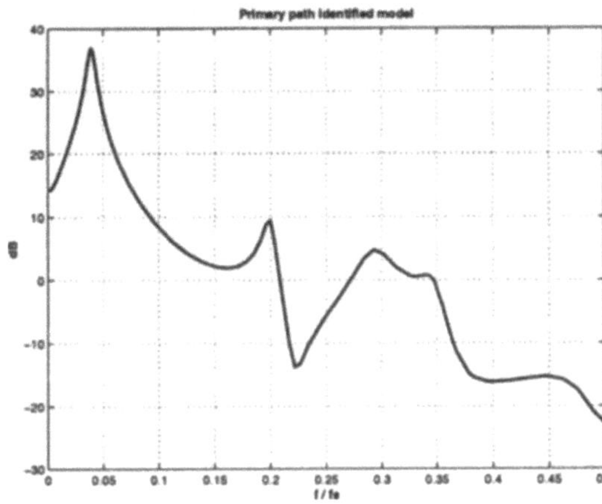

Fig. 4.9. The frequency characteristic of the primary path model (input: shaker's signal, output: residual acceleration)

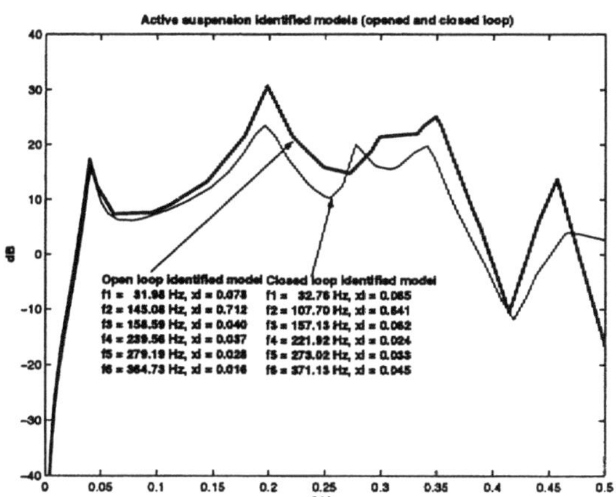

Fig. 4.10. The frequency characteristics of the secondary path model (input: piston displacement, output: residual acceleration)

Note that the reduced controller K_2 corresponds to the complexity of the pole placement controller with the fixed part H_R. Controller K_3 has a complexity lower than that corresponding to the pole placement.

The simulated output and input sensitivity functions (computed with the estimated model $\widehat{G}$) are shown in Fig. 4.10 and Fig. 4.11. The various ν-gap between the nominal and reduced order controllers and between the nominal simulated sensitivity functions and reduced simulated sensitivity functions are sumarized in Table 2.

Controller	K_n $n_R = 27$ $n_S = 28$	K_1 $n_R = 17$ $n_S = 18$	K_2 $n_R = 12$ $n_S = 13$	K_3 $n_R = 9$ $n_S = 10$
$\delta_\nu(K_n, K_i)$	0	0.7300	0.7629	0.7859
$\delta_\nu(S_{yp}^n, S_{yp}^i)$	0	0.1591	0.1752	0.1738
$\delta_\nu(S_{up}^n, S_{up}^i)$	0	0.7159	0.7585	0.7867
C.L. error variance	0.0091	0.0113	0.0131	0.0161

Table 4.2. Comparison of the various controllers (controller reduction using CLOM

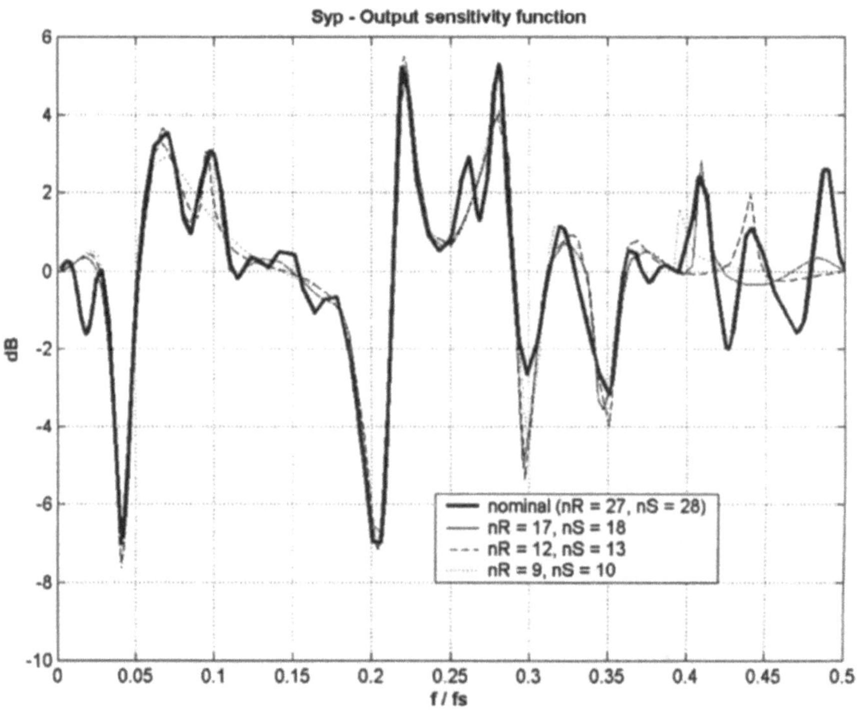

Fig. 4.11. Output sensitivity for the active suspension (controller reduction using CLOM algorithm on real data)

To illustrate the performances of the resulting controllers (K_n, K_1, K_2 and K_3) on the real system, the spectral density of the residual acceleration in open and in closed loop is shown in Fig. 4.13. The primary source of vibration (shaker's excitation) is a PRBS. The characteristics obtained in closed loop operation with the four controllers are compared with those corresponding to the open loop operation (the interesting frequency range is 0 to $0.25 f_s$ (200 hz)). One can see that the performances of the reduced order controllers are

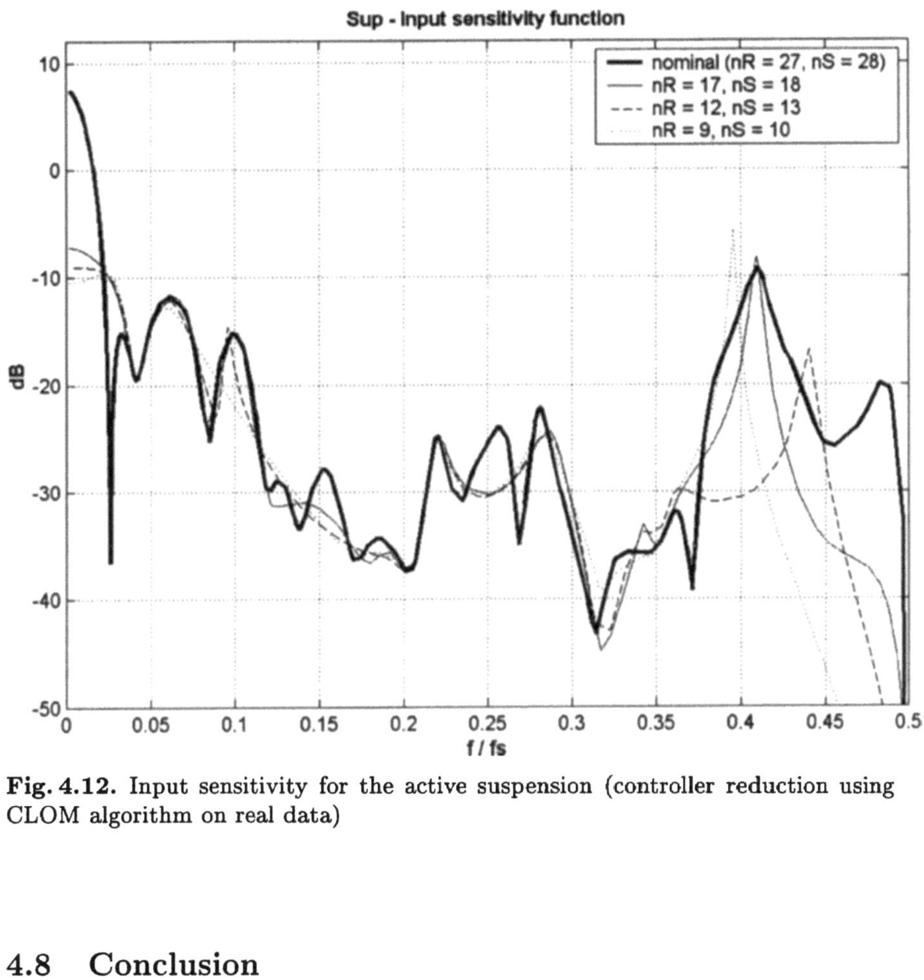

Fig. 4.12. Input sensitivity for the active suspension (controller reduction using CLOM algorithm on real data)

4.8 Conclusion

The first objective of this paper was to show theoretically and experimentally that algorithms for direct identification in closed loop of reduced order controllers can take advantage of using real data. By doing so, the resulting controller takes in account the discrepancy between the true plant model and the model used for design and controller reduction.

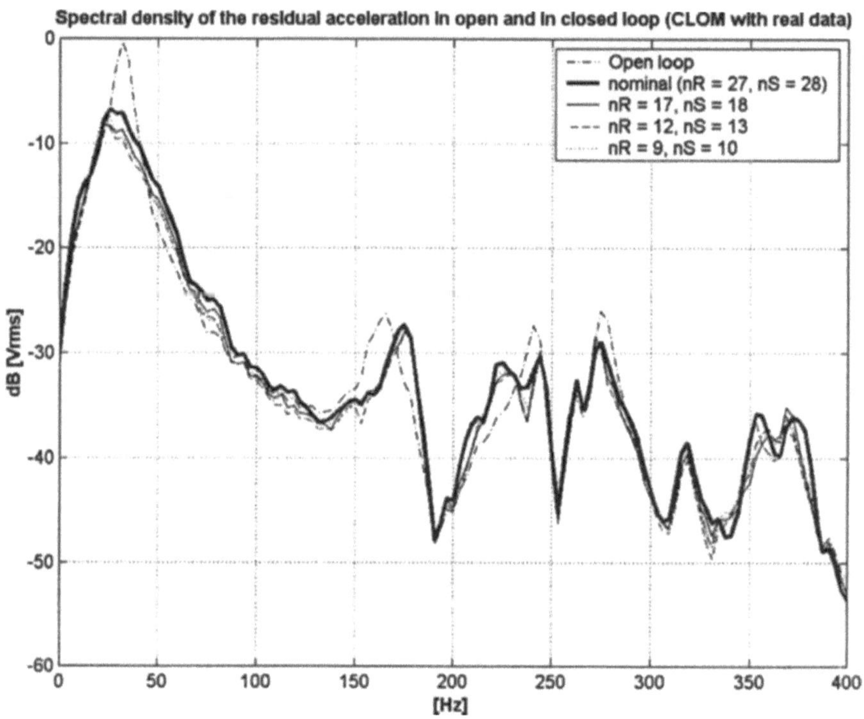

Fig. 4.13. Spectral density of the residual acceleration in open and in closed loop (controller reduction using CLOM algorithm on real data)

Acknowledgement

The author acknowledge the contribution of A. Karimi and A. Constantinescu to this work.

References

1. B.D.O. Anderson. A system theory criterion for positive real matrices. *SIAM J. of Control*, 5(2):171–182, 1967.
2. B.D.O. Anderson. A simplified viewpoint on hyperstability. *IEEE Trans. on Aut. Control*, Ac-13:292–294, 1968.
3. B.D.O. Anderson. Controller design: Moving from theory to practice. *IEEE Control Magazine*, August 1993.
4. B.D.O. Anderson and Y. Liu. Controller reduction: Concepts and approaches. *IEEE Transactions on AC*, 34(8):802–812, August 1989.
5. P. Bendotti, B. Cordons, C.M. Falinower, and M. Gevers. Control oriented low order modeling of a complex pwr plant: a comparison between open loop and closed loop methods. *Proc, IEEE-CDC 1998 (FAO 7-2)*, December 1998.
6. A. Karimi and I.D. Landau. Controller reduction by direct closed loop identification (output matching). *Proc. IFAC-ROCOND 2000 Symp. Prague, Czech Republic*, June 2000.
7. I.D. Landau and A. Karimi. Recursive algorithms for identification in closed loop - a unified approach and evaluation. *Automatica*, 33(8):1499–1523, 1997.
8. I.D. Landau and A. Karimi. Direct closed loop identification of reduced order controllers (input matching). *Proc. IFAC-SYSID 2000 Symp. S. Barbara, USA*, June 2000.
9. I.D. Landau and A. Karimi. Model estimation and controller reduction: dual closed loop identification problems. *submitted to IEEE-CDC 2000, Sydney, Australia*, December 2000.
10. I.D. Landau, R. Lozano, and M. M'Saad. *Adaptive Control*. Springer, London, 1997.
11. I.D. Landau, D. Rey, A. Karimi, A. Voda, and A. Franco. A flexible transmission system as a benchmark for robust digital control. *European Journal of Control*, 1(2):77–96, 1995.
12. J. Langer and I.D. Landau. Combined pole placement/sensitivity function shaping method using convex optimization criteria. *Automatica*, 35:1111–1120, 1999.
13. L. Ljung. *System identification. Theory for the user*. Prentice Hall, N.Y., 2 edition, 1999.
14. K. Zhou and J. Doyle. *Essentials of robust control*. Prentice-Hall, N.Y., 1998.

5 Making a Nonlinear Observer Adaptive *

Ramon Costa[1], Petar Kokotović[2], and Iven Mareels[3]

[1] Department of Electrical Engineering
COPPE/UFRJ, Rio de Janeiro, Brazil
e-mail: ramon@coep.ufrj.br

[2] Center for Control Engineering and Computation
University of California, Santa Barbara CA 93106
e-mail: petar@seidel.ece.ucsb.edu

[3] Department of Electrical Engineering
University of Melbourne, Melbourne, Australia
e-mail: i.mareels@unimelb.edu.au

Dedicated to Brian Anderson for his 60th birthday.

Abstract. For systems with monotonic nonlinearities, the Arcak observer provides globally convergent state estimates. In the presence of unknown parameters this convergence property is lost and our goal is to recover it by adaptation. Here we present some preliminary results in this direction.

Keywords: Nonlinear observer, adaptive schemes.

5.1 Introduction

Brian Anderson's creative path is marked by many seminal results in diverse areas of systems and control theory. Prominent among those are his results on passivity and adaptive systems. They have influenced much of current research in constructive nonlinear and adaptive control including output feedback control with nonlinear observers. An example is a recent nonlinear observer, referred to as Arcak observer [2], for systems with monotone nonlinear functions of unmeasured states. In the observer error equation these functions give rise to time-varying sector nonlinearities, which, in turn, make a passivity (circle criterion) design feasible.

This paper reports on two attempts to make the Arcak observer adaptive for a representative example consisting of a van der Pol oscillator with an unknown parameter multiplying the unmeasured cubic nonlinearity. While the simulation evidence abundantly confirms that in both adaptive schemes the observation error converges to zero, the proof of the fact is still an open problem.

* Research supported in part by the National Science Foundation under grant ECS-9812346, the Air Force Office of Scientific Research under grant F49620-95-1-0409, and CAPES-Brazil.

5.2 Arcak observer

The systems for which Arcak observers [2] can be designed are of the form

$$\dot{x} = Ax + G\gamma(Hx) + \varrho(y,u)\,, \qquad (5.1)$$
$$y = Cx\,,$$

where $x \in \mathbb{R}^n$ is the state, $y \in \mathbb{R}^p$ is the measured output, $u \in \mathbb{R}^m$ is the control input, the pair (A, C) is detectable, and $\gamma(\cdot)$ is a locally Lipschitz r-dimensional nonlinearity.

For the Arcak observer

$$\dot{\hat{x}} = A\hat{x} + L(C\hat{x} - y) + G\,\gamma(H\hat{x} + K(C\hat{x} - y)) + \varrho(y,u)\,, \qquad (5.2)$$

the error $e = x - \hat{x}$ is governed by

$$\dot{e} = (A + LC)e + G\,[\gamma(v) - \gamma(\omega)]\,, \qquad (5.3)$$

where $v := Hx$, and $\omega := H\hat{x} + K(C\hat{x} - y)$. Introducing

$$z := v - \omega = (H + KC)e\,, \qquad (5.4)$$
$$\varphi(t,z) := \gamma(v) - \gamma(\omega)\,, \qquad (5.5)$$

where the time dependence in $\varphi(t, z)$ is due to $v(t) = Hx(t)$, we rewrite the observer error system (5.3) as

$$\dot{e} = (A + LC)e + G\varphi(t,z)\,, \qquad (5.6)$$
$$z = (H + KC)e\,.$$

It is easy to see that this error system represents a feedback interconnection of a linear block defined by the triple $(A + LC,\ G,\ H + KC)$ and the time-varying nonlinearity $\varphi(t, z)$. The design task is to find matrices $K \in \mathbb{R}^{r\times p}$ and $L \in \mathbb{R}^{n\times p}$ to render the linear block passive, and to impose restrictions on $\gamma(\cdot)$ to assure that $\varphi(t, z)$ is a sector nonlinearity,

$$z_i\varphi_i(t,z_i) \geq 0\,, \quad \forall z_i \in \mathbb{R}\,. \qquad (5.7)$$

Then the multivariable circle criterion can be applied to guarantee the exponential convergence of the state estimation error $e(t)$ to zero.

Theorem 1. *[2] Assume that each component $\gamma_i(\cdot)$ of $\gamma(\cdot)$ is nondecreasing, $(a - b)[\gamma_i(a) - \gamma_i(b)] \geq 0$ for all $a, b \in \mathbb{R}$, and suppose that $x(t)$ exists for all $t \geq 0$. If a matrix $P = P^T > 0$, a constant $\nu > 0$, and a diagonal matrix $\Lambda > 0$ can be found such that*

$$\begin{bmatrix} (A+LC)^TP + P(A+LC) + \nu I & PG + (H+KC)^T\Lambda \\ G^TP + \Lambda(H+KC) & 0 \end{bmatrix} \leq 0, \qquad (5.8)$$

then the observer error $e(t)$ *satisfies, for all* $t \geq 0$,

$$|e(t)| \leq \kappa |e(0)| \exp(-\beta t)\,, \tag{5.9}$$

where $\kappa = \sqrt{\frac{\lambda_{max}(P)}{\lambda_{min}(P)}}$, $\beta = \frac{\nu}{2\lambda_{max}(P)}$.

With Theorem 1, the observer design for system (5.1) is reduced to solving (5.8), a linear matrix inequality (LMI) for P, PL, Λ, ΛK and ν. Efficient numerical tools are available for LMI's to determine whether the problem is feasible and, if so, to compute K and L.

5.3 Van der Pol oscillator with a parameter

5.3.1 Observer design

In the remainder of this paper we study the adaptive observer problem for the van der Pol oscillator

$$\begin{aligned} \dot{x}_1 &= x_2\,, \\ \dot{x}_2 &= -x_1 + x_2 + \theta^* x_2^3\,, \\ y &= x_1\,, \end{aligned} \tag{5.10}$$

which is in the form (5.1) with $\varrho(y,u) = 0$ and

$$A = \begin{bmatrix} 0 & 1 \\ -1 & 1 \end{bmatrix}, \quad G = \begin{bmatrix} 0 \\ \theta^* \end{bmatrix}, \quad C = \begin{bmatrix} 1 & 0 \end{bmatrix}.$$

If the parameter θ^* is known and negative, $\theta^* < 0$, the Arcak observer is applicable,

$$\begin{aligned} \dot{\hat{x}}_1 &= \hat{x}_2 - l_1(x_1 - \hat{x}_1)\,, \\ \dot{\hat{x}}_2 &= -x_1 + \hat{x}_2 - l_2(x_1 - \hat{x}_1) + \theta^* \omega^3\,. \end{aligned} \tag{5.11}$$

and the observer error system is

$$\dot{e} = \bar{A}e + G(x_2^3 - \omega^3)\,, \tag{5.12}$$

where

$$\bar{A} = A + LC = \begin{bmatrix} l_1 & 1 \\ l_2 & 1 \end{bmatrix}, \quad L = \begin{bmatrix} l_1 \\ l_2 \end{bmatrix}.$$

The observer design consists in finding l_1, l_2 and constructing an implementable signal ω to achieve GAS of (5.12). Evaluating the derivative of $2V(e) = e^T P e$ along the trajectories of (5.12) we get

$$2\dot{V} = e^T(P\bar{A} + \bar{A}^T P)e + 2e^T PG(x_2^3 - \omega^3)\,. \tag{5.13}$$

We select L such that $\bar{A}$ is Hurwitz and $P\bar{A}+\bar{A}^TP = -2Q$, $Q = Q^T > 0$. Letting

$$P = \begin{bmatrix} p & k \\ k & m \end{bmatrix}, \tag{5.14}$$

we rewrite (5.13) as

$$\dot{V} = -e^TQe + \theta^*(ke_1 + me_2)(x_2^3 - \omega^3) \tag{5.15}$$

and recall the inequality

$$(x_2 - \omega)(x_2^3 - \omega^3) \geq 0, \tag{5.16}$$

which holds for all $x_2, \omega \in \mathbb{R}$. Because, by assumption, $\theta^* < 0$, we can construct ω to satisfy

$$ke_1 + me_2 = x_2 - \omega. \tag{5.17}$$

We select $m = 1$ and obtain

$$\omega = \hat{x}_2 - ke_1. \tag{5.18}$$

Denoting

$$z = ke_1 + e_2, \tag{5.19}$$

$$\sigma(t,z) = z^2 + 3\omega z + 3\omega^2 \geq 0, \tag{5.20}$$

using (5.17), and $(x_2^3 - \omega^3) = z\sigma(t,z)$, we establish that for $e \neq 0$ the derivative $\dot{V}$ in (5.15) is negative,

$$\dot{V} = -e^TQe + \theta^*\sigma z^2 < 0, \tag{5.21}$$

which proves GAS of the observer error system (5.12). We point out that ω constructed in (5.18) is implementable because it depends only on e_1 and $\hat{x}_2$, which are available. Likewise, the LMI (5.8), with the restriction $m = 1$ imposed on matrix P, is feasible. A particular solution is used for simulations in Section 5.6.

5.3.2 An ISS property

When θ^* is unknown but negative, $\theta^* < 0$, the *certainty-equivalence* form of the observer (5.11) is

$$\dot{\hat{x}}_1 = \hat{x}_2 - l_1e_1, \tag{5.22}$$

$$\dot{\hat{x}}_2 = -x_1 + \hat{x}_2 - l_2e_1 + \theta\omega^3,$$

in which θ^* is replaced by an estimate θ, restricted by $\theta < 0$. The observer error system is

$$\dot{e} = \bar{A}e + b[\theta\sigma z - \tilde{\theta}x_2^3], \tag{5.23}$$

where $b = \begin{bmatrix}0 & 1\end{bmatrix}^T$ and

$$\tilde{\theta} = \theta - \theta^* \tag{5.24}$$

is the parameter error.

To clearly see an input-to-output stability (ISS) property of (5.23) we perform the coordinate transformation

$$Te = \begin{bmatrix}1 & 0\\ k & 1\end{bmatrix} e = \begin{bmatrix}e_1\\ z\end{bmatrix}, \tag{5.25}$$

resulting in

$$\tilde{A} = T\bar{A}T^{-1} = \begin{bmatrix} l_1 - k & 1\\ -k^2 + (l_1 - 1)k + l_2 & k + 1\end{bmatrix}, \quad \tilde{b} = Tb = \begin{bmatrix}0\\ 1\end{bmatrix}. \tag{5.26}$$

We also take advantage of the freedom to select k and l_2 to render $\tilde{A}$ upper-triangular. Then, the observer error system becomes a cascade of a linear and a nonlinear subsystem with the input $-\tilde{\theta}x_2^3$,

$$\begin{aligned} \dot{e}_1 &= \lambda_1 e_1 + z, \\ \dot{z} &= (\lambda_2 + \theta\sigma)z - \tilde{\theta}x_2^3. \end{aligned} \tag{5.27}$$

where $\lambda_1 < 0$ and $\lambda_2 < 0$ are the eigenvalues of $\tilde{A}$. Now the time derivative of $\frac{1}{2}z^2$ along the trajectories of the z-subsystem of (5.27) is

$$\begin{aligned} \frac{d}{dt}\left(\frac{1}{2}z^2\right) &= \lambda_2 z^2 + z[\theta\sigma z - \tilde{\theta}x_2^3] \\ &= \lambda_2 z^2 + \theta\sigma z^2 - \theta z x_2^3 + \theta^* z x_2^3 \\ &\le \lambda_2 z^2 - |\theta||z|\left(\sigma|z| - |x_2^3|\right) + |\theta^*||z||x_2^3|. \end{aligned} \tag{5.28}$$

From (5.20) we find that the minimum of σ is $\sigma_{min} = \frac{1}{4}z^2$, which yields

$$\frac{d}{dt}\left(\frac{1}{2}z^2\right) \le -|z|\left(|\lambda_2||z| - |\theta^*||x_2^3|\right) - |\theta||z|\left(\frac{1}{4}|z^3| - |x_2^3|\right). \tag{5.29}$$

It follows that if

$$|z| \ge \max\left\{4^{1/3}|x_2|,\ 2\frac{|\theta^*||x_2^3|}{|\lambda_2|}\right\},$$

then $\frac{d}{dt}(z^2) \le \frac{1}{2}\lambda_2 z^2$. This means that the z-subsystem is ISS with respect to its input $-\tilde{\theta}x_2^3$, and that the same is true for the whole error system (5.27).

Because, by assumption, x_2 and θ^* are bounded, the observer error e is bounded for all bounded estimates θ.

5.4 Adaptation: a MRAC scheme

Our search for an update law for θ begins by verifying that the derivative of $\frac{1}{2}e^T Pe + \frac{1}{2\mu}\tilde{\theta}^2$ along the trajectories of the observer error system (5.23) is rendered negative semi-definite by $\dot{\theta} = \mu z\omega^3$, where $\mu > 0$ is the adaptation gain. This is so because the transfer function from $[\theta\sigma z - \tilde{\theta}x_2^3]$ to z is SPR. Unfortunately, this update law cannot be implemented because z is not an available signal.

On the other hand, the transfer function from $[\theta\sigma z - \tilde{\theta}x_2^3]$ to e_1, which is an available signal, is $W(s) = C(sI - \bar{A})^{-1}b$ and has relative degree 2. Following Monopoli [4] we let $L(s) = s + \alpha$ and select α such that $W(s)L(s)$ is SPR. With the signal $\eta = L^{-1}(s)\left[\omega^3\right]$, we modify the certainty equivalence observer (5.22) by adding to it $\dot{\theta}\eta$,

$$\begin{aligned} \dot{\hat{x}}_1 &= \hat{x}_2 - l_1 e_1 \,, \\ \dot{\hat{x}}_2 &= -x_1 + \hat{x}_2 - l_2 e_1 + \theta\omega^3 + \dot{\theta}\eta \,. \end{aligned} \tag{5.30}$$

The modified observer error system is

$$\begin{aligned} \dot{\epsilon} &= \bar{A}\epsilon + \bar{b}\left[\theta^* L^{-1}(s)\left[\sigma z\right] - \tilde{\theta}\eta\right] \,, \\ e_1 &= C\epsilon \,, \end{aligned} \tag{5.31}$$

where $\bar{b}$ is determined from $W(s)L(s) = C(sI - \bar{A})^{-1}\bar{b}$. We assume that an upper bound for θ^* is known, $\theta^* < -\varepsilon < 0$. The Monopoli update law with a projection to assure $\theta(t) \leq -\varepsilon < 0$ is

$$\dot{\theta}(t) = \begin{cases} 0 \,, & \text{if } \theta \geq -\varepsilon \text{ and } \mu e_1\eta \geq 0 \,, \\ \mu e_1 \eta \,, & \text{otherwise} \,. \end{cases} \tag{5.32}$$

Since $W(s)L(s)$ is SPR, then there exist matrices $P_2 > 0$ and $Q_2 > 0$ such that $P_2\bar{A} + \bar{A}^T P_2 = -2Q_2$ and $P_2\bar{b} = C^T$. Therefore, the time derivative of

$$2V_2(\epsilon, \tilde{\theta}) = \epsilon^T P_2 \epsilon + \mu^{-1}\tilde{\theta}^2 \,, \tag{5.33}$$

along the trajectories of (5.31), yields

$$\begin{aligned} \dot{V}_2 &= -\epsilon^T Q_2 \epsilon + e_1\theta^* L^{-1}(s)\left[\sigma z\right] + \mu^{-1}\tilde{\theta}\left[-\mu e_1\eta + \dot{\tilde{\theta}}\right] \\ &\leq -\epsilon^T Q_2\epsilon + \theta^* e_1 L^{-1}(s)\left[\sigma z\right] \,, \end{aligned} \tag{5.34}$$

where the inequality is due to the fact that when $\dot{\theta} = 0$, then $\mu e_1\eta \geq 0$, and, hence, $-\tilde{\theta}e_1\eta \leq 0$.

While the modified observer (5.30) and the update law (5.32) are both implementable, the difficulty is now in the analysis. The presence of $L^{-1}(s)[\sigma z]$ in (5.34) has hindered our efforts to establish the boundedness and convergence properties of this adaptive scheme, clearly shown by simulations.

5.5 Adaptation: a gradient scheme

For an alternative update law, we return to an old gradient scheme [3] using the sensitivity filters implemented with parameter estimate θ treated as a constant *(MIT rule)*. To minimize $\frac{1}{2}e_1^2$ while keeping $\theta(t) \leq -\varepsilon < 0$ we employ the update law

$$\dot{\theta}(t) = \begin{cases} 0, & \text{if } \theta \geq -\varepsilon \text{ and } -\mu e_1 \frac{\partial e_1}{\partial \theta} \geq 0 \ , \\ -\mu e_1 \frac{\partial e_1}{\partial \theta} , & \text{otherwise.} \end{cases} \tag{5.35}$$

Differentiating the observer error system (5.23) with respect to θ, we obtain the sensitivity filter

$$\begin{aligned} \dot{\xi}_1 &= \lambda_1 \xi_1 + \xi_2 \, , & \xi_1(0) &= 0 \, , \\ \dot{\xi}_2 &= \lambda_2 \xi_2 - \omega^3 - \theta \frac{\partial}{\partial \theta}(\omega^3) \, , & \xi_2(0) &= 0 \, . \end{aligned} \tag{5.36}$$

If $\theta = const$, the standard ODE theorems establish that $\xi_1 = \frac{\partial e_1}{\partial \theta}$, $\xi_2 = \frac{\partial z}{\partial \theta}$. To be close to this situation, $\theta(t)$ should vary slowly, that is, μ in (5.35) should be sufficiently small.

In the van der Pol oscillator (5.10), x_2 is independent of the estimate θ. Also, $\omega = x_2 - z$, and, hence $\frac{\partial}{\partial \theta}(\omega^3) = -3\omega^2 \xi_2$. Thus, the final form of the sensitivity filter is

$$\begin{aligned} \dot{\xi}_1 &= \lambda_1 \xi_1 + \xi_2 \, , \\ \dot{\xi}_2 &= (\lambda_2 + 3\theta\omega^2)\xi_2 - \omega^3 \, . \end{aligned} \tag{5.37}$$

This filter is implementable. Its ξ_1-equation (5.37) is clearly ISS with respect to ξ_2. We need to show that the ξ_2-equation is ISS with respect to ω^3, which is known to be bounded from the preceding analysis. Because $\lambda_2 < 0$ and $\theta(t) < 0$ for all t, the homogeneous ξ_2-equation is exponentially stable, and, hence, it is ISS.

In this way we have shown that all the signals in the adaptive system remain bounded, provided $\theta(t)$ is bounded.

5.6 Simulations and discussion

While our analysis is incomplete, especially for the Monopoli scheme, the extensive simulations show excellent convergence properties of both adaptive schemes. In Figures 5.1 and 5.2 we present typical results obtained with $l_1 = -6$, $l_2 = -12$, $k = -4$, and $\varepsilon = 0.1$, for medium size parameter error $\tilde{\theta}(0)$. The performance remains good for larger $\tilde{\theta}(0)$.

During the first 30 *sec* of each simulation, the Arcak observer operated without adaptation, using a constant parameter estimate. Significant observer

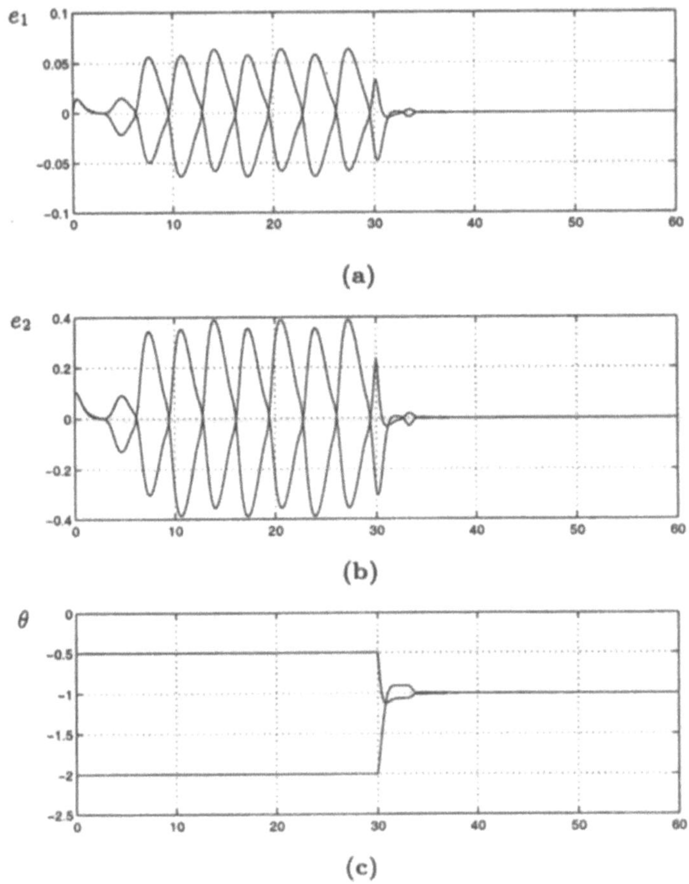

Fig. 5.1. MRAC scheme with $\mu = 50$ and $\theta^* = -1$. First run : $\theta(0) = -0.5$. Second run : $\theta(0) = -2$.

error $e(t)$ is shown in plots (a,b). With this estimate as the initial guess $\theta(0)$, the adaptation is switched on at $t = 30$ *sec*. The convergence of $e(t)$, in the plots (a,b), and $\tilde{\theta}(t)$, in plots (c), is fast and its speed can be further increased by increasing the adaptation gain μ.

It is remarkable that the gradient scheme with sensitivity filters actually benefits from the increase of the adaptation speed, a fact not predicted by the averaging theory in [1]. Because of the strong ISS property derived in Section 5.5, it is likely that a mechanism other than the simple gradient is at work in this scheme.

The prospect of making the Arcak observer adaptive is highly promising, leaving the proof of convergence a challenge for future research.

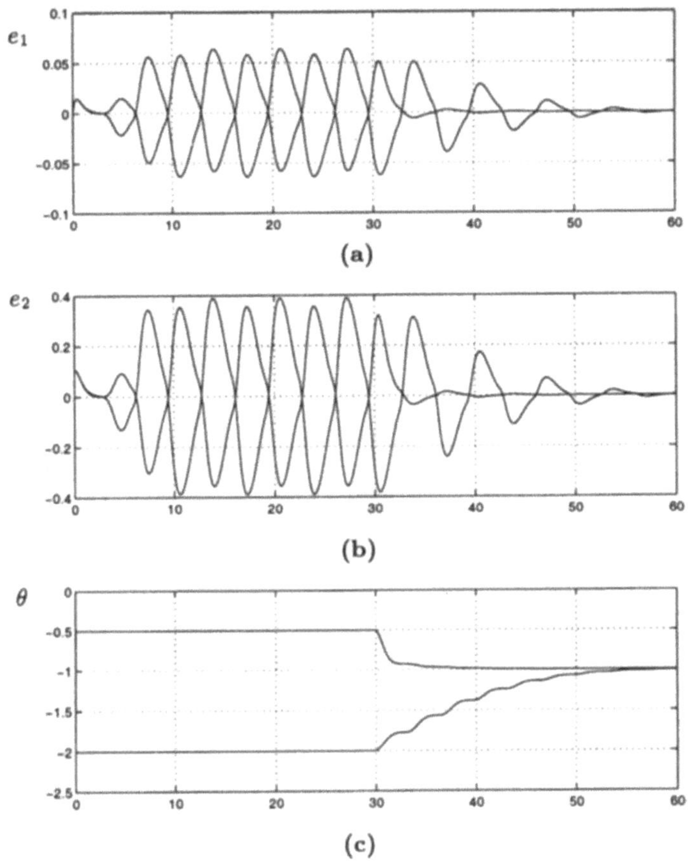

Fig. 5.2. Gradient scheme with $\mu = 50$ and $\theta^* = -1$. First run : $\theta(0) = -0.5$. Second run : $\theta(0) = -2$.

Acknowledgement The authors are grateful to Murat Arcak for helpful comments.

References

1. B. D. O. Anderson, R. R. Bitmead, C. R. JohnsonJr., P. K. Kokotović, R. L. Kosut, I. M. Y. Mareels, L. Praly, and B. D. Riedle (1986). Stability of Adaptive Systems. MIT Press.
2. M. Arcak and P. V. Kokotović (1999). Nonlinear Observers: A circle criterion design. Proceedings of the 38th IEEE Conference on Decision and Control, Phoenix, AZ. 4872-4876.
3. K. J. Åström and B. Wittenmark (1995). Adaptive Control (Second edition). Addison–Wesley.
4. P. A. Ioannou and J. Sun (1996). Robust Adaptive Control. Prentice-Hall.

6 Sensor Adaptive Target Tracking over Variable Bandwidth Networks

Robin Evans, Vikram Krishnamurthy, and Girish Nair

Centre of Expertise in Networked Decision Systems
Department of Electrical and Electronic Engineering
University of Melbourne
Australia

6.1 Introduction

Tracking a maneuvering target is often formulated as a problem of estimating the state of a partially observed jump Markov linear system. For example, consider a radar track-while-scan system where a stream of range and bearing (and perhaps doppler) measurements are obtained at regular intervals (the radar scan rate) on all targets (aircraft, clouds) within the field of view. The purpose of the tracking system is to search through this data in real time for trails of measurements which correspond to targets of interest (aircraft) and then to follow these targets and estimate their kinematic parameters (range, speed, heading, etc). False tracks can arise from consistent trails of noise and clutter and an important measure of a tracking algorithm's performance is its ability to minimize the number and duration of false tracks while reliably following real tracks.

A typical multitarget tracking system [7] consists of three sub-systems: (i) a *track initiation procedure* which detects tracks and initializes the track state estimator, (ii) a *track state estimator* which generates estimates of the target kinematic parameters based on the available noisy measurements, and (iii) a *data association process* that determines which data arise from which track.

Increasingly, tracking systems attempt to combine the above processes into a single integrated procedure. Real tracks are described by a Markov chain which models track existence, a Markov sequence which models track kinematics, and a measurement model for both real target measurements and false (clutter) measurements. The optimal estimator for this complete track model is then approximated in some computationally feasible fashion. Algorithms such as IPDA [25], MHT [29] and HMM-based trackers [31] integrate these tracking system functions to various degrees.

Here we consider a further enhancement and integration of tracking system functionality by extending existing algorithms to handle flexible multimode measurement sensors and variable bandwidth data links between the sensors and the tracker. Flexible multi-mode sensors are capable of rapidly switching between various measurement modes so that the tracking system can select which mode to use for the next measurement. Significant increases in tracker performance have been demonstrated for such systems [15]. Similarly when the sensors are connected to the tracker via a varying bandwidth data link, the tracker must dynamically decide which sensor data is the best to transfer using the available bandwidth during each measurement interval. We distinguish two cases. If the data link is only one-way from the sensor to the tracker then the optimal use of the available bandwidth must be determined by the sensor without knowledge of the current tracker state [27]. Alternatively, if there is a two-way link then the tracker can control the sensor and instruct the sensor as to what data to send, based on current tracking performance. This situation is clearly very similar to the flexible sensor situation.

Below we focus on the flexible sensor and two-way data link problem whereby the tracker must control the sensor or the data link. This is a stochastic control problem. Furthermore, since the tracker attempts to control tracking estimation error performance without direct measurement of the tracking error, this is a partially observed stochastic control problem. We model the sensor adaptive tracking problem as a control problem for a jump Markov linear system (JMLS). JMLS have widely been used in areas such as target tracking [17], [11], CDMA spread spectrum telecommunication systems [19] and pulse-train de-interleaving [21]. JMLS are a significant generalization of the concept of Hidden Markov Models in that they relax the conditional independence assumption of the observations given the state which is a key requirement for standard Hidden Markov Models (HMM). Moreover, they are also a generalization of the linear Gaussian state space model since they allow the parameters to the system to evolve randomly according to the sample path of a Markov chain. Unfortunately, unlike a HMM or Gaussian state space model where optimal state estimation can be performed via the HMM filter and Kalman filter, respectively, computing the optimal (conditional mean) state estimate of a JMLS requires exponential complexity (see [21], [10] and [9] for an exposition of optimal and suboptimal state estimation algorithms for JMLS). Yet, the optimal filtered state estimate is a key ingredient (called the "information state") required in the optimal solution of the sensor adaptive stochastic tracking problem.

This paper presents optimal and practically implementable sub-optimal algorithms which integrate sensor and/or data link control into the tracker. The optimal sensor adaptive algorithm is presented in terms of stochastic dynamic programming for a finite horizon cost function. The suboptimal sensor adaptive algorithms we present consist of two approximations: (i) instead of optimizing a finite horizon cost function, we show that satisfactory results can be obtained by optimizing the expected instantaneous cost at each time. (ii) approximating the optimal state filter by the Interacting Multiple Model (IMM) algorithm (or some other sub-optimal algorithm).

The remainder of this paper is organised as follows. In Section 2, the sensor adaptive tracking problem is formulated. Section 3 discusses two illustrative examples of the sensor adaptive tracking problem. Section 4 presents special cases where optimal processing is possible. In Section 5 a suboptimal one-step ahead algorithm is presented and a new Sensor Adaptive Interacting Multiple Model (SA-IMM) tracker is described. Section 6 presents two numerical examples. Finally, in Section 7 we briefly consider a new Sensor Adaptive Probabilistic Data Association (SA-PDA) tracker to handle false measurents and analyse the case of a one-way, bandwidth-limited communication link.

6.2 Problem Formulation

6.2.1 Target and Sensor Model

Let $t = 1, 2 \ldots$ denote discrete time. The signal model consists of the following ingredients.

a. Target Mode: In order to model target maneuvers the underlying target trajectory model jumps between a finite set of known modes representing various trajectory models such as constant speed and heading, constant heading with linear acceleration, co-ordinated turn etc.. This mode jumping can be modelled by a finite state Markov chain r_t. Let r_t denote a discrete-time, time-homogeneous, s-state, first-order Markov chain with state space $\{e_1, \ldots, e_s\}$ where e_i denotes the unit s dimensional vector with 1 in the i-th entry. Denote the transition probabilities as $p_{m,n} = \Pr\{r_{t+1} = n | r_t = m\}$ for any $m, n \in S$ where $S = \{1, 2, ..., s\}$. The transition probability matrix $[p_{m,n}]$, is thus an $s \times s$ stochastic matrix with elements satisfying $p_{m,n} \geq 0$ and $\sum_{n=1}^{s} p_{m,n} = 1$, for each $m \in S$. Denote the initial probability distribution by $p_m = \Pr\{r_1 = m\}$, for $m \in S$, such that $p_m \geq 0, \forall m \in S$ and $\sum_{m=1}^{s} p_m = 1$.

b. Target Kinematic State: Target trajectories of interest are modelled by a process $\{x_t\}$ where typically x_t is the target kinematic state (position, velocity) in Cartesian coordinates. Because the target model jumps according to the modes r_t, the target state x_t can be modelled as the following Jump Markov linear system (JMLS):

$$x_{t+1} = A(r_{t+1}) x_t + B(r_{t+1}) w_t \tag{6.1}$$

where $x_t \in \mathbb{R}^{n_x}$ is the system state, and $w_t \overset{i.i.d.}{\sim} \mathcal{N}(0, I_{n_w}) \in \mathbb{R}^{n_w}$. $A(\cdot)$ and $B(\cdot)$ are functions of the Markov chain state r_t, *i.e.* $(A(\cdot), B(\cdot)) \subset \{(A(m), B(m)); m \in S\}$, and they evolve according to the realisation of the finite state Markov chain r_t. We assume that $x_0 \sim \mathcal{N}(\hat{x}_0, P_0)$ where $P_0 > 0$ and let x_0 and w_t be mutually independent for all t.

c. Sensor Choice and Observation process: Assume that at time t there are L_t possible sensors available which can be used to give measurements of the target's state x_t. However, at each time instant only one of these

L_t possible sensors can be chosen. The variable $u_t \in \{1, \ldots L_t\}$ denotes the sensor selected at time t from among the L_t possible sensors. u_t will be chosen to maximize a cost function as described later. For notational simplicity we will ignore the time dependence t of L_t.

Let the vector $y_t(u_t) \in \mathbb{R}^{n_v}$ denote the observation vector at time t if sensor $u_t \in \{1, \ldots, L\}$ was chosen. In tracking problems, as well as the "true" observation $y_t(u_t)$ several false measurements are also obtained by sensor u_t. These false measurements are typically modeled as detections which are uniformly distributed throughout the observation span of a particular sensor. For example, a range measuring sensor might detect a real target present at a particular range with a certain detection probability and will report this range with an additive measurement error. The sensor will also report other detections (e.g., clutter, noise) at other ranges. These false measurements are assumed to be uniformly distributed over the full measurement range of the sensor and the probability of seeing a false-range measurement at any particular range is governed by a Poisson arrival process. In this paper we concentrate mostly on the no false measurements case and only briefly consider false measurements in Section 7.

In general the real target observation probability density $p(y_t|x_t, r_t, u_t)$ of a given sensor at time t could be non-Gaussian. However, to simplify our presentation, in this paper we will use the following linear Gaussian sensor model

$$y_t(u_t) = C(r_t, u_t)\, x_t + D(r_t, u_t)\, v_t \tag{6.2}$$

where $v_t \overset{i.i.d.}{\sim} \mathcal{N}(0, I_{n_v}) \in \mathbb{R}^{n_v}$ are i.i.d. Gaussian sequences and $D(i,j)D(i,j)^{\mathrm{T}} > 0, \forall i \in S, j \in \{1, 2, \ldots, L\}$, i.e. positive definite. We assume that x_0, w_t and v_t are mutually independent for all t.

Equations (6.1) and (6.2) denote a controlled partially observed jump Markov linear system. The model parameters

$$\lambda = \{p_m, p_{mn}, A(m), B(m), C(m, u_t), D(m, u_t), \hat{x}_0, P_0; m, n \in S\}$$

are assumed known. This is a reasonable assumption in many tracking problems.

Note that if the realization of the Markov chain sequence r_t is known (or if $S = 1$) and there is only one sensor, then (6.1) and (6.2) reduce to a linear Gaussian state space model and the Kalman filter yields optimal (conditional mean) estimates of the the target state. On the other hand if $L = 1$, $A(r_t) = I$, $B(r_t) = 0$, then y_t in (6.2) specializes to a standard Hidden Markov Model – i.e., the finite state Markov chain r_t observed in i.i.d. noise.

6.2.2 The Sensor Adaptive Tracking Problem

Let $Y_t = \{u_1, u_2, \ldots, u_t, y_1(u_1), y_2(u_2), \ldots, y_t(u_t)\}$ so that Y_t represents the information available at time t upon which to base state estimates and sensor selection decisions. As depicted in Fig.6.1, the sensor adaptive tracking problem proceeds in three stages for each $t = 0, 1, \ldots, N-1$, where N is a fixed positive integer

1) Selection: Based on Y_t we generate $u_{t+1} = \mu_{t+1}(Y_t)$ which determines which sensor is to be used at the next measurement time.

2) Observation: We then observe $y_{t+1}(u_{t+1})$ where u_{t+1} is the sensor selected in the previous stage.

3) Estimation: After observing $y_{t+1}(u_{t+1})$ we generate our best estimate $\hat{x}_{t+1}$ of the state of the Markov chain x_{t+1} as $\hat{x}_{t+1} = \mathbf{E}\{x_{t+1}|Y_{t+1}\}$. Note that the state estimate $\hat{x}_{t+1}$ is dependent on the sequence of sensors selected from time 1 to $t+1$, i.e. $u_1, \ldots, u_{t+1}$ (since it depends on Y_{t+1}).

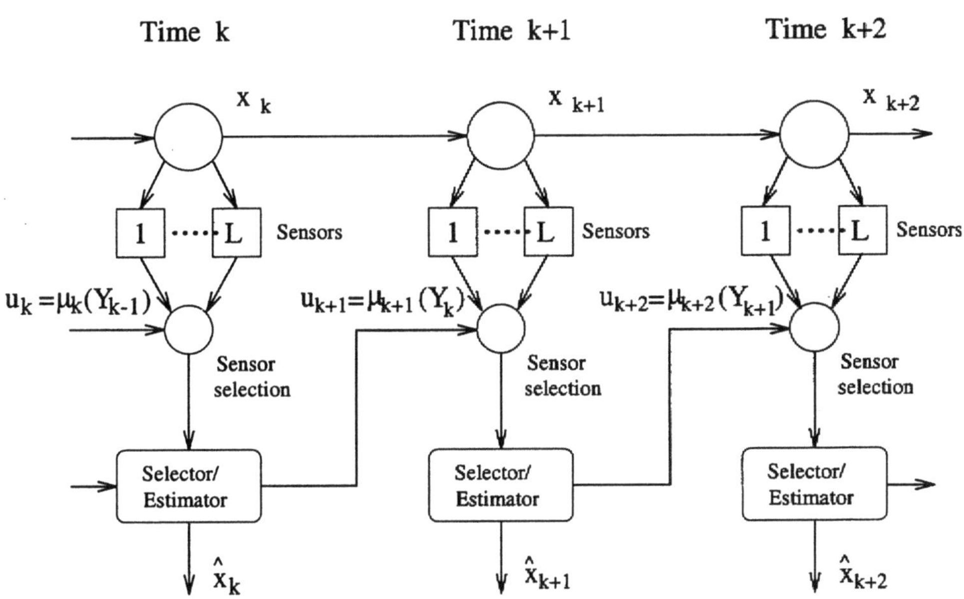

Fig. 6.1. The Sensor Selection and Track Estimation Problem

We define the sensor sequence

$$\mu = \{\mu_1, \mu_2, \ldots, \mu_N\}$$

and say that sequences are admissible if $u_{t+1} = \mu_{t+1}(Y_t)$ where

$$\mu_{t+1} : Y_t \to \{1, \ldots, M\}$$

is a sequence of functions.

We assume the following cost is associated with estimation errors and with the particular sensor sequence chosen. If based on the observation at time t, the decision is made to choose $u_{t+1} = l$ (i.e. to choose the l-th sensor at time $t+1$ where $l \in \{1, \dots, L\}$), then the instantaneous cost incurred at time t is

$$\alpha_t(l)\, \|x_t - \hat{x}_t\|^2 + \beta_t(l) \|r_t - \hat{r}_t\|^2 + \gamma_t(x_t, r_t, l). \tag{6.3}$$

Here $\alpha_t(l)$ and $\beta_t(l)$, $l = 1, 2 \dots, L$ are known non-negative scalar weights. $\hat{x}_t = \mathbf{E}\{x_t|Y_t\}$ and $\hat{r}_t = \mathbf{E}\{r_t|Y_t\}$ denote the optimal (conditional mean) estimates of x_t and r_t given the information sequence Y_t – note that these depend on the sequence μ. $\|x_t - \hat{x}_t\|^2$ denotes the mean square state estimation error in the target trajectory at time t due to the choice of sensor sequence $u_1, \dots, u_t$. Finally, $\gamma_t(x_t, r_t, u_t) \geq 0$ denotes the instantaneous cost of using the sensor u_{t+1} when the target 's mode is r_t and the target's state is x_t.

Our aim is to find the sensor sequence which minimizes the total accumulated cost J_μ from time 1 to N over the set of admissible sensor selection rules:

$$J_\mu = \mathbf{E}\left\{ \sum_{t=1}^{N} \alpha_t(u_{t+1}) \|x_t - \hat{x}_t\|^2 + \beta_t(u_{t+1}) \|r_t - \hat{r}_t\|^2 + \sum_{t=0}^{N-1} \gamma_t(x_t, r_t, u_{t+1}) \right\} \tag{6.4}$$

where $u_{t+1} = \mu_{t+1}(Y_t)$.

The terms $\alpha_t(l)$ and $\beta_t(l)$ allow different sensors $l \in \{1, 2, \dots, L\}$ to be weighted differently and the time index in α_t allows us to weight the state estimate errors over time. The final summation term reflects the cost involved in using a particular sensor (i.e. the unit time sensor charge) when the target is in state x_t and mode r_t.

The above problem is a finite horizon partially observed stochastic control problem. Infinite horizon problems with discounted cost can be formulated similarly.

6.2.3 Information State Formulation

Define the information state $\pi_t^{x_t, r_t}$ as the conditional density of x_t, r_t given the history Y_t, i.e.,

$$\pi_t^{x_t, r_t}(x, i) \triangleq p(x_t = x, r_t = e_i | Y_t) \text{ where } x \in \mathbb{R}^{n_x}, \quad i \in \{1, \dots, S\} \tag{6.5}$$

To simplify notation in the sequel define the probability density functions

$$\pi_t^{x_t}(x) \triangleq p(x_t = x|Y_t) = \sum_{i=1}^{s} \pi_t^{x_t,r_t}(x,i)$$

$$\pi_t^{r_t}(i) \triangleq P(r_t = e_i|Y_t) = \int_{\mathbb{R}^{n_x}} \pi_t^{x_t,r_t}(x,i)dx, \quad \pi_t^{r_t} \triangleq \begin{bmatrix} P(r_t = e_1|Y_t) \\ \vdots \\ P(r_t = e_s|Y_t) \end{bmatrix}$$

Using Bayes' rule it can be shown that the information state evolves as

$$\begin{aligned} \pi_{t+1}^{x_{t+1},r_{t+1}} &= T_t(\pi_t^{x_t,r_t}, y_{t+1}, u_{t+1}) \\ &= \frac{1}{K_t} p(y_{t+1}|r_{t+1}, x_{t+1}, u_{t+1}) \int_{\mathbb{R}^{n_x}} \sum_{m=1}^{s} p(x_{t+1}, r_{t+1}|x_t = x, r_t = m) \\ &\qquad \pi^{x_t,r_t}(x,m)dx \end{aligned} \tag{6.6}$$

where K_t denotes a normalization term. Defining $Q(r_t) = C(r_t)C'(r_t)$ and $\Sigma(r_t, u_t) = D(r_t, u_t)D'(r_t, u_t)$ for the JMLS (6.1) and (6.2), the various terms in (6.6) are

$$\begin{aligned} p(y_t|r_t, x_t, u_t) &\sim \mathcal{N}\left(C(r_t, u_t), \Sigma(r_t, u_t)\right) \\ p(x_{t+1}, r_{t+1}|x_t = x, r_t = m) &= p_{m,r_{t+1}} \mathcal{N}\left(A(r_{t+1})x, Q(r_{t+1})\right) \end{aligned} \tag{6.7}$$

The information state is a sufficient statistic for Y_t. Thus the admissible policies (sensor selection rules) satisfy $\mu_{t+1} : \pi^{x_t,r_t} \to \{1, \ldots, L\}$.

In terms of the information state, the cost function can be re-expressed as

$$\begin{aligned} J_\mu &= \mathbf{E}\Bigg\{\sum_{t=1}^{N} \alpha_t(u_{t+1}) \left[\langle x^2, \pi_t^{x_t}\rangle - \langle x, \pi_t^{x_t}\rangle^2\right] + \sum_{t=1}^{N} \beta_t(u_{t+1}) \left(1 - \pi_t^{r_t\prime} \pi_t^{r_t}\right) \\ &\quad + \sum_{t=0}^{N-1} \int_{\mathbb{R}^{n_x}} \sum_{i=1}^{S} \gamma_t(x, i, u_{t+1}) \pi_t^{x_t,r_t}(x,i)dx\Bigg\} \\ &\triangleq \mathbf{E}\left\{\sum_{t=0}^{N} \epsilon_t(u_{t+1}, \pi_t^{x_t,r_t})\right\} \end{aligned} \tag{6.8}$$

For the JMLS (6.1), (6.2) the information state $\pi_t^{x_t,r_t}$ is a Gaussian mixture with s^t components. Computing $\pi_t^{x_t,r_t}$ and hence also $\mathbf{E}\{x_t|Y_t\}$ or $\mathbf{E}\{s_t|Y_t\}$ are NP hard problems which involve prohibitive computational cost that is exponential in t. Thus in general the solution to the sensor adaptive tracking problem for a JMLS target model is not practical without some sort of approximation.

6.3 Illustrative Applications

In this section two examples of the above problem formulation are presented: tracking over bandlimited data links and tracking using an electronically steerable array.

6.3.1 Tracking Over Bandlimited Data Links

The architecture we have in mind is shown in Figure 6.2. Several sensors are connected to a multiplexer which in turn is connected via a limited bandwidth channel to a computer which performs target tracking based on observations provided by the sensors. Clearly if the channel bandwidth is adequate to allow all the sensor data to reach the estimation computer in a timely manner then we have a straightforward multisensor estimation problem which can be solved using standard methods such as the Kalman filter. However, if the channel bandwidth is limited then we must decide how best to allocate the sensor data to fit the available capacity so that optimal estimation is achieved. For example consider a scenario where 20 sensors each capable of making 16 bit measurements every 1 millisecond (i.e. 320 kbits/sec sensor data) are connected to a multiplexer which is connected to the estimation computer with a 32 kbits/sec link. The required 10:1 data rate reduction can be achieved by any number of static or dynamic assignment strategies such as only sending the most significant bit from each sensor (20 kbits/sec) or sending 16 bit data from 2 different sensors during each 1 millisecond time slot (32 kbits/sec) and so on.

We describe an approach to this problem based on the sensor selection ideas discussed above. We create a set of *virtual sensors* each supplying information at the required data rate. The bandwidth constrained estimation problem then becomes a problem of switching at each time slot between virtual sensors, i.e. a sensor selection problem. Continuing with the example above we might create virtual sensor 1 by taking the two most significant bits from real sensors 1 and 2, taking the 11 most significant bits from real sensor 6 and taking the most significant bit from the other 17 real sensors. Thus virtual sensor 1 consist of 32 bits and meets the 32 kbits/sec limit. Proceeding in this way we can create a set (possibly very large) of virtual sensors each supplying data at an acceptable rate. The problem then is for the tracker to dynamically choose the virtual sensors in an optimal fashion and transmit this choice to each sensor over a feedback channel with the same data rate.

This approach centralises decisions at the tracker and is suitable when only limited computational power can be located at the sensors. If there are no such computational constraints then clearly a better approach is to perform additional processing at the sensors, using the actual observations as well as the feedback bits received from the tracker. In the single sensor case feedback from the tracker is redundant, since all the data available to

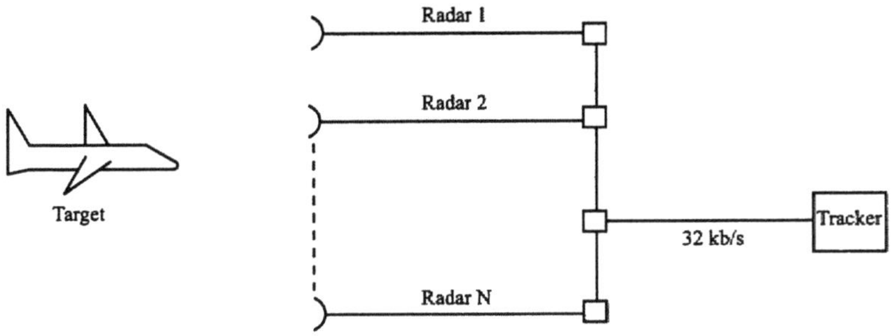

Fig. 6.2. Bandlimited Tracking

it originates at that sensor in the first place, and hence the optimal scheme locates all decisions at the sensor. This is further explored in Section 6.7.2.

6.3.2 Tracking Using an Electronically Steerable Array

Consider the following target tracking problem: P targets (e.g. aircraft) are being tracked by an electronically scanned array (ESA) with only one steerable beam. The coordinates of each target $x_t^{(p)}$, $p = 1, 2 \ldots, P$ evolve independently according to a JMLS and the conditional mean estimate of all target states is required. Since there is only one steerable beam, we can obtain noisy measurements $y_t^{(p)}$ of **only one target** at any given time instant. *Which single target should the tracker choose to observe at each time instant in order to optimize some specified cost function?* This is the flexible sensor situation described above and this question is an integral part of any agile beam tracking system [2,6,7]. Practical implementation of such systems is quite complex and beam scheduling must handle a variety of search, identification and tracking functions. Figure 6.3 shows a schematic representation of the problem. As shown in [18], the above problem is a special case of a sensor selection problem for JMLS with a multi-arm bandit structure. It is well known that the multi-arm bandit problem has a rich structure which results in the selection problem decoupling into P independent optimization problems. Indeed, it turns out that the optimal policy has an *indexable rule* [20], [32]: for each target p there is a function $\iota^{(p)}(x_t^{(p)})$ called the *Gittins index*, which is a function only of p and the information state $x_t^{(p)}$, and the optimal beem steering policy at time t is to steer the beam towards the target with the smallest Gittins index, i.e.,

$$\text{steer beam towards target } q \text{ where } q = \min_{p \in \{1, \ldots, P\}} \left\{ \iota^{(p)}(x_t^{(p)}) \right\}. \tag{6.9}$$

For a proof of this index rule for general multi-arm bandit problems see [32].

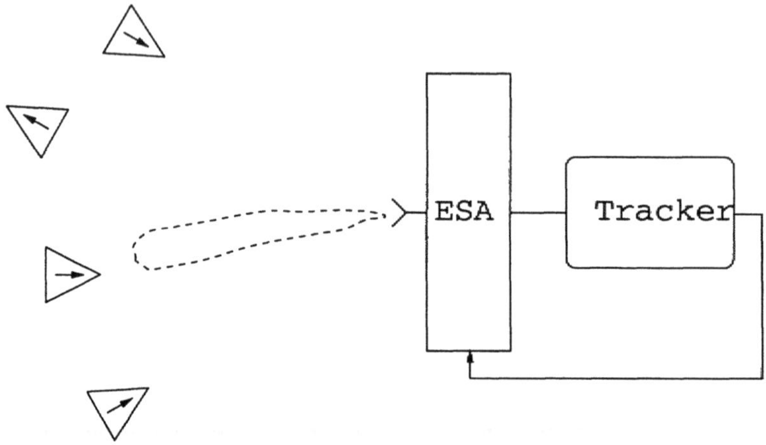

Fig. 6.3. Multi-target tracking with one multi-mode sensor.

6.4 Optimal Algorithms

6.4.1 Stochastic Dynamic Programming

Based on the above formulation, the solution to the optimal sensor selection problem is obtained using the Dynamic Programming algorithm [5] which proceeds backwards in time from $t = N$ to $t = 0$:

$$V_N(\pi^{x_t,r_t}) = \epsilon_N(\pi^{x_N,r_N})$$

and for $t = N-1, N-2, \dots, 0$

$$V_t(\pi^{x_t,r_t}) = \min_{u \in \{1,\dots,L\}} \left[\epsilon_t(u, \pi_t^{x_t,r_t}) + \mathbf{E}_{y_{t+1}} \left\{V_{t+1}\left(T(\pi_t^{x_t,r_t}, y_{t+1}, u)\right)\right\}\right] \tag{6.10}$$

The optimal cost starting from the initial condition π_0 is given by $V_0(\pi_0)$ and if $u_{t+1}^* = \mu_{t+1}^*(\pi_t)$ minimises the right hand side of (6.10) for each t and each π_t, the selection policy

$$\mu^* = \{\mu_1^*, \mu_2^*, \dots, \mu_N^*\}$$

is optimal.

The optimality of the above DP algorithm (6.10) follows from Theorem 3.2.1 in [14]. In particular, the measurable selection condition 3.3.2 in [14] is trivially satisfied (trivial because the control space is finite) as long as γ_t in (6.4) is lower semi-continuous.

In general, the above DP recursion (6.10) does not translate directly into practical solution methodologies. The fundamental problem with (6.10) is that at each iteration t, one needs to compute $V_t(\pi^{x_t,r_t})$ over an uncountable infinite set of states π^{x_t,r_t}. However, in two special cases, $V_t(\pi^{x_t,r_t})$ has a finite dimensional characterization.

6.4.2 Linear Quadratic Gaussian Sensor Selection

If the sequence $\{r_t\}$ is known (or $s = 1$) and $\gamma_t(x_t, r_t, u_{t+1}) = c_t(u_{t+1})$ then computing the optimal policy μ is equivalent to the following nonlinear deterministic control problem [23]. Minimize

$$J_\mu = \sum_{t=1}^{N} \mathrm{tr}[P_{t|t}] + \sum_{t=0}^{N-1} r_t(u_{t+1}) \tag{6.11}$$

subject to the constraint that

$$\begin{aligned} P_{t|t} &= P_{t|t-1} - P_{t|t-1}C'(r_t,u_t) \\ &\quad \times \left[C(r_t,u_t)P_{t|t-1}C'(r_t,u_t) + D(r_t,u_t)D'(r_t,u_t)\right]^{-1} C(r_t,u_t)P_{t|t-1} \\ P_{t|t-1} &= B(r_t)B'(r_t) + A(r_t)P_{t-1|t-1}A'(r_t), \qquad t = 1,\dots,N \end{aligned} \tag{6.12}$$

The above equations are merely the covariance update equation of a Kalman filter. Thus in the linear Gaussian quadratic case, the optimal sensor selection can be computed off-line independently of the data. This is not surprising; since the Kalman filter covariance is independent of the observation sequence.

6.4.3 Hidden Markov Model Sensor Selection

If $A(r_t) = I$ and $B = 0$, then (6.1), (6.2) reduces to a partially observed Markov decision process (POMDP). Suppose further that v_t is finite state i.i.d noise so that the measured output y_t belongs to a finite number of symbols, say $\{O_1,\dots,O_M\}$. Then in the case $\alpha = \beta = 0$, the optimal sensor sequence can be exactly determined by a finite dimensional algorithm. This finite dimensional characterization was first derived in [30], where it was shown that the value function $V_t(\pi)$ is piecewise linear and convex. Several linear programming based algorithms have been presented in the operations research literature for computing these piecewise linear segments. These include Sondik's algorithm [30], Monahan's algorithm [24], Cheng's algorithm [22], and the Witness algorithm [8]. We refer the reader to the website `http://www.cs.brown.edu/research/ai/pomdp/index.html` for an excellent tutorial exposition with graphics of these various algorithms.

6.5 Sub-optimal Algorithms

In practice the algorithms presented above are computationally too demanding for real-time on-line implementation. As mentioned above optimal state estimation for a JMLS is an NP hard problem and moreover, it is difficult to include false measurements. In this section we describe a one-step ahead sub-optimal solution to the sensor selection problem with multiple jumping models. The idea behind the one-step ahead algorithm is to compute the expected posterior density for the target state for each sensor based on the current posterior density and the known measurement models, then make a measurement using the sensor which gave the best predicted cost. The idea is illustrated below for the no false measurement case. The false measurement case is briefly described in Section 7.1.

6.5.1 One-step Ahead Algorithm

To simplify our notation, we will assume that the weights α and β in (6.4) are constants. The one-step ahead algorithm proceeds recursively as follows: Assume that $\pi_{t-1}^{x_{t-1},r_{t-1}}$ (or a suitable approximate value) has been computed at time $t-1$.

Step 1: Minimize at time t the instantaneous cost defined in (6.3) which can be re-expressed in terms of the information state as

$$J_{\mu_t} = \alpha\left[\langle x^2, \pi_t^{x_t}\rangle - \langle x, \pi_t^{x_t}\rangle^2\right] + \beta\left(1 - \pi_t^{r_t\prime}\pi_t^{r_t}\right) + \langle \gamma_t'(\cdot,\cdot,u_t), \pi_t^{x_t,r_t}\rangle \tag{6.13}$$

subject to $\pi_t^{x_t,r_t} = T(\pi_{t-1}^{x_{t-1},r_{t-1}}, y_t, u_t)$ where $T(\cdot)$ is defined in (6.6).

Using the dynamic programming recursion (6.10), the above cost function is straightforwardly minimized by

$$\begin{aligned} V_t(\pi^{x_t,r_t}) &= \alpha\left[\langle x^2, \pi_t^{x_t}\rangle - \langle x, \pi_t^{x_t}\rangle^2\right] + \beta\left(1 - \pi_t^{r_t\prime}\pi_t^{r_t}\right) \\ u_t^* &= \arg\min_u \left[\langle \gamma_{t-1}'(\cdot,\cdot,u_t), \pi_{t-1}^{x_{t-1},r_{t-1}}\rangle + \right. \\ &\qquad\qquad \left. \mathbf{E}_{y_t}\left\{V_t(T(\pi_{t-1}^{x_{t-1},r_{t-1}}, y_t, u))\right\}\right] \end{aligned} \tag{6.14}$$

Step 2: Set $\pi_t^{x_t,r_t} = T(\pi_{t-1}^{x_{t-1},r_{t-1}}, y_t, u_t^*)$
Step 3: $t \to t+1$

For a JMLS, the information state $\pi_t^{x_t,r_t}$ is a Gaussian mixture with s^t components. So in principle (although the computational complexity grows exponentially with t), the above one-step ahead algorithm provides an algorithm for the sensor adaptive tracking problem. Below we discuss how an approximation of $\pi_t^{x_t,r_t}$ can be efficiently computed using the IMM algorithm – the resulting algorithm will be termed the Sensor Adaptive IMM Algorithm (SA-IMM). But first, we briefly consider the linear Gaussian and HMM cases where π_t is easily computable.

Linear Quadratic Sensor Selection: Consider the same setup as in Sec. 6.4.2. Given $P_{t-1|t-1}$, it follows from (6.14) that the one-step ahead algorithm at time t for selecting the optimal sensor is

$$u_t^* = \arg\min_{u\in\{1,\dots,L\}} \left[P_{t|t}(u) + \gamma_{t-1}(u)\right]$$

where $P_{t|t}(u_t)$ evolves according to (6.12). Unlike the optimal algorithm of Sec. 6.4.2 the one-step ahead algorithm only requires L Kalman covariances to be evaluated at each time.

HMM Sensor Scheduling: Assume the same conditions as Sec. 6.4.3 but with β non-zero. The information state evolves according to the well known HMM filter:

$$\pi^{r_t} = T(\pi_{t-1}^{r_{t-1},y_t,u_t}) = \frac{\phi(y_t(u_t))P'\pi_{t-1}^{r_{t-1}}}{\mathbf{1}'\phi(y_t(u_t))P'\pi_{t-1}^{r_{t-1}}}$$

where $\phi(y_t(u_t)) = \operatorname{diag}[p(y_t|r_t = e_1, u_t), \ldots, p(y_t|r_t = e_s, u_t)]$. Given $\pi_{t-1}^{r_{t-1}}$, the one-step ahead adaptive sensor solution for HMMs becomes:

$$u_t^* = \arg \min_{u \in \{1,\ldots,L\}} \gamma_{t-1}'(\cdot, u)\pi_{t-1}^{r_{t-1}} + \beta \sum_{m=1}^{M} \left[1 - \frac{\pi_{t-1}^{r_{t-1}\prime} P\phi^2(y_t(u_t))P'\pi_{t-1}^{r_{t-1}}}{(\mathbf{1}'\phi(y_t(u_t))P'\pi_{t-1}^{r_{t-1}})^2} \right]$$

6.5.2 Sensor Adaptive IMM (SA-IMM) Algorithm

The IMM algorithm computes the approximate information state for a JMLS in a computational efficient manner by approximating the information state π^{x_t,r_t} (s^t component Gaussian mixture) by an s-component Gaussian mixture at each time instant t. It is worthwhile mentioning that there are several other sub-optimal algorithms for approximating the information state of a JMLS – in particular, particle filter based sequential sampling methods appear to be extremely promising, see [9,10] for details.

Given an approximate computation of the state $\pi^{x_{t-1},r_{t-1}}$, the SA-IMM algorithm operates recursively via the steps below.

Step 1 – IMM Algorithm: Compute an approximation of the updated information state $T(\pi^{x_{t-1},r_{t-1}}, y_t, u_t)$ as follows. We have $\pi^{x_t,r_t} = \pi^{r_t}p(x_t|r_t, Y_t)$. Furthermore π^{r_t} and $p(x_t|r_t, Y_t)$ can be expressed recursively according to the following two equations:

(i) By elementary application of Bayes' rule we have

$$\begin{aligned}
p(x_t|r_t, Y_t) &= p(y_t|x_t, r_t, u_t)p(x_t|r_t, Y_{t-1}) \\
&= p(y_t|x_t, r_t, u_t) \int_{\mathbb{R}^{n_x}} p(x_t|x_{t-1}, r_t)p(x_{t-1}|r_t, Y_{t-1})dx_{t-1}
\end{aligned}$$

$$= p(y_t|x_t, r_t, u_t) \int_{\mathbb{R}^{n_x}} p(x_t|x_{t-1}, r_t) \times \left[\frac{\sum_{i=1}^{s} p_{i,r_t} p(x_{t-1}|r_{t-1} = e_i, Y_{t-1})p(r_{t-1} = e_i|Y_{t-1})}{\sum_{i=1}^{S} a_{i,r_t} p(r_{t-1} = i|Y_{t-1})} \right] dx_{t-1} \tag{6.15}$$

(ii) Modal update

$$\pi^{r_t}(j) = \sum_{i=1}^{s} p_{i,j}\pi^{r_{t-1}}(i) \int_{\mathbb{R}^{n_x}} p(y_t|x_t, r_t = e_j, u_t)p(x_t|r_t = e_j, Y_{t-1})dx_t$$

Consider the term $p(x_t|r_t = e_j, Y_{t-1})$ enclosed in square brackets on the last line of (6.15). The IMM algorithm approximates this by a single Gaussian density with the same mean and variance for each $j = 1, \ldots, s$. Note that the densities $p(y_t|x_t, r_t, u_t)$, $p(x_t|x_{t-1}, r_t)$ are Gaussian as described in (6.7). Thus with the IMM approximation, π^{x_t,r_t} is a Gaussian mixture with s components.

Step 2 – Sensor Adaptive Step: Compute u_t^* according to (6.14).

Step 3 – IMM State Update: Update the information state π^{x_t,r_t} using the new observation $y_t(u_t^*)$ according to the IMM algorithm. Each of the components $p(x_t|r_t = e_j, Y_t)$, $j = 1, \ldots, s$ can be computed via a Kalman filter – hence the SA-IMM algorithm involves a bank of s parallel Kalman filters, see [3] for details.

6.6 Numerical Examples

We present two numerical examples. The first involves a multisensor tracking situation where 4 sensors comprising 2 range sensors and 2 bearing sensors are able to communicate with a tracking system via variable bandwidth data links. The target undergoes a maneuver simultaneously but there is no clutter. The bandwidth is such that all sensors cannot send their data at full accuracy to the tracker. The simulation results in Figures 6.4 and 6.5 show typical tracking performance for SA-IMM at data rates of 10 and 22 bits per update interval. A total bandwidth of 44 bits/interval is needed to communicate all the data from the four sensors.

The second numerical example concerns an unattended ground sensor network (UGSN) where 100 bearing measurement sensors are located at known locations and a target moves through this field of sensors. The tracking system has only enough bandwidth to obtain a bearing measurement from one sensor in each track update time. The simulation results in Figure 6.6 show typical tracking performance of SA-IMM when the bearing measurement error variance for each sensor is 1 radian squared.

6.7 Extensions

6.7.1 Sensor Adaptive PDA (SA-PDA)

When false measurements (clutter) are present the usual procedure is to identify which measurements are 'close' enough to the predicted measurement for a particular target track then use this set of 'validated' measurements in the track update procedure. The nearest-neighbour rule simply selects the validated measurement closest to the predicted measurement position whereas the PDA algorithm uses all of the validated measurements. Basically the track is updated with each measurement in the validated set and the updated estimates are combined in a weighted averaging process to form

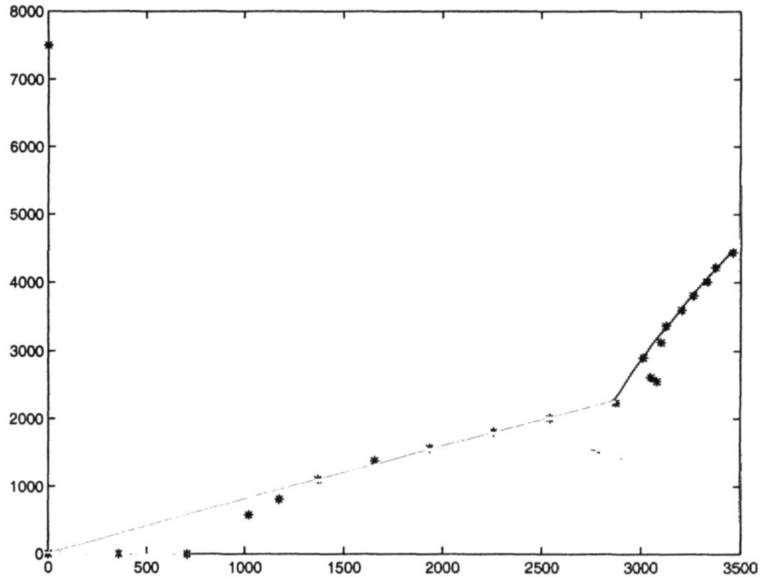

Fig. 6.4. SA-IMM Tracking at 10 bits/interval

the new estimate. The weighting is proportional to the likelihood of each measurement scenario. In the SA-PDA algorithm we must decide on which sensor to use next, based only on the current state estimate and known statistical properties of each sensor and the clutter.

Let f_t be the false measurements at time t and Ω_t the sequence of all real and false measurements and sensor choices up to and including time t. The SA-PDA algorithm computes the expected posterior density if sensor l was used by first computing the posterior density if real measurement y_t and clutter f_t were obtained and then averaging it over y_t and f_t, with x_t, Ω_{t-1} and u_t fixed, to yield

$$\begin{aligned}
&\bar{p}(x_t|\,\Omega_{t-1}, u_t = l)\\
&\triangleq \mathbf{E}\{p(x_t|y_t, f_t, u_t = l, \Omega_{t-1})|\,x_t, \Omega_{t-1}, u_t = l\},\\
&= \int\int \frac{p(y_t, f_t|x_t, u_t = l)p(x_t|\Omega_{t-1})}{p(y_t, f_t|\Omega_{t-1}, u_t = l)} p(y_t, f_t|x_t, u_t = l)dy_t df_t\\
&= p(x_t|\Omega_{t-1}) \int\int \frac{p(y_t, f_t|x_t, u_t = l)^2}{p(y_t, f_t|\Omega_{t-1}, u_t = l)} dy_t df_t. \qquad (6.16)
\end{aligned}$$

The sensor l^* which minimises a specific cost related to this expected density after normalisation is chosen and a standard PDA or IPDA update follows.

Instead of minimising a cost of the expected posterior density, another approach is to minimise the expected cost of the actual posterior density. For instance, we could compute the variance of the posterior density given

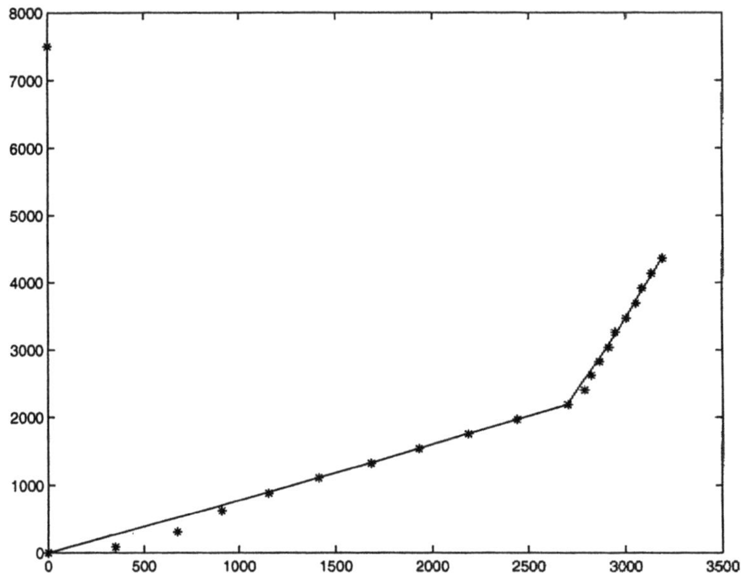

Fig. 6.5. SA-IMM Tracking at 22 bits/interval

$y_t, f_t, u_t = l$ and Ω_{t-1}, average it over y_t and f_t to yield

$$\mathbf{E}\{\text{var}\{x_t|y_t, f_t, \Omega_{t-1}, u_t = l\}|\, \Omega_{t-1}, u_t = l\}$$

and select the minimising sensor. In the linear Gaussian case with uniformly distributed false measurements, this is equivalent to computing a modified matrix Riccati equation update for each sensor, similar to that used in [16].

6.7.2 One-Way Communication Links

Consider again the scenario in which communication between a number of sensors and a tracker occurs over finite bandwidth links, as discussed in section 3.1. In the sensor selection approach, the tracker uses its information state at each time instant to dynamically select the next batch of virtual sensors. If the feedback channel from the tracker to each sensor is absent then the nature of the problem changes significantly, for instead of passively implementing the selections of the tracker each sensor must make its own bit allocation decisions, based on local information. Conversely the task of the tracker is now simply to filter the data it receives from the sensors, without providing any feedback. Thus this is no longer a centralised stochastic control problem.

In order to describe the essential characteristics of this problem, we focus here on the single link case without target mode-jumping. Let the target and sensor be governed by equations (6.1) and (6.2), with the mode r_t and control

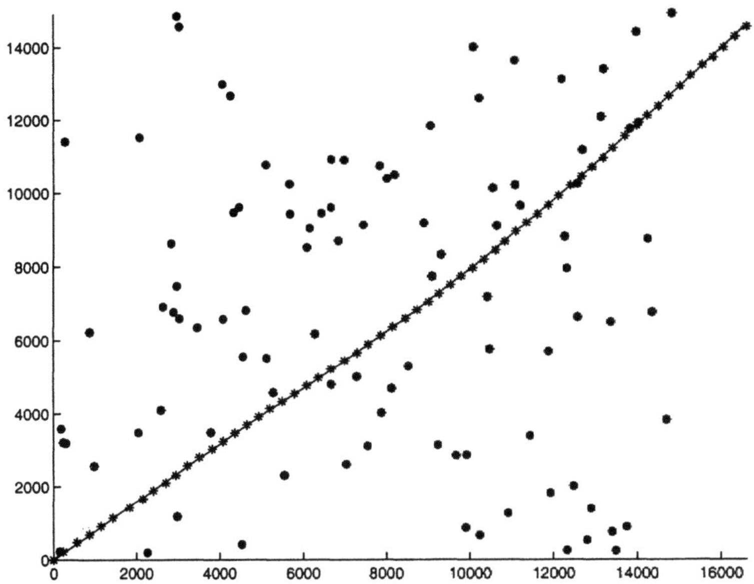

Fig. 6.6. SA-IMM Tracking with a Ground Sensor Network

variable u_t suppressed. Suppose that at each time t only one symbol ψ_t from the finite set $\mathbb{Z}_M \triangleq \{0, 1, \ldots, M-1\}$ can be transmitted from the sensor to the tracker, where $R \triangleq \log_2 M$ is the corresponding bit rate. Due to the finite rate, each symbol arrives at the tracker at time $t+1$. The coder and estimator are represented by the equations

$$\psi_t = \gamma_t(Y_t, \Psi_{t-1}) \in \mathbb{Z}_M,$$
$$\hat{x}_t = \eta_t(\Psi_{t-1}) \in \mathbb{R}^{n_x}, \quad t = 1, 2, \ldots,$$

where $\Psi_{t-1} \triangleq (\psi_1, \ldots, \psi_{t-1})$. The aim is to find a *coder-estimator* $(\gamma, \eta) \triangleq (\{\gamma_t\}, \{\eta_t\})$ which minimizes the finite horizon cost

$$J \triangleq \mathbf{E}\left\{\sum_{t=1}^{N} \alpha_t \|x_t - \hat{x}_t\|^2\right\}.$$

Note that since the information available to the tracker is a subset of that available to the coder, stochastic dynamic programming cannot be applied to minimize this. Nevertheless, several important structural properties may be stated.

Firstly, it can be shown that the minimum cost may be decomposed as

$$\min_{\gamma,\eta} J = \mathbf{E}\left\{\sum_{t=1}^{N} \alpha_t \|x_t - \hat{x}_{t|t-1}\|^2\right\} + \min_{\gamma,\eta} \mathbf{E}\left\{\sum_{t=1}^{N} \alpha_t \|\hat{x}_{t|t-1} - \hat{x}_t\|^2\right\}, \tag{6.17}$$

where $\hat{x}_{t|t-1} \triangleq \mathbf{E}\{x_t | Y_{t-1}\}$. As such the problem of coding and estimating $\{x_t\}$ under a quadratic cost is exactly equivalent to coding and estimating $\{\hat{x}_{t|t-1}\}$. In other words, as the coding step is where information is lost, the optimal coder first applies a Kalman filter to extract the best possible state estimates and then encodes them rather than the raw measurements. Note that the first term on the right hand side of (6.17) is just the cost that would be obtained without a data rate constraint. As expected, it is a lower bound for the rate-constrained cost.

Secondly, for a given estimator η the optimal coder is

$$\psi_t = \arg\min_{\psi \in \mathbf{Z}_M} f_{t+1}\left(\hat{x}_{t+1|t}, (\Psi_{t-1}, \psi)\right), \qquad \forall t \in [1, 2, \ldots, N-1],$$

where the functions $f_t : \mathbb{R}^{n_x} \times \mathbf{Z}_M^t \to [0, \infty)$ are defined by the downward recursion

$$f_t(\hat{x}_{t|t-1}, \Psi_{t-1}) \triangleq \alpha_t \|\hat{x}_{t|t-1} - \eta_t(\Psi_{t-1}\|^2 + \mathbf{E}\left\{ \min_{\psi \in \mathbf{Z}_M} f_{t+1}(\hat{x}_{t+1|t}, (\Psi_{t-1}, \psi)) \middle| \, \hat{x}_{t|t-1}, \Psi_{t-1} \right\} \quad (6.18)$$

with $f_{N+1} \triangleq 0$ and $\min_{\gamma,\eta} J = \min_\eta f_1$. Hence the symbol transmitted by an optimal coder at time t depends only on the current Kalman filter output $\hat{x}_{t+1|t}$ and the symbols previously transmitted. This is not surprising, since by virtue of the Gaussian assumption $\{\hat{x}_{t+1|t}\}$ is Markovian and thus $\hat{x}_{t+1|t}$ is a sufficient statistic for the future evolution of the filter process. Past filter outputs affect the current choice of symbol only through the symbols already transmitted. See [26] for more details.

In general it is impossible to derive explicit formulae for optimal coder-estimators, even for the case of a one-dimensional target model. The Generalized Lloyd Algorithm [13] could be applied or alternatively f_1 could be approximated numerically for fixed horizon, data rate and target and sensor models, but the entire procedure would have to be repeated if any of these changed. Nevertheless, the structural properties outlined above provide useful guidelines for the construction of suboptimal coder-estimators, as discussed in [27].

Finally, consider the following question: for given target and sensor models, what is the minimum data rate for which there exists a coder-estimator that produces bounded mean square estimation errors? In the special case of an autoregressive, linear time-invariant target model with no process or measurement noise, it has been shown that for there to exist a coder-estimator that takes the mean square error to zero it is necessary and sufficient that

$$R > \log_2 |\lambda|,$$

where λ is the model pole with largest magnitude [28]. We conjecture that the same condition is necessary and sufficient for bounded mean square errors in the stochastic case.

References

1. M. Athans. On the determination of optimal costly measurement strategies for linear stochastic systems. *Automatica*, 8:397–412, 1972.
2. Y. Bar-Shalom, editor. *Multitarget Multisensor Tracking: Applications and Adcances*, volume 2, chapter 8: Multitarget Tracking with an Agile Beam Radar. Artech House, 1993.
3. Y. Bar-Shalom and X. Li. *Estimation and Tracking: Principles, Techniques and Software*. Artech House, Boston, 1993.
4. J. S. Baras and A. Bensoussan. Optimal sensor scheduling in nonlinear filtering of diffusion processes. *SIAM Journal Control and Optimization*, 27(4):786–813, July 1989.
5. D. P. Bertsekas. *Dynamic Programming and Optimal Control*, volume 1 and 2. Athena Scientific, Belmont, Massachusetts, 1995.
6. D. R. Billetter. *Multifunction Array Radar*. Artech House, 1989.
7. S. Blackman and R. Popoli. *Design and Analysis of Modern Tracking Systems*. Artech House, 1999.
8. A. R. Cassandra. *Exact and Approximate Algorithms for Partially Observed Markov Decision Process*. PhD thesis, Brown University, 1998.
9. A. Doucet, N. Gordon, and V. Krishnamurthy. Particle filters for state estimation of jump Markov linear systems. *IEEE Trans. Signal Proc.*, 2000. to appear.
10. A. Doucet, A. Logothetis, and V. Krishnamurthy. Stochastic sampling algorithms for state estimation of jump Markov linear systems. *IEEE Trans. Auto. Control*, 45(2):188–202, Feb 2000.
11. J. S. Evans and R. J. Evans. Image-enhanced multiple model tracking. *Automatica*, 35(11):1769–1786, Nov. 1999.
12. J. S. Evans and V. Krishnamurthy. Optimal sensor scheduling for hidden Markov models. In *IEEE International Conference on Acoustics, Speech and Signal Processing, ICASSP98*, pages 2161–2164, Seattle, May 1998.
13. A. Gersho and R. M. Gray. *Vector Quantization and Signal Compression*. Kluwer, Boston, 1993.
14. O. Hernández-Lerma and J. B. Laserre. *Discrete-Time Markov Control Processes: Basic Optimality Criteria*. Springer-Verlag, New York, 1996.
15. D. Kershaw and R. Evans. Optimal waveform selection for tracking systems. *IEEE Transactions on Information Theory*, 40(5):1536–1550, Sep. 1994.
16. D. Kershaw and R. Evans. Waveform selective probabilistic data association. *IEEE Transactions on Aerospace and Electronic Systems*, 33(4):1180–1188, Oct. 1997.
17. V. Krishnamurthy and R. J. Elliott. Filters for estimating Markov modulated poisson processes and image based tracking. *Automatica*, 33(5):821–833, May 1997.
18. V. Krishnamurthy and R. J. Evans. Beam scheduling for electronically scanned array tracking systems using multi-arm bandits. In *Proceedings of 38th IEEE Conf. on Decision and Control*, pages 1039–1044, Phoenix, Arizona, Dec 1999.
19. V. Krishnamurthy and A. Logothetis. Adaptive nonlinear filters for narrowband interference suppression in spread spectrum CDMA systems. *IEEE Transactions on Communications*, 47(5):742–753, May 1999.

20. P. R. Kumar and P. Varaiya. *Stochastic systems – Estimation, Identification and Adaptive Control.* Prentice-Hall, New Jersey, 1986.
21. A. Logothetis and V. Krishnamurthy. Expectation maximization algorithms for MAP estimation of jump Markov linear systems. *IEEE Transactions on Signal Processing*, 47(8):2139–2156, August 1999.
22. W. S. Lovejoy. A survey of algorithmic methods for partially observed Markov decision processes. *Annals of Operations Research*, 28:47–66, 1991.
23. L. Meier, J. Perschon, and R. M. Dressler. Optimal control of measurement systems. *IEEE Transactions on Automatic Control*, 12(5):528–536, October 1967.
24. G. E. Monahan. A survey of partially observable Markov decision processes: theory, models and algorithms. *Management Science*, 28(1), January 1982.
25. D. Musicki, R. Evans, and S. Stankovic. Integrated probabilistic data association. *IEEE Transactions on Automatic Control*, 39(6):1237–1241, June 1994.
26. G. N. Nair and R. J. Evans. Structural results for finite bit-rate state estimation. In *Proceedings of 1999 Information, Decision and Control*, pages 47–52, Adelaide, Australia, Feb. 1999.
27. G. N. Nair and R. J. Evans. A finite-dimensional coder-estimator for rate-constrained state estimation. In *Proceedings of 14th IFAC World Congress*, volume I, pages 19–24, Beijing, P.R. China, July 1999.
28. G. N. Nair and R. J. Evans. Stabilization with data-rate-limited feedback: tightest attainable bounds. *Systems and Control Letters*, 41(1):49–56, Aug. 2000.
29. D. Reid. An algorithm for tracking multiple targets. *IEEE Transactions on Automatic Control*, 24(6):843–854, Dec. 1979.
30. R. D. Smallwood and E. J. Sondik. Optimal control of partially observable Markov processes over a finite horizon. *Operations Research*, 21:1071–1088, 1973.
31. X. Xianya and R. Evans. Multiple target tracking and multiple frequency line tracking using Hidden Markov Models. *IEEE Transactions on Signal Processing*, 39(12):2659–2676, December 1991.
32. P. Whittle. Multi-armed bandits and the Gittins index. *J. R. Statist. Soc. B*, 42(2):143–149, 1980.

Part III

Illustrative Applications of Identification and Control

7 Adaptive Servo Control of Large Antenna Structures

R.J. Evans[1], I.M.Y. Mareels[1], L. J. Sciacca[2], D.N.Cooper[3], R.H. Middleton[4], R.E. Betz[4], and R. A. Kennedy[5]

[1] Department of Electrical and Electronic Engineering, The University of Melbourne, VIC 3010, http://www.ee.mu.oz.au
[2] Tenix Defence Systems, Mawson Lakes, SA, 5095
[3] Formerly Division of Telecomms and Industrial Physics, CSIRO, Epping, NSW 2121, http://www.tip.csiro.au
[4] Department of Electrical and Computer Engineering, The University of Newcastle, NSW 2308, http://www.ee.newcastle.edu.au
[5] Department of Telecommunications, Research School of Information Sciences and Engineering, The Australian National University , ACT 2600, http://online.anu.edu.au/RSISE/teleng

Prelude

It is a pleasure to contribute to this volume in honour of Prof. Brian Anderson's sixtieth birthday. Prof. Brian Anderson taught and influenced us all. The paper illustrates the strength of some simple ideas from adaptive control in a difficult practical control context. We build on some of the earliest work in the area of adaptive control to realise a robust engineering solution to servo control of large antenna structures.

Abstract

This contribution considers the design of servo tracking loops for large antennae structures which exhibit lightly damped flexible modes. A cascaded loop control structure employing performance oriented adaptive feedforward in the outer loop is developed. The procedures for designing the inner current (motor drive), rate loop and position loop as well as the adaptive feedforward compensator are presented. It is demonstrated that this constitutes a practical and robust solution to the problem of controlling servo tracking systems in the presence of non trivial resonances. The control structure provides excellent tracking of parabolic trajectories as well as disturbance rejection. The conclusions are supported by analysis, simulation, and experimental results on 22m, 26m and 64m radio-astronomy and telecommunication antennae across Australia.

Fig. 7.1. Some of the antennae in the Australia Telescope Compact Array, Culgoora, New South Wales, Australia.

7.1 Introduction

We distill here the control engineering considerations from our involvement over a decade in a number of controller designs for large antennae used in radioastronomy and satellite tracking applications in Australia. These antennae require precision tracking and are fully steerable in azimuth and elevation. Typically the tracking drive systems and the antenna structure exhibit multiple, poorly damped, resonant modes. Figure 7.1 shows some of the antennae of the Australia Telescope Compact Array (ATCA) (for more information we refer to http://wwwnar.atnf.csiro.au/). The antennae in this array were used for some of the experimental results described below.

The problem of controlling large steerable antennae structures has been the subject of practical and theoretical studies for over 40 years. In an early survey of antenna control techniques, Wilson [4] points out that the key design issues revolve around the placement of the drive mechanical and structural resonances. He advocates the use of high gear ratios and velocity feedback to manipulate drive train resonances and thus improving control stability. The design of controllers for antenna systems is complicated by the fact that no two antenna designs are identical in structural, mechanical and operational characteristics. This leads Wilson to observe that antenna controllers will invariably require tuning and servo modifications during the commissioning phase of the antenna.

During the late 50's and early 60's, when many large radio-telescopes were being built, the standard approach to antenna control was to employ large gear ratios (of the order of 40,000:1) and Ward-Leonard and Metadyne generator systems to provide speed and torque control. Backlash, a significant inherent problem with large gear trains, was largely eliminated through the use of dual motors per axis and a torque-share-bias drive. All the antennae we worked with employ this mechanism.

In order to provide position feedback and coordinate translation from celestial to azimuth/elevation coordinates, a large antenna was often slaved to a small precision optical-based equatorial telescope. The Parkes 64m radio-telescope antenna used an ingenious error detection system mounted behind the main dish at the azimuth and elevation axes giving accurate feedback of the 64m dish position to hundredths of a degree [5], [6]. In these early systems, human operators were required to provide regular manual adjustments to the speed reference of the Ward-Leonard motor sets. The Parkes 64m radio-telescope was built in 1961. Detailed computer studies were carried out to design the servo controls for this complex mechanical and structural system [7]. To avoid exciting the low frequency resonance modes of the structure a low bandwidth servo system was used [6]. This consisted of high gear ratios, a low-pass filter in the error feedback loop and velocity feedback in the motor servo loop. As part of the commissioning trials, experiments were performed to determine the exact structural and mechanical resonances [8]. The careful design of mechanical and structural components of Parkes ensured the antenna remained one of the world's most accurate large radio telescope antennae for over two decades.

During the late 1970's, computers replaced the operators to provide true digital closed-loop position control of the antennae. With the advent of reliable power transistors, high bandwidth variable speed DC drives replaced the Ward-Leonard motor drive systems.

During the mid 1980's the drive system was upgraded with DC servo controllers and new high performance DC motors. A distributed computing system was used for digital position control. At about the same time the Australia Telescope, an array of 22m radio-telescopes, was being planned and Cooper, Gouth and Evans [9,10] were instrumental in designing the structure, drive, computing and control systems for the array. Lessons learnt from Parkes were employed, particularly in the drive system. Accurate position sensors (22 bit angle encoders resolve angle position to within 0.3 seconds of arc) and computers enabled digital real-time coordinate transformation to be carried out. Greater emphasis was placed on the design of the control feedback loops to reduce resonance effects and improve drive performance in the presence of disturbances.

Following the Australia Telescope experience we designed several more large antenna control systems including a number of 26m and 30m telecommunication antennae for satellite tracking. As a result of our work on these large antennae, we have designed a range of adaptive satellite tracking systems and adaptive position tracking systems and gained insight into some of the key factors influencing the design of control systems for large antennae. Here we describe our approach to some of the major control problems including the use of adaptive position feedback and cascaded feedback control for the motor drive system to suppress resonances and improve disturbance rejection.

The main conclusion is that a properly designed cascade of a high bandwidth current-loop, surrounded by a rate-loop, and finally enclosed by an adaptive position loop is a robust, cost-effective and high performance solution that ensures accurate tracking and effective disturbance rejection. We have examined a number of alternative control designs but finally concluded that provided the rate and position loops are designed in an appropriate fashion then the three loop cascade is a satisfactory engineering solution. In this way, the motor rate and current loops are tuned to dampen resonance effects and compensate for disturbances while the position loop focuses on meeting the tracking and low frequency disturbance rejection objectives. Our approach can be summarized as follows. The bandwidth of the inner current-loop is chosen high enough to ensure that the motor closely approximates a torque source. This is enclosed by a rate-loop designed to achieve maximum damping of the fixed-rotor resonance subject to acceptable reduction in stiction effects. Finally, the outer position-loop has a low bandwidth, in order not to excite drive train and structural resonances yet sufficient to reject wind disturbances. Furthermore, the loop must be capable of accurately tracking constant ramp and parabolic trajectories while still possessing good transient response. Other requirements include having the capability of rapidly slewing through large angles then smoothly entering precision tracking mode. In order to meet these competing requirements our controller incorporates an adaptive reference feed forward system [1,2] and resonance filtering. This approach allows a reduction in the number of closed-loop integrators (required to track the parabolic trajectories) thus allowing an increase in the loop-gain. The tracking requirements are met by the adaptive feedforward. The basic idea is that the feedforward control input is related to the amount of integral error in the position-loop which has the further advantage of an improved step response because of low integrator storage. A simple discrete event supervisor handles bumpless transfer between the control modes (including precision tracking, slewing, stowing, stopping etc).

The remainder of this contribution is organised as follows. Section 2 formulates the servo tracking problem and briefly describes the structural and mechanical as well as mathematical models. In Section 3 we discuss the servo control objectives, and consider solutions to the servo controller design prob-

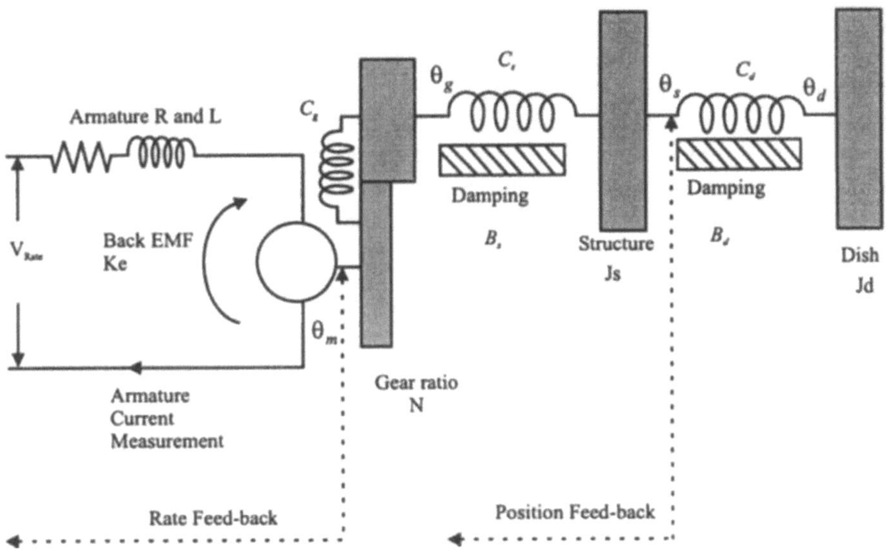

Fig. 7.2. Structural model of azimuth axis

lem. In Section 4 we discuss the current and tacho loops. Section 5 discusses the position servo loop in detail and presents an adaptive reference feedforward compensator. The adaptive law is a modification of the well known MIT-Rule. The resulting controller achieves the design objectives. We conclude in Section 6.

The design philosophy expounded here has been used to implement servo controllers for many large antennae. The experimental results presented throughout were taken from these systems.

7.2 System description

In this section the basic dynamic models used in the controller design are presented. The important non-linear effects are also described.

Essentially the antennae of interest are low compliance, large inertia structures. Figure 7.2 shows a simplified three mass/spring/damper model of an antenna. In our discussions we restrict ourselves to the azimuth axis as this axis has the highest inertia and lowest resonance and is thus considered the most difficult to control. We show experimentally that it possesses characteristics that vary depending on the elevation axis position. This can easily be appreciated by observing that the moment of inertia as reflected to the azimuth axis varies with the elevation angle of the dish.

For practical reasons the position encoder is placed at the structural support. This implies that the actual dish bore sight position is unknown and cannot be controlled directly. Some form of (dynamic) feed-forward compensation based on calibration against well known celestial objects must be employed to correct for the systematic deviation between encoder measurement and actual bore sight position, which has to be controlled. This feed-forward compensation also corrects for structural alignment, antenna seating misalignment, and dish distortions. These bias corrections are of considerable size and may be orders of magnitude greater than the required position loop accuracy. A detailed and elaborate calibration procedure is employed to largely remove these errors. It is also clear that any control implementation must be robust with respect to encoder placement variations. In the sequel we assume that this calibration is performed, and we make no further distinction between measurement position and desired bore sight position.

A detailed analysis was carried out on the gear train components taking into account the individual contributions of inertia, compliance and viscous friction of each gear and mesh section. The results showed a complex arrangement of poorly damped higher order resonance frequencies. However, for the purposes of controller design it turns out that a single lumped mass and compliance model is sufficiently accurate. Simulations show close similarity with the real plant response provided of course the closed loop bandwidth is low. Many of our preliminary controller designs were based on the gear train having the dominant inertia with the compliance reflected through the gears and lumped with the structural compliance.

An accurate model of the gear backlash is further complicated by the several gear trains that are present. The high gear ratios involved (in the range 15,000:1 to 40,000:1) mean that the final backlash may be several tens of degrees when reflected to the motor side. To counter the effects of backlash, a torque-share arrangement employing two DC motors was used to keep gear teeth in contact even in low torque situations (see Figure 7.10). This system works well and essentially allows backlash to be ignored for control system design purposes [11].

Friction effects, coulomb friction as well as stiction, in the gear and motor assembly and bearings are important in determining the achievable control loop performance. Figure 7.3 shows part of the friction model used for the motor system. A similar model is used to describe bearing (or wheel-on-track) stiction. For the antennae we have studied, motor stiction torque amounts to about one sixth of peak motor torque. Moreover, in star tracking applications motor speed is often near zero and control under stiction conditions is therefore of critical importance. The main source of stiction is found in the azimuth bearing. The effect of stiction on position error is illustrated in the Figure 7.4 which shows stiction induced limit cycles in the azimuth position when an Australia Telescope antenna is in holding position. (Observe the angle quantisation from the encoders.)

Large antenna structures have a multitude of resonances that must not be excited by the control actions. These resonances depend on the antenna construction, and for the azimuth control loop they depend on the elevation angle. The resonances are determined by the antenna structure, dish and

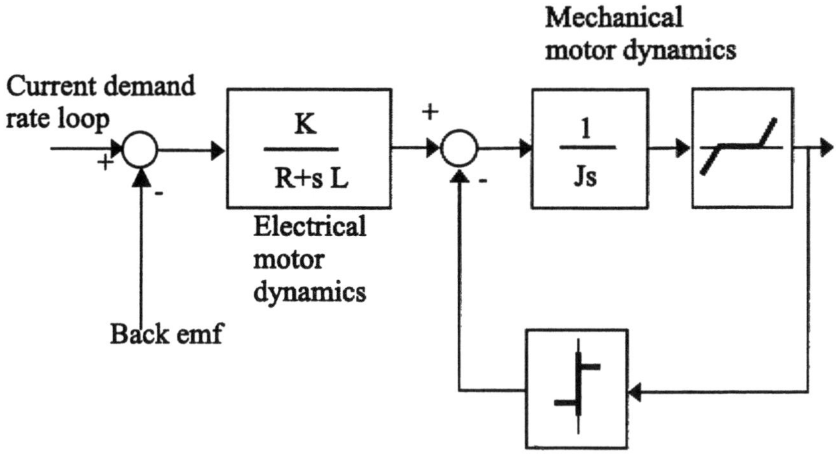

Fig. 7.3. Motor dynamics including stiction

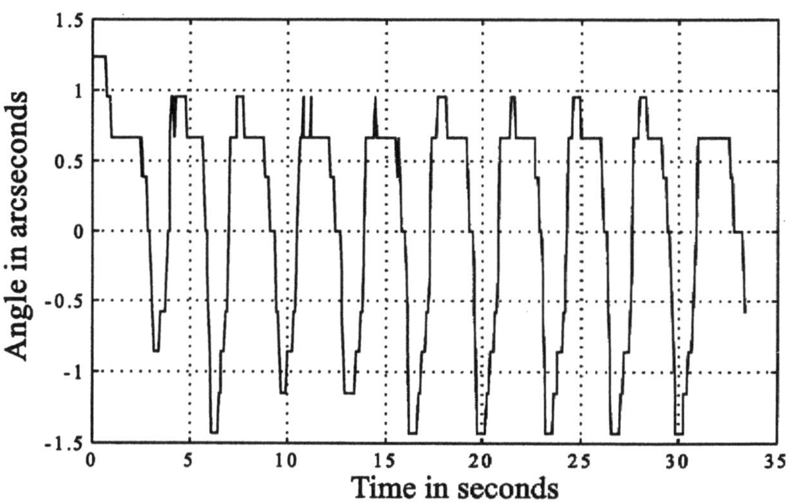

Fig. 7.4. Position limit cycles due to stiction in an ATCA antenna

major axes but also other structural features such as the supporting structure for the feed or feed reflector. For example, the feed tripod structure on the 64m antenna located in Parkes weighs several tonnes and the 1 tonne feed cabin is over 40m from the elevation drive system. This extended structure contributes a resonant mode which is reflected back into the servo loops.

We have performed resonance tests on many of the antennae we have studied in order to understand the multitude of resonances involved and aid in the servo loop design process. The resonance tests performed to date do not fully quantify the effects of wind disturbances on the structure. There are also higher frequency resonance peaks casued by the dish framework, but we found that these have reasonable damping, and are not normally excited by the controller or by wind disturbances. Figures 7.5 and 7.6 display some results of resonance tests performed on an Australia Telescope Antenna (22m diameter dish) and a (26m diameter dish) telecommunication antenna respectively. Figure 7.5 clearly shows that the resonance modes vary with elevation angle and that there are multiple significant resonant peaks.

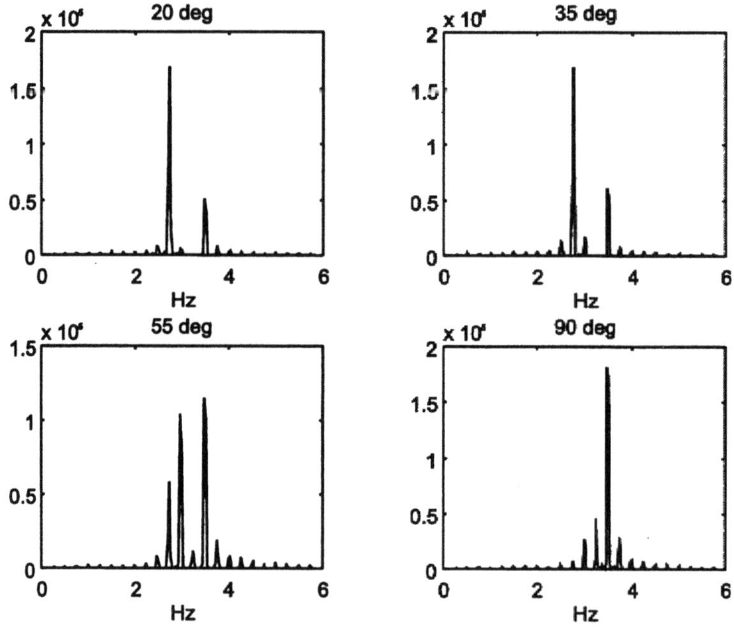

Fig. 7.5. Resonance tests on 22m Australia Telescope antenna, for different elevations

Identification experiments were carried out on several antennae to better model the relevant characteristics of the structure and drive system. Many significant non-linearities became apparent during these experiments inlcuding stiction in the gear/motor assembly, DC offsets and drift in the analog drive control circuitry. Various standard techniques from linear system identification [16] were employed to model the resonances. Figure 7.7 shows the Bode plot for a high order ARMAX model obtained for one of the antennae in the ATCA with an elevation angle of 55 degrees. The azimuth drive system was excited in open-loop with step inputs into the rate loop in order to show the dominant structural resonances.

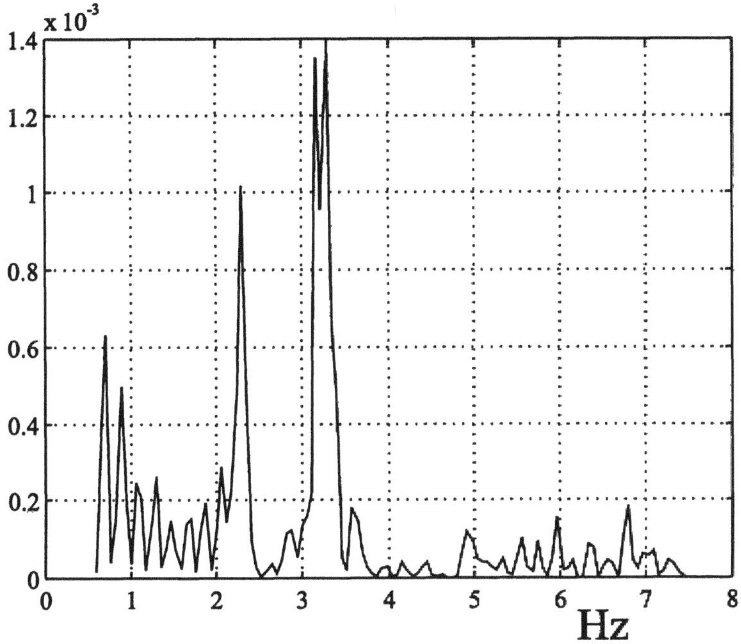

Fig. 7.6. Resonance test on 26m telecommunications antenna, frequency in Hz

Detailed models of the antennae, including the relevant nonlinearities, were implemented using MATRIXX, Easy5 and MATLAB. These models were used for simulation validation of the controller designs, before experimentation.

Initial studies (for the design of the gear train drives) considered the individual gear train resonance dynamics in great detail. For the position loop design however we focussed on the main lumped inertia and compliance models based on known structural data. Figure 7.8 shows one such model including some non-linear effects.

Partly pre-empting our servo objective considerations, and assuming the controlled system bandwidth to be less than the locked-rotor resonance, we can restrict ourselves to a model of the low frequency dynamics of the antenna. Then provided the control action is small, to avoid nonlinear considerations, the system transfer function from motor control action to azimuth angle position can be well approximated by:

$$\frac{G}{s(s^2 + 2\zeta\omega_2 s + \omega_2^2)}.$$

Here ω_2 represents the dominant structural resonance, which as discussed above depends on the elevation angle. This transfer function representation

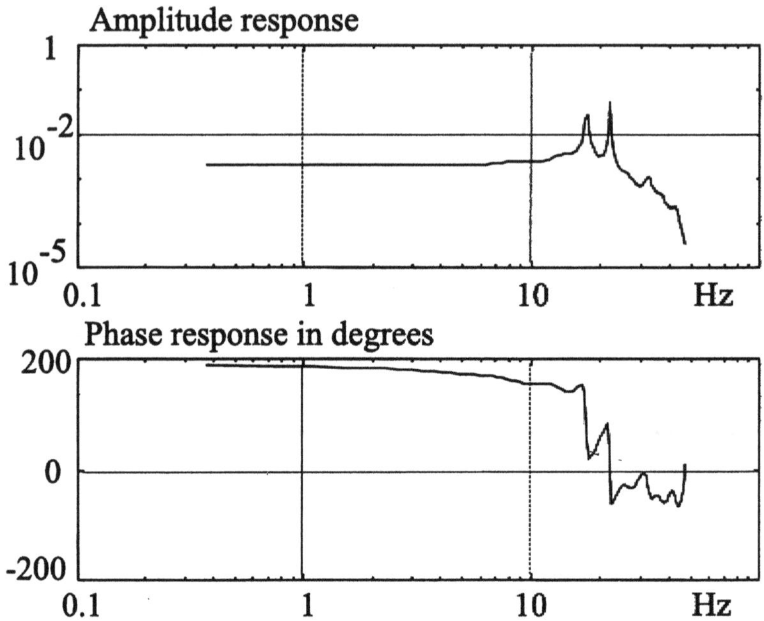

Fig. 7.7. Bodeplot, 22m antenna, azimuth rateloop, elevation 55 degrees

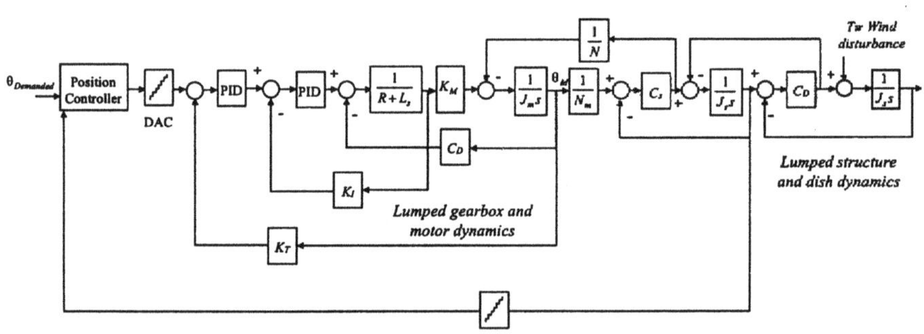

Fig. 7.8. Antenna model

ignores all nonlinear effects, assumes no friction, and is only valid if we ignore the presence of wind gusts that may excite frequency components above the locked rotor resonances. Essentially it only takes into account the lumped inertia of the motor armature and load, and the lumped compliance of the structure, gear boxes, and dish. Nevertheless it serves control design well.

7.3 Servo controller design objectives

The typical functions to be performed by the large antennae determine the servo objectives.

7.3.1 Ability to track stars

This is clearly the main aim for radio-astronomy applications, but it is also useful for antenna calibration and alignment in satellite tracking applications. In addition to tracking a single star, it is often desirable to be able to rapidly move from tracking one star to tracking another. Therefore the control system should be able to accurately track parabolic azimuth and elevation reference trajectories, and switch from one parabolic trajectory to another, at the maximum antenna speed, and with as low a settling time as possible. Typical star tracking rates range from 0 to 200 arcsec/s and accelerations are in the order of 0.3 arcsec/s^2. As indicated before stringent position accuracy requirements are imposed, angle position errors have to be less than 3 arcsec. This error limit must include all residual errors that remain after the calibration process, where errors of the order of 50 arcsec or more have to be removed. Observe that this is indeed a stringent requirement given that the allowed error is only marginally larger than the amplitude of the stiction induced limit cycles (amplitude of 1 arcsec see Figure 7.4 and not far removed from the position encoder's resolution accuracy (22 bit encoding achieves an angle resolution of 0.3 arcsec).

From a linear control perspective, tracking quadratic time trajectories and fast settling time are somewhat contradictory objectives. Indeed the tracking objective requires two or more integrators in the feedback loop, which is known to adversely affect settling time.

7.3.2 Ability to track geosynchronous and low earth orbit satellites

One method for tracking geosynchronous satellites is the so-called step track technique. The details of this technique are beyond the scope of this paper. Let it suffice to say that a rapid settling of the response to small step changes in the set point is required. This imposes a specification on the step response similar to that described above, except that the step changes in this case are generally much smaller in magnitude than those used when switching between stars.

Low earth orbit satellites, on the other hand, require faster tracking rates than what is required for star tracking. The rates are of the order of several degrees per second and often require tracking at full motor torque to achieve the desired antenna slewing rate and acceleration. This places a heavy demand on the position loop bandwidth. Again integrators are required in the position feedback loop to achieve zero steady-state error for parabolic trajectories.

7.3.3 Disturbance rejection

The impact of wind on an antenna can cause significant tracking errors. The torques created on the antenna by a wind of given velocity can normally be accurately predicted from the mechanical design. However, it is much harder to characterise the more typical random variations in wind speed. Stochastic models incorporating non-Gaussian distributions and non-white spectral properties are used, however it remains near impossible to account realistically for local ground effects and turbulence. For control design it suffices to approximate the wind effects as a combination of step torque disturbances or sinusoidal torque disturbances on the motor drive.

Of course in conditions of high wind the antenna will be stowed, but normal operations proceed in conditions of winds up to 15 m/s and typical wind spectra (under acceptable working conditions) have bandwidths of the order of 0.5 Hz [20]. Observe that this is only a factor of 4 below the lowest dominant resonance frequency for the ATCA.

7.3.4 Robustness

Most controller designs are based on a simplified model of the antenna that only captures the lowest frequency resonance and neglects all higher order resonant modes. In addition, there is significant uncertainty in the frequency and damping of the dominant resonance. Thus, there is considerable unstructured uncertainty in the antenna transfer function above the frequency of the first resonance. We rely on a simple small gain argument to ensure robust stability, by requiring that the feedback gain is insignificant in the region above the dominant resonance mode. This is completely in line with the safety requirements. Safety of personnel and equipment demands that under no circumstances the control action excite any resonance frequency. This necessitates the controller bandwidth limitation as determined by the location of the resonances.

7.3.5 Maintainability and tunability

One of the key aspects of any engineering system is the ease with which it can be installed, commissioned and maintained. In the case of a control system, this is related to the number of parameters that can be varied, and the level of expertise needed to be able to tune these parameters well. It is essential that all controller parameters be tunable in a straightforward and logical manner. As the radio telescope antennae are essentially laboratory equipment they are constantly being upgraded and changed. The control system is at the heart of the accuracy of these instruments, and it must therefore be reliable, robust and easy to tune and calibrate to achieve near best performance.

7.3.6 Operational modes

The above operational specifications suggest that an overall control strategy should consider different operational modes for the antenna. Typical implemented control modes are

- accurate tracking mode (single object tracking against parabolic trajectories),
- slewing mode (fast response to step change in position),
- fast tracking mode (higher velocity, reduced tracking accuracy requirements),
- docking mode (hold safety position in high wind situations),
- start up and shut down modes,
- manual mode (reference position with manual command).

In the sequel we focus solely on the accurate tracking mode. The different modes have their own specialised servo controllers, which are simply switched as required. The switching strategy must ensure a smooth control command signal transition, this is essential from a wear and tear perspective as well.

7.3.7 Design alternatives

In this subsection we briefly consider some of the design approaches we have trialled and subsequently rejected for the servo controller.

Pole placement
The R-S-T pole-placement method described in [12] was investigated in some depth [14]. A discretization method was devised which maintained robustness [15]. Satisfactory robust performance was observed in simulation studies. Practical trials, however, proved unsuccessful due to robustness issues associated with unmodelled dynamics, in particular the variable nature of the dominant resonance due to the changing elevation angle. Furthermore, the pole-placement controller contained 14 tunable and closely coupled coefficients, which makes it near impossible to gather intuition on how to tune these parameters. After installing the controller, an unsatisfactory controller response was very difficult to diagnose and correct. The poor maintainability/tunability properties make this controller strategy unmanageable in practice.

Adaptive pole placement
In an attempt to solve the tuning problem, alluded to before, gain scheduling and adaptive pole placement were briefly considered. Gain scheduling involves a major design and evaluation phase, and proved unmangeable and unreliable. The major difficulty encountered with adaptive pole-placement in this application was the lack of sufficient excitation. Typical reference trajectories are parabolic, hence contain only sufficient excitation for the indentification

of a single parameter. This coupled with disturbances and unmodelled dynamics lead to well known stability problems [17]. The complexity of the adaptive mechanism did little to alleviate the maintainability concerns. This control strategy was therefore abandoned.

7.4 Classical cascaded feedback

Our preferred solution for the accurate tracking mode antenna servo design involves the use of classical feedback design, based on three cascaded single input single output servo loops, together with an adaptive feedforward control scheme. The cascade provides the required robustness and the adaptive feedforward ensures tracking performance despite variations in the resonances, gains and reference trajectories.

The perceived advantages are

- easy to understand, simple classical design procedures can satisfactorily meet the disturbance rejection and safety requirements.
- the cascaded structure allows controller tuning to be broken up into a sequence of simple tuning operations.
- the adaptive feedforward is structured so that safety, stability and robustness are uncompromised, while ensuring essentially zero parabolic tracking error.

The cascaded feedback involves three control loops:

- the inner loop is a high bandwidth current feedback loop, using measurement of the DC motor armature current,
- the intermediate loop is a medium bandwidth velocity feedback loop, using tachometer measurement of the DC motor velocity,
- the outer loop is a low bandwidth position feedback loop, using the angle encoders.

For the ATCA antennae, the respective bandwidths of the cascade are chosen to be 90Hz, 20Hz and 1.8Hz.

High bandwidth motor current feedback loop
To reduce the sensitivity to motor parameters and to improve rejection of torque disturbances caused by wind on the dish, a high bandwidth current loop is placed around the motor as shown in Figure 7.9. A lead-lag compensator with integral action is used to ensures that the motor is essentially a torque source.

This current control loop ensures that the the motor torque T_m will be $T_m \approx k_m I_a^*$ where k_m is the motor back emf constant and I_a^* is the desired current from the rate loop. The current loop makes the DC motor look like a torque source and removes the effects of back-emf.

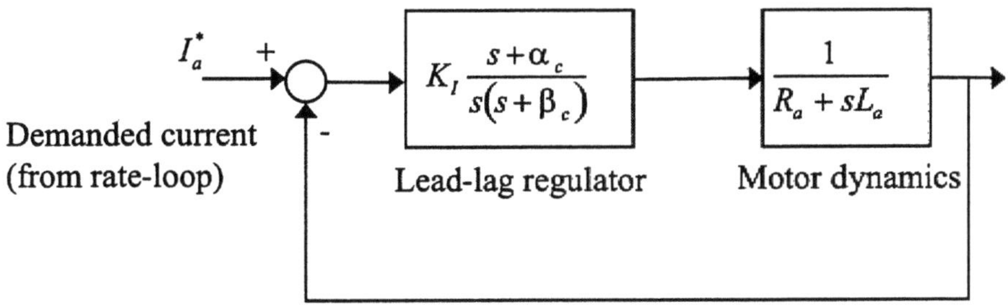

Fig. 7.9. Current loop controller

The current loop bandwidth is chosen as large as possible consistent with the type of DC chopper used. Current-loop bandwidths of around 60 Hz to 100 Hz are easily achievable using standard techniques.

There are two other advantages to using a high bandwidth current loop. Firstly, by using a high bandwidth current loop, we effectively render the rate loop to be passive. Indeed, the control variable for the rate-loop is up to a positive scalar the motor torque, and the measured variable is the motor speed. Thus, regardless of the number/frequency/damping of resonances, this system is passive, and is thus easy to control (high gain is feasible). A second advantage of a high bandwidth current loop is that it pushes the resonant frequency towards the free rotor resonant frequency. This frequency is higher than the locked rotor resonant frequency which results if a low bandwidth current loop is used. Unfortunately, because of the enormous gear ratio, the effective motor inertia reflected through the gear train is larger than the equivalent antenna inertia. As a consequence, the difference between the free and locked rotor resonant frequencies is small, and this second benefit is therefore minor in some cases.

The rate loop

The rate-loop serves the following main functions:

- It improves the damping of the main resonance.
- It assists in (wind) disturbance rejection.
- It reduces the number of integrators seen by the position loop, and hence assists the position loop to achieve a shorter settling time.
- It reduces the effects of nonlinear friction (stiction).

Here we focus on the effect of velocity feedback in improving the damping of the main resonance and assisting in disturbance rejection. For example, consider a wind disturbance that hits the dish and supporting structure. (Typical wind disturbance bandwidth is less 0.5 Hz.) This disturbance is reflected through the gears to the motor rotor. As the current loop ensures that in the first instance the motor reacts as a constant torque source, this disturbance results in a change in motor speed. This change in speed is then compensated for using the rate loop.

In our design, the rate loop's bandwidth is smaller than the current loop's bandwidth, by roughly a factor of 4. A bandwidth of 15Hz to 25Hz is sufficient to guarantee wind disturbance rejection.

The tacho voltage V_T to current loop reference I_a transfer function is approximately given by [19],

$$\frac{I_a}{V_T} = \frac{K\left(s^2 + \omega_0^2\right)}{s\left(s^2 + \omega_1^2\right)},$$

where K represents the gain, and is essentially determined by the inertia reflected back to the load side of the gear train, ω_0 is the locked rotor resonant frequency and ω_1 is the free rotor resonant frequency ($\omega_0 < \omega_1$). For example for a typical ATCA antenna, this transfer function is given by

$$\frac{I_a}{V_T} = \frac{1.42\left(s^2 + 355\right)}{s\left(s^2 + 361\right)}.$$

The main structural resonance ω_n is typically just above the locked rotor resonance ω_1. Again for the ATCA antenna, typical values are $\omega_1 = 19\frac{\mathrm{rad}}{\mathrm{s}}$ and $\omega_n = 19.1\frac{\mathrm{rad}}{\mathrm{s}}$ and $\omega_0 = 18.8\frac{\mathrm{rad}}{\mathrm{s}}$.

A PI regulator suffices to control the rate-loop system. Gains are selected to achieve a bandwidth of about 15 to 25 Hz and to maximise the damping of the closed loop poles. For the ATCA antenna this leads to a PI controller with transfer function $20.4\frac{(s+9.4)}{s}$. A further roll-off filter to limit the control action beyond the 25 Hz bandwidth is recommended.

One may be tempted to choose a rate loop with low bandwidth, to dampen the lowest resonant frequency, unfortunately, this is not compatible with the wind disturbance rejection requirement, and a higher bandwidth rate loop (as selected) is necessary.

The angle position loop

Finally the position loop is closed via a PI controller detuned with a low pass filter to ensure that no structural resonances can be excited. The overall bandwidth for the ATCA antennae is set at 1.8 Hz, just below the 2.9 Hz locked rotor resonance.

The main disturbance to be rejected is the wind disturbance which is most easily modelled as a load disturbance on the dish side of the dynamical model. The servo loop design considers the effect of wind as a torque disturbance reflected through the gears and affecting the motor speed as described earlier. Quantative analysis of the position error due to wind loading is difficult and simulations based on known wind gust profiles are used to help tune the position loop controller. However significant insight can be gained by considering a simple sinusoidal wind torque disturbance [10]. Ensuring that the RMS error due to a maximum wind velocity and sinusoidal disturbance remains inside the allowable position error budget allows one to determine the required P action of the position servo. Similarly, the limits of the integral action in the position loop can be estimated by ensuring that the maximum steady wind action on the antenna does not violate the position accuracy requirements.

The cascaded feedback system, together with the anti-backlash torque share arrangement is represented in the Figure 7.10.

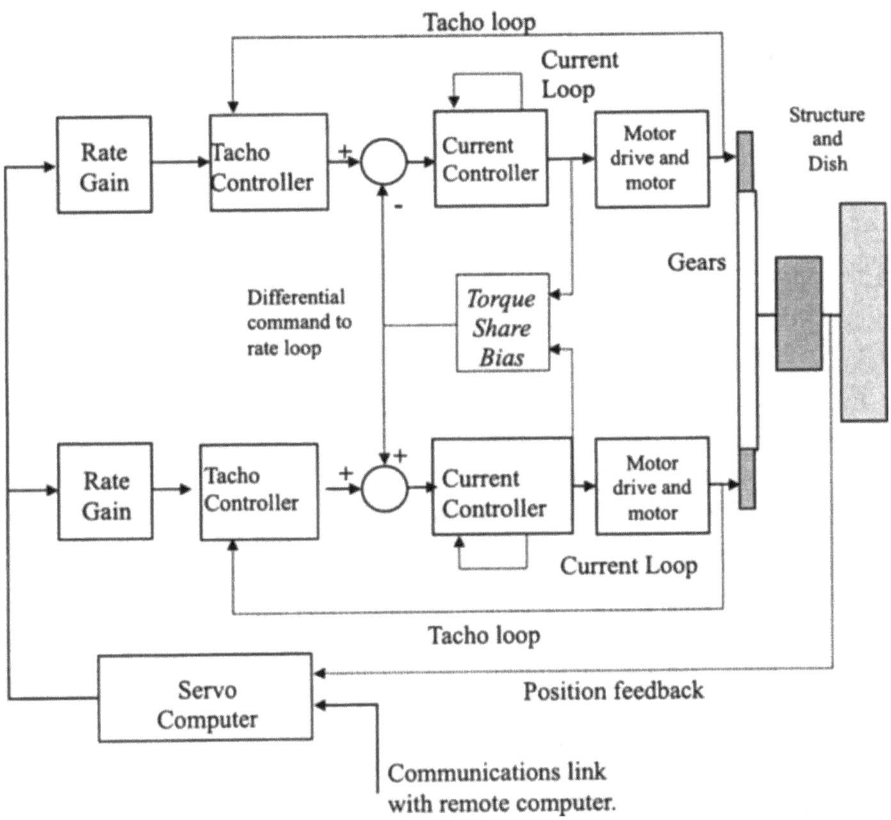

Fig. 7.10. Feedback controller and torque share arrangement

7.5 Adaptive position control

This section discusses in greater detail the design of the outer position loop, as it applies to the accurate tracking mode. In particular we describe the adaptive feedforward action.

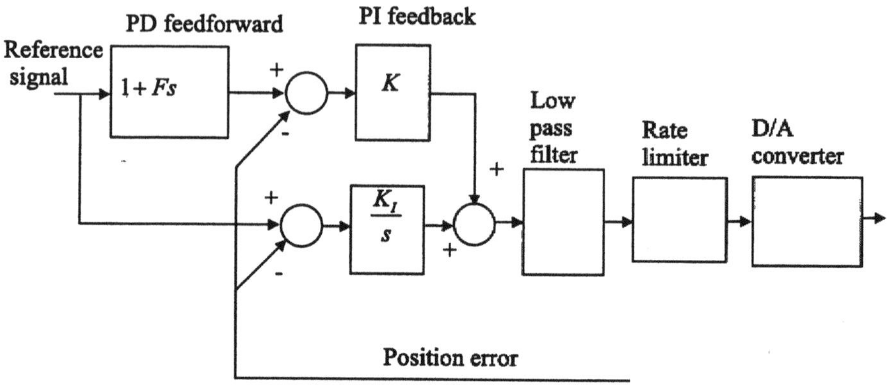

Fig. 7.11. PD feedforward with PI feedback in position loop

7.5.1 Non-adaptive position control

Lead-lag controllers have been advocated and successfully used in antenna position control [4]. Integral action, with two integrators in the feedback loop, enables the antenna to track ramps and parabolas and compensates for steady-state wind disturbances. Finally, a low pass filter with a roll-off below the lowest antenna resonant frequency is used to avoid exciting the resonant modes of the structure.

However, during commissioning of the Australia Telescope a fixed proportional derivative (PD) feed-forward combined with proportional and integral (PI) feed-back loop was used as shown in Figure 7.11 [10]. Observe the rate limiter, which is included to protect the motor and gear train from excessive wear.

The rationale for this somewhat unconventional approach is that in order to speed up the response one of the integrators is removed from the feedback loop and replaced by derivative action in the feedforward [3]. This was necessary as the two integrators in the feedback path solution was too sluggish to be able to achieve both a satisfactory settle time as well as sufficient wind disturbance rejection. The problem with this alternative design is that it achieves the desired tracking accuracy only for a narrow range of parabolic trajectories. Moreover as the feedforward action depends on the actual dynamics in the feedback loop, it must be re-tuned whenever this loop changes, e.g. when the motor drive units are replaced, or updated, a not infrequent event.

Figure 7.12 depicts an unsatisfactory position angle response for the case of the non-adaptive PD feed-forward together with a PI feedback loop. The D action is not well calibrated in this case. Figure 7.12 illustrates the response

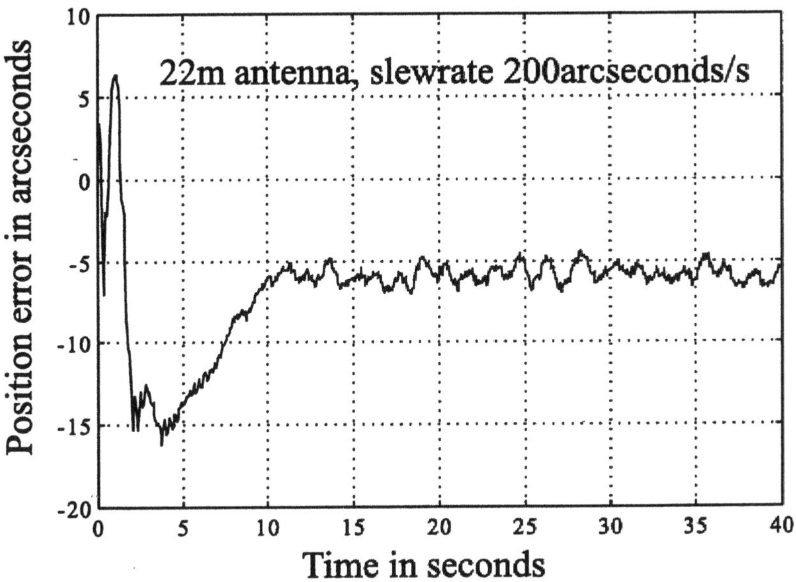

Fig. 7.12. Unsatisfactory non-adaptive control response

of one of the ATCA antennae when a maximum rate motion of 200 arcsec/s is imposed. The steady state error is of the order of 5 arcsec, which does not meet the design specifications.

7.5.2 Adaptive feedforward control

The position loop philosophy expounded in the previous section suggest that adaptively tuning of the feedforward derivative action may be an option. Indeed robustness and stability can be guaranteed by the feedback action, whilst the required tracking accuracy can act as a guide to tune the feedforward action. Such a configuration is of course reminiscent of the MIT rule for adaptive control [18,21]. This philosophy was adopted in this context, and a simple tuning algorithm, essentially a gradient method on an appropriate position error signal, was developed [2]. It was inspired by the work done on the dynamics of the MIT rule [18]. The proposed method has the additional advantage that it is reasonably easy to explain to the antenna users, who are not necessarily conversant in control theory.

To illustrate the ideas, consider the system represented in Figure 7.13. The model system's response is selected such that the model output y_m equals the reference signal r ($y_m = r$) in steady state for any reference signal $r = \alpha + \beta t + \gamma t^2$, with α, β, γ constant. The transient response of the model is the desired design response for the antenna.

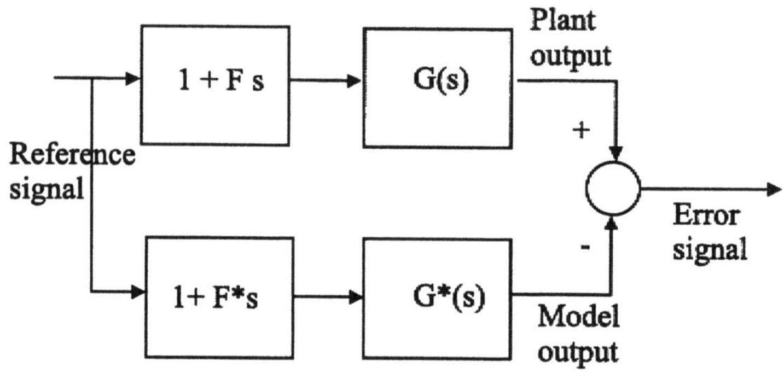

Fig. 7.13. Adaptive feedforward reference control

In order to tune the feedforward derivative action, we adjust its gain to minimise the error between the actual response y_p and the desired model system response y_m, $e = y_p - y_m$. This leads to an update law of the form (with standard abuse of notation)

$$\dot{F} \propto -e\frac{\partial e}{\partial F}.$$

Here F is the adaptively adjusted gain of the derivative action of the feedforward controller. The sensitivity of the output error with respect to this gain is given by

$$\frac{\partial e}{\partial F} = G\dot{r}.$$

G represent the actual closed loop dynamics of the controlled antenna. It is assumed that the reference trajectory r is sufficiently smooth, which is indeed ensured. Also, because the reference trajectory is generated by the user, an accurate $\dot{r}$ is directly available. Moreover, given we are only interested in tracking reference signals that are quadratic functions of time, well within the pass band of the closed loop G, we can approximate $G\dot{r} \approx \dot{r}$. This leads to the adaptive law

$$\dot{F} = -\mu e\frac{\dot{r}}{(\eta + |\dot{r}|)^2}.$$

The normalisation helps to make the adaptation law to be less sensitive to variations in $\dot{r}$, $\eta > 0$ and small compared to typical values of $|\dot{r}|$. This form of adaptation was tested successfully, but a significant performance improvement can be obtained with a different error signal, namely

$$\epsilon(t) = \int_0^t e(\tau)d\tau + \nu e(t).$$

Here $\nu > 0$, it reflects the relative importance attached to the error versus its integral.

This error signal ϵ is selected such that it steers the closed loop behavior towards a step response which is independent of the desired reference trajectory. The practical motivation stems from the desire to minimise the integral action required in the feedback loop. If ϵ were identically zero, the output error $e(t)$ would converge to zero as $e^{-\nu t}$. This goes a long way towards explaining why $\epsilon(t)$ is an acceptable choice for adaptation. We refer the reader to the concept of ***tunability*** in [17] for a more complete analysis.

The actually implemented adaptation law is given by

$$e = y_p - y_m,$$

$$\epsilon(t) = \int_0^t e(\tau)d\tau + \nu e(t), \tag{7.1}$$

$$\dot{F} = -\mu \text{sat}\left(\epsilon(t)\frac{\dot{r}}{(\eta+|\dot{r}|)^2}\right).$$

Here sat is a symmetric saturation function that limits the rate of adaptation. Moreover, the adaptive feedforward gain $F(t)$ is itself restrained to reside in $F_{min} \leq F(t) \leq F_{max}$. The limits F_{min} and F_{max} can be precomputed from the knowledge that the possible reference trajectories are within a certain prespecified range, $\dot{r}_{min} \leq \dot{r}(t) \leq \dot{r}_{max}$.

In order to guarantee that the adaptation response itself is independent of the reference signal, and to ensure that the adaptation speed is well inside the the bandwidth of the controlled loop and does not conflict with the feedback action of the controller, the following choice is appropriate $\nu = \frac{2}{\sqrt{\mu}}$. With this selection the adaptation time constant is determined by $\sqrt{\mu}$, actually more precisely the adaptation response is well approximated by a critically damped second order response. Our choice is $\sqrt{\mu} = 0.5$s, which makes the adaptation response compatible with the overall response bandwidth of 1.9Hz. The above choices can be justified using an averaging analysis, where the adaptation is considered fast as compared to the reference signals that have to be tracked. Such an analysis completed for the MIT rule can be found in [18].

The performance of the above adaptive feedforward action, combined with the PI feedback loop is illustrated in Figure 7.14 for a 22m ATCA antenna and in Figure 7.15. for a 26m telecommunications antenna. In both instances the achieved position responses are well within specifications despite the fact that the reference trajectories were outside the normal high accuracy operational conditions. The adaptation response is valid across the entire range of trajectory tracking operation. Moreover using the above implementation (7.1) the adaptation response is indeed almost independent of the external reference signal. In Figure 7.14 one can observe the effect of a rather severe wind gust that struck the antenna around $t = 25$s. The disturbance response was excellent, and the adaptation law did not respond in any significant way. It may also be observed by comparing Figure 7.12 with 7.14 that the controller with the adaptive feedforward results in a closed loop which is about twice as fast when compared to the non-adaptive loop under similar experimental conditions.

It is our experience, justified by extensive field trials, and several years of active implementation, that the proposed adaptation law, achieves excellent tracking accuracy across the whole operating range for parabolic trajectories, despite the fact that the closed loop only contains a single integrator. This implies improved settling time and disturbance rejection response as compared to the more classical internal model principle controller. By setting the adaptation bandwidth below the typical disturbance bandwidth, and due to the fact that it only responds to rate errors, the adaptation is not aversely affected by external disturbances. Adaptation is almost entirely restricted to react to plant mismatch, and reference trajectory changes. This is important, as often adaptive control fails because the adaptation law cannot sufficiently distinguish between plant changes (for which adaptation should take place) and signal environment disturbances (against which the controller should respond without changing its parameters). In this practical circumstance, thanks to the clean separation of bandwidth between the reference signal and plant variations (which require adaptation) as opposed to the wind disturbance (which do not require adaptation) we were able to effectively implement an adaptive feedforward law. This frequency domain interpretation about adaptive control is expounded in [22]..

Because in practice, any changes in reference trajectories are smoothly implemented, tracking performance can often be maintained even during the change over. In the extensive field experiments the adaptive feedforward has performed well over the entire tracking regime. It is now a fixture in all the antenna controllers we have installed.

7.6 Conclusion

Large antennae used in radio astronomy or telecommunications can be adequately controlled using an adaptive PD feedforward controller combined with a three tier cascade of feedback loops. The feedback cascade consists of an inner high bandwidth current loop, an intermediate medium bandwidth tacho loop and an outer low bandwidth position loop. The feedback loops can be closed using classical PI controllers complemented with detuning low pass filters. The adaptation of the feedforward action ensures high performance in tracking accuracy whilst the feedback action ensures robustness, fast settling and disturbance rejection. Ultimately the control performance is limited by the stiffness of the mechanical structure and the stiction effects in the gear train; which is precisely the goal of automatic control.

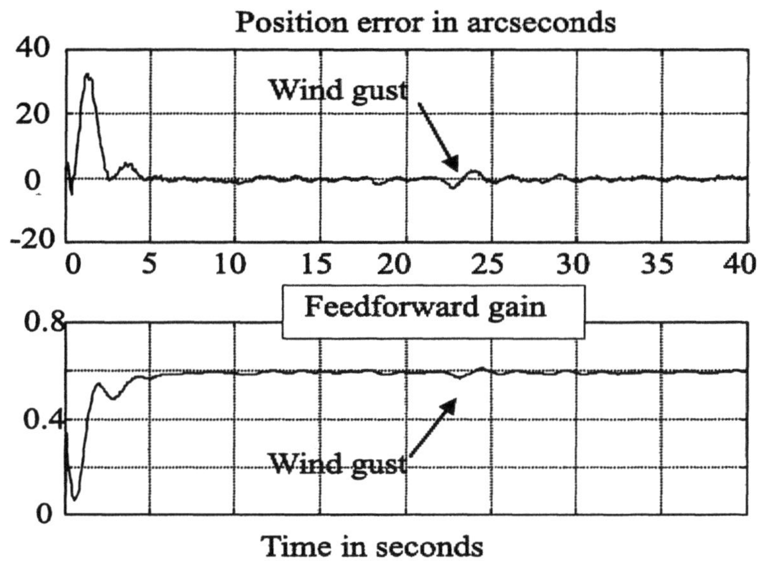

Fig. 7.14. Adaptive feedforward with PI feedback, tracking error response, 22m antenna, slewrate 200 arcseconds/s

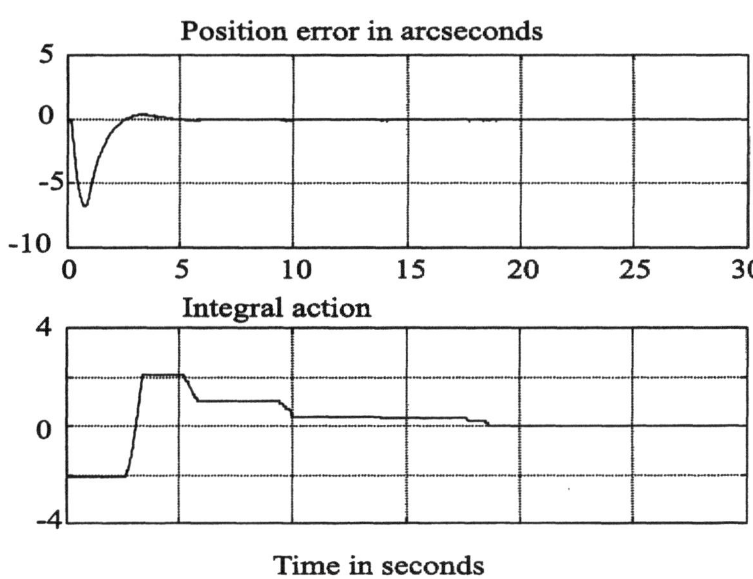

Fig. 7.15. Adaptive feedforward with PI feedback, tracking error response, 26m antenna, maximum slewrate

References

1. L.J. Sciacca, *Experiments in Adaptive Control and System Identification - A Case Study*, Technical Report EE9044, University of Newcastle, Department of Electrical and Computer Engineering, May 1990.
2. L.J. Sciacca, *Adaptive Servo Control for Tracking Systems*, Masters dissertation, Department of Electrical and Computer Engineering, August 1989.
3. L.J. Sciacca, R. J. Evans, R. E. Betz, D. N. Cooper, An Adaptive Controller for the Australia Telescope, *Proceedings of the 11th IFAC World Congress*, Tallinn, Estonia, August 1990.
4. D.R. Wilson, *Modern Practice in Servo Design* , Pergamon Press, New York, 1969.
5. B. Wallis, Improvement in Telescope Mounting, Brit. Pat. App. No. 29248/1955, 11 January 1957.
6. J. Rothwell, The Control of a Radio-Telescope, *AEI Engineering*, Jan/Feb, pp. 15-21, 1963.
7. H. C. Minnett, Design Study of Servo Drive System for an Alt-azimuth Radio Telescope, *Freeman, Fox and Partners*, London, Appendix C of Report Proposed Radio Telescope, September, 1957.
8. J.A. Macinante, The Dynamic Behaviour of a Large Steerable Radio Telescope, *Mechanical and Chemical Transactions*, I.E. Aust, November, pp. 147-155, 1967.
9. D.N. Cooper, N. Gouth, R. J. Evans, The Australia Telescope Antennas: Structural, Mechanical and Servomechanism Features IREECON-85, Melbourne, October 1985.
10. R.J. Evans, *Australia Telescope Project : Servo Design : Preliminary Considerations* Technical Report ECE 8311, May 1983, Department of Electrical and Computer Engineering, University of Newcastle.
11. R.J. Evans, D. N. Cooper, R. A. Kennedy, Servo System Design for the Australia Telescope, *I.E. Aust. Third Conference on Control Engineering*, Sydney, 1986.
12. K.J. Astrom, B. Wittenmark, *Computer Control Systems-Theory and Design*, Prentice-Hall, 1984
13. K.J. Astrom, B. Wittenmark, *Adaptive Control* Addison-Wesley, 1988.
14. R.A. Kennedy, *Digital Control of a Radio-Telescope*, Masters of Engineering, University of Newcastle, Department of Electrical Engineering and Computer Science, 1985.
15. R.A. Kennedy, R.J. Evans, Digital redesign of continuous controllers based on closed loop performance *Proceedings of the 29th IEEE Conf Dec Contr*, pp1898-1901, Hawaii, 1990.
16. L. Ljung, *System Identification - Theory for the User*, Prentice-Hall, N. J., 1987.
17. I.M.Y. Mareels, J.W. Polderman, *Adaptive Systems: An Introduction*, Birkhauser, 1994.
18. I.M.Y. Mareels, *Dynamics of adaptive systems*, PhD thesis, Department of Systems Engineering, Australian National University, August 1986.
19. R.H. Middleton, R. J. Evans, R. E. Betz, Servo-regulator designs for robotic systems with resonances, *Proceedings International Robotics Conference*, Sydney, pp.1200-1217, Nov 1998.

20. C. Scruton, *Introduction to wind effects on structures*, Oxford University Press, 1980.
21. P.V. Osburn, H.P. Whitaker, A. Kezer, New developments in the design of model reference adaptive control systems, *IAS*, paper No 61-39, Feb 1961.
22. B.D.O. Anderson, R.R. Bitmead, C.R. Johnson, P.V. Kokotovic, R. Kosut, I.M.Y. Mareels, L. Praly, B.D. Riedle, *Stability of Adaptive Systems: Averaging and Passivity Analysis*, MIT Press, Boston, MA, 1986.

8 Quantizer Optimization: Application of Passivity in Telecommunications

Rodney A. Kennedy[1], Zhi Ding[2], and Shazia Hasnie[3]

[1] Telecommunications Engineering, RSISE, The Australian National University, Canberra, ACT 0200, Australia
[2] Department of Electrical and Computer Engineering, University of Iowa, Iowa City, IA 52242, USA
[3] Asia Pacific Institute of Information Technology, Karachi, Pakistan

Abstract. Passivity theory is used to examine the stability of a generalized decision feedback structure which incorporates a general memoryless non-linearity. This non-linearity is considered in lieu of the nearest neighbor quantizer conventionally used for generating data estimates that are fed back. By associating error propagation recovery in a finite time with stability, we determine an optimal class of memoryless non-linearities. This class includes the saturation characteristic. Among constrained quantization rules, we also determine the optimal size of a quantizer deadband region – a problem that has been previously been raised in the literature on erasures in decision feedback equalization. These new general structures are shown theoretically to have superior error propagation suppression characteristics to the conventional decision feedback equalizer. The theoretical results are corroborated by a set of simulations.

8.1 Introduction

Passivity theory has been developed over a number of decades as an effective tool for studying the stability of non-linear systems. Conceptually it arose out of circuit theory and from the familiar linear circuit definition of passivity [1]. Passivity ties in with the notion of stability in the sense that a passive network tends to dissipate energy and, therefore, various signals in a passive network have a tendency to go to zero (which may be regarded as an equilibrium condition)[1]. Beyond circuit theory, passivity finds strongest expression in control theory including adaptive control and in power systems [2]. However, more recently, passivity has been shown to be a useful tool in studying telecommunication problems related to equalization [3] and speech coding [4]. This work carries the application of passivity in telecommunications further. As in earlier work, the application domain is decision feedback equalization – a non-linear strategy to mitigate the temporal distortions of a linear communication channel.

What is the notion of stability for decision feedback equalization that lends itself to the use of passivity? To answer this question we need to clarify

[1] Whenever inputs to the network are finite energy signals.

what is an appropriate definition of the equilibrium state of the decision feedback equalizer (DFE). As will be shown later the structure of a DFE consists of a finite impulse response filter in feedback around a quantizer. The quantizer output of the DFE forms a nearest neighbor estimate of the transmitted discrete level data. Within the DFE these estimates are fedback to aid in compensating for distortions in the linear communication channel due to past data symbols. However, if the data estimates are in error then the channel compensation is imperfect leading to the increased likelihood of future errors – referred to as *error propagation*. So an equilibrium of the DFE can be associated with an error free mode, a perturbation can be associated with some initial error state and stability can be associated with the ability of the DFE to flush the errors and return to the equilibrium state. The degree of stability can be associated with the speed at which error recovery is made from some initial error condition.

To date, stability investigations for the DFE (primarily those based on passivity) have restricted their attention to a fixed structure where the parameters of the DFE are tuned to optimal values [5,3]. For these systems passivity can inform us which communication channels pose stability problems (leading to poor performance) and which channels are well suited to the use of a DFE because error propagation is greatly reduced. In the current work we consider a different problem in equalization but the methodology will be similar. This work uses passivity to inform us of the best way to quantize the fedback data estimates to best suppress error propagation. We derive an optimal memoryless non-linear quantizer to act in lieu of the usual nearest neighbor quantizer of the conventional DFE. This leads to a surprising result that one of the best ways to suppress error propagation is to use no quantizer at all. Alternatively the theory can be used to design optimal deadband regions for the quantizer known as "erasures" in the literature [6]. With the new structures, which differ only in the quantizer characteristic for fedback decisions, we get a strict enlarging of the stability region relative to the conventional DFE, i.e., an enlarging of the region in the parameter space of the channel coefficients where error propagation is suppressed.

In section 8.2 we give background on the DFE and the philosophy behind its operation. We also look at the channel and equalizer models. This leads to the important error system towards which our analysis is primarily directed. In section 8.3 we review needed passivity definitions and derive the key technical results concerning Passivity Theory and Loop Transformations. In section 8.4 we make a formal connection between the concept of error propagation and stability of the DFE. Our main results, which result from the application of Passivity Theory, are given in section 8.5. These results give the optimal class of memoryless non-linearities that yield the greatest stability. The conclusions are given in section 8.6.

8.2 Problem Formulation

8.2.1 Background on Decision Feedback

Decision feedback is an equalizer design philosophy which greatly simplifies practical implementation while maintaining reasonable performance relative to more optimal structures [7]. The philosophy expounds that estimates of past data symbols can be used as if correct in determining current and future symbol estimates [8]. This seems a rather strong assumption. However, the alternative is to use a symbol sequence estimation scheme, such as the Viterbi algorithm, which is typically of high complexity and impractical in many cases of practical interest unless the channel memory is short [7]. The simplicity of decision feedback structures is further emphasized by the recognition that, in terms of implementation, they are of similar or less complexity to transversal filter structures (in the form of a linear equalizer)[2].

The notion of decision feedback is usually associated with the decision feedback equalizer [9]. However, decision feedback has become prominent as a device in generating simplified reduced state sequence estimation schemes which attempt to approximate the optimal Viterbi algorithm[10,11]. These schemes implicitly utilize multiple past symbol estimates (decisions) to reduce the complexity and dimensionality of the trellis. Thus it can be said that decision feedback is an effective tool in producing implementable and practical receiver structures in telecommunications. The ideas in this work which are formulated in terms of the simpler decision feedback equalizer should have applicability also to reduced state Viterbi schemes.

The symbol error rate or the bit error rate are central measures to evaluate the performance of any communications receiver structure. For structures which utilize decision feedback, symbol errors are made in two ways: i) *primary errors* which can be thought of as being generated by channel noise; and ii) *secondary errors* due to error propagation triggered by a primary error. The distinction need not be so explicit but it is convenient for our work. Primary errors tend to be straightforward to understand and tend to be unavoidable in any communication system subject to noise. Error propagation is well recognized but less well understood [3,8,12–15]. This difficulty in understanding arises since decision feedback is inherently non-linear and recursive.

The analysis of decision feedback equalizers has concentrated on the study of primary errors. To make such an analysis authors make the assumption that past decisions are correct. The details are not important but this has the rather idealistic effect of permitting linear systems analysis tools to be used. Necessarily this analysis glosses over error propagation and researchers' only resort is to use simulations to characterize it.

---

[2] The complexity can be less since the decided symbols require fewer bits in their representation and the multiplications can be accomplished via shifting.

The study of error propagation has concentrated on two approaches: i) finite state machine analysis [8,12–14]; and ii) non-linear stability analysis using passivity theory [3]. This work uses the latter approach to devise a new structure what has diminished error propagation yet retains the essential qualitative operation of the DFE.

This work considers the question about what is the best way to generate (preliminary) decisions given that they are fed back. It is convenient to refer to the conventional nearest neighbor decision device as a *hard* decision device. A decision device which admits the possibility that more levels than the symbol alphabet are possible is referred to as *soft*.

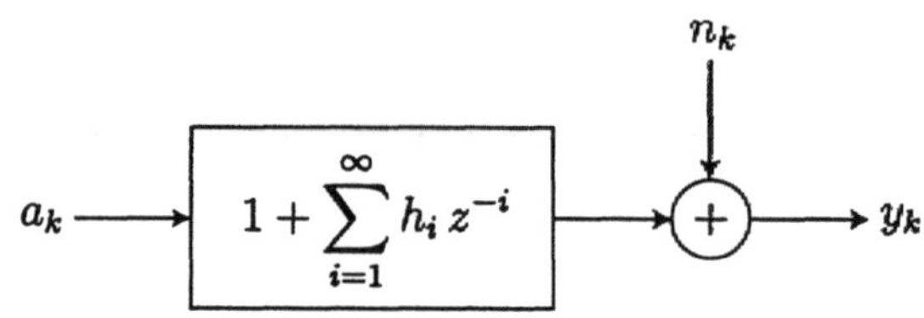

Fig. 8.1. Linear baseband equivalent discrete time channel model driven by discrete data a_k and subject to channel noise n_k

8.2.2 Channel and Equalizer Models

A communication channel, shown in Fig. 8.1, subject to intersymbol interference (ISI) is represented by the following input-output equivalent baseband discrete time representation:

$$y_k = a_k + \sum_{i=1}^{\infty} h_i a_{k-i} + n_k \tag{8.1}$$

where k denotes time, a_k is the discrete data belonging to some alphabet $\mathbb{A}$, $\{h_i\}$ are the channel parameters, n_k is additive noise and y_k is the noisy channel output. Note that in (8.1), the leading channel coefficient (which could be represented by h_0) is defined to be unity without loss of generality. The only condition we impose on the channel is that the coefficients are absolutely summable[3].

With regards to correcting for the intersymbol interference found on this channel, the basic DFE takes the form in Fig. 8.2. It uses the quantized signal $\widehat{a}_k$, which estimates the discrete transmitted data a_k, to drive an FIR filter which models the tail of the channel (noting that the coefficients are identical

[3] The channel coefficients, given by the $\{h_i\}$, are absolutely summable if and only if $\|h\|_1 \triangleq \sum_{i=0}^{\infty} |h_i| < \infty$. We can write this also as $h \in l_1$.

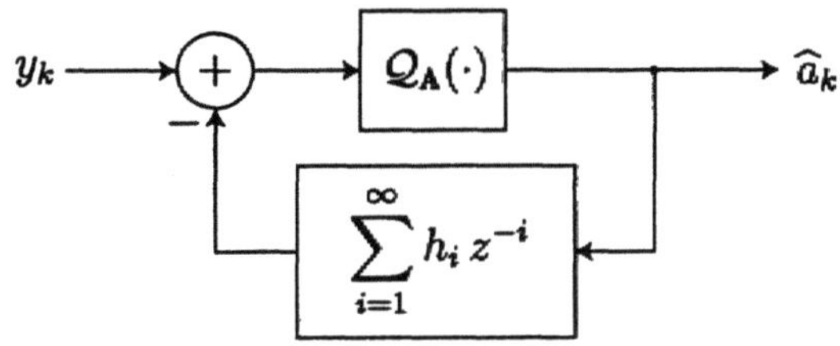

Fig. 8.2. Conventional decision feedback equalizer where the output decisions, $\widehat{a}_k$, are generated by the $\mathcal{Q}_{\mathbb{A}}(\cdot)$ nearest neighbor quantizer

to those in the channel, referred to as a zero-forcing criterion). The quantizer used is a nearest neighbor rule given by

$$\mathcal{Q}_{\mathbb{A}}(x) \triangleq \arg\min_{a \in \mathbb{A}} \left\{ (x-a)^2 \right\}, \quad x \in \mathbb{R} \tag{8.2}$$

which maps a continuous value to an alphabet symbol in a nearest neighbor fashion. In this way the quantized signal $\widehat{a}_k$ is used both for the fedback signal and the ultimate decisions (equalizer output).

Example 1. In this work we largely focus on the classical M-ary constellation $\mathbb{A} = \{\pm 1, \pm 3, \ldots, \pm(M-1)\}$. In this case the $(M-1)$ decision thresholds for the nearest neighbor rule (8.2) lie at $\mathbb{D} = \{0, \pm 2, \ldots, \pm(M-2)\}$. This leads to the expression

$$\mathcal{Q}_{\mathbb{A}}(x) = \sum_{j \in \mathbb{D}} \operatorname{sgn}(x-j), \quad x \in \mathbb{R} \tag{8.3}$$

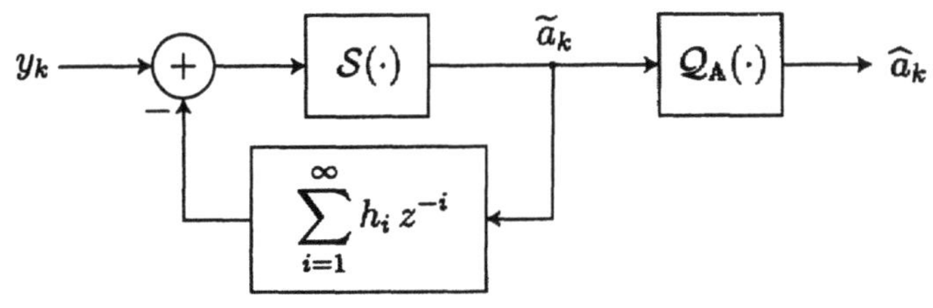

Fig. 8.3. Variant on decision feedback equalizer where the fedback decisions, $\widetilde{a}_k$, are generated by a soft quantizer $\mathcal{S}(\cdot)$ which can differ from the output decisions, $\widehat{a}_k$, generated by the $\mathcal{Q}_{\mathbb{A}}(\cdot)$ nearest neighbor quantizer

This work considers a variant to the DFE in Fig. 8.2 to generalize it to the structure in Fig. 8.3. Now we separate the role of fedback decision and equalizer output. The problem we investigate concerns the optimum choice, in the appropriate sense, of a suitable memoryless non-linearity (in lieu of the nearest neighbor quantizer), which we call the soft decision function or quantizer,

$$\mathcal{S}(\cdot):\ \mathbb{R} \mapsto \mathbb{R}, \tag{8.4}$$

that is placed inside the feedback loop, see Fig. 8.3. Naturally, we ultimately need to generate estimates of the sent data and, thereby, force the estimates at the output of the equalizer to take on one of the alphabet values. This justifies the use of the terminating nearest neighbor quantizer $\mathcal{Q}_{\mathbb{A}}(\cdot)$ seen in Fig. 8.3. We refer to the output of the first non-linearity as the soft decision and denote it by the symbol $\widetilde{a}_k$. The hard decisions are denoted by $\widehat{a}_k$ and the two decisions are related via

$$\widehat{a}_k = \mathcal{Q}_{\mathbb{A}}(\widetilde{a}_k). \tag{8.5}$$

The defining recursive equation for the new non-linear equalizer structure from (8.1) and Fig. 8.3 is given by

$$\widetilde{a}_k = \mathcal{S}(a_k + r_k) \tag{8.6}$$

where

$$r_k = \sum_{i=1}^{\infty} h_i (a_{k-i} - \widetilde{a}_{k-i}) \tag{8.7}$$

is the residual intersymbol interference (which ideally should be zero or small relative to unity). In this equation we have set the noise, n_k, to zero. The intention in our work is to characterize the effect of an initial error state and subsequent error propagation. Noise is not needed to make this study except perhaps to give a mechanism for an initial non-zero error state but equally well such an initial state could arise from a cold start (power up).

Finally, we present some important but straightforward constraints that soft decision quantizer $\mathcal{S}(\cdot)$ should satisfy if the non-linear equalizer is to behave as expected. That is, because of (8.5), if the residual ISI in (8.7) is zero or sufficiently close to zero then the soft decisions should be such that the related hard decisions are correct, i.e., $\widehat{a}_k = a_k$. These conditions are expressed mathematically as the invariance property:

$$\mathcal{S}(a) = a, \quad \forall a \in \mathbb{A}. \tag{8.8}$$

and the continuity property:

$$\mathcal{S}(\cdot) \text{ is continuous at } a, \quad \forall a \in \mathbb{A}. \tag{8.9}$$

8.2.3 Error System

This work focuses on means to mitigate the detrimental effects of error propagation in a non-linear feedback structure by optimal selection of the soft decision function $\mathcal{S}(\cdot)$. Towards this goal we define two error signals

$$\widetilde{e}_k \triangleq a_k - \widetilde{a}_k, \tag{8.10}$$

the soft error, and

$$\widehat{e}_k \triangleq a_k - \widehat{a}_k, \tag{8.11}$$

the hard or equalizer output decision error. An equation that then captures the soft decision feedback recursion but expressed in terms of the soft error signals, $\widetilde{e}_k$, is given by

$$\widetilde{e}_k = a_k - \mathcal{S}(a_k + r_k) \tag{8.12}$$

where from (8.7) and (8.10), the residual ISI is given by

$$r_k = \sum_{i=1}^{\infty} h_i \widetilde{e}_{k-i}. \tag{8.13}$$

We can then express the new error signal, $\widetilde{e}_k$, as a memoryless non-linear function of the signals a_k and r_k. In what follows it is convenient to write

$$w_k \triangleq -\widetilde{e}_k \tag{8.14a}$$

$$= \mathcal{S}(a_k + r_k) - a_k. \tag{8.14b}$$

to permit representation the error system in a conventional negative feedback control form [1]. It is also useful because the new signal w_k tends to have the same sign as r_k.

A crucial but simple task is to relate geometrically the non-linear, memoryless mapping which takes r_k to w_k, and is parametrized by a_k, in terms of the memoryless non-linearity $\mathcal{S}(\cdot)$. We write this non-linear, memoryless mapping as the operator

$$\mathcal{P}_{a_k}: \; r_k \mapsto w_k. \tag{8.15}$$

The nature of $\mathcal{P}_{a_k}(\cdot)$ can be directly inferred from (8.14b). For a frozen a_k, $\mathcal{P}_{a_k}(\cdot)$ is identical to $\mathcal{S}(\cdot)$ apart from a shift in the input and an equal shift in magnitude and sign at the output.

To characterize $\mathcal{P}_{a_k}(\cdot)$ completely it remains to take into account the finite variation of a_k over $\mathbb{A}$. This equivalent non-linear operator – which we denote as $\mathcal{P}_{\mathbb{A}}(\cdot)$ – may be represented by the superposition of the translated non-linear mappings for the various a_k ranging over $\mathbb{A}$. Since only one input a_k is operative at any time instant k then the mapping is truly a function (one-to-one). However, it is convenient to informally regard the mapping from r_k to w_k as a multi-valued non-linearity which is independent of a_k because, as will be seen later, the effect of the variation in a_k is relatively minor.

Example 1 (Conventional Binary DFE). In Fig. 8.4 we show the case for the conventional binary DFE which uses $\mathcal{S}(\cdot) = \mathcal{Q}_{\mathbf{A}}(\cdot) = \text{sgn}(\cdot)$. In this case, $\mathcal{P}_{\mathbf{A}}(\cdot)$ is the overlay of two offset sgn($\cdot$) functions given by $\mathcal{P}_{\pm 1}(x) = \text{sgn}(x \mp 1) \pm 1$.

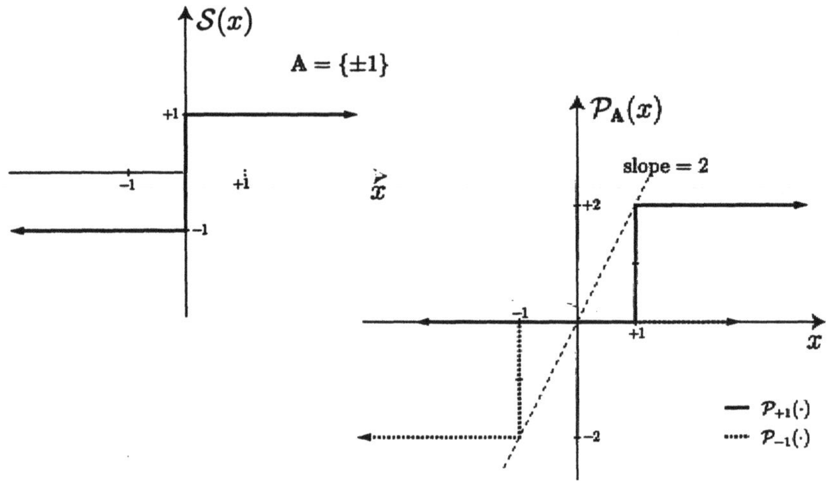

Fig. 8.4. Example of $\mathcal{S}(\cdot)$ and $\mathcal{P}_{\mathbf{A}}(\cdot)$ for the conventional DFE where $M = 2$, $\mathbf{A} = \{\pm 1\}$ (binary), $\mathcal{S}(\cdot) = \mathcal{Q}_{\mathbf{A}}(\cdot) = \text{sgn}(\cdot)$ and $\mathcal{P}_{\pm 1}(x) = \text{sgn}(x \mp 1) \pm 1$. For the constituents of $\mathcal{P}_{\mathbf{A}}(\cdot)$: $\mathcal{P}_{+1}(\cdot)$ is shown as a solid line and $\mathcal{P}_{-1}(\cdot)$ is shown as a dotted line. Note that $\mathcal{P}_{\mathbf{A}}(\cdot)$ is sector bounded in the first and third quadrants tightly bounded between a line of slope 2 and a line of slope 0

Example 2 (Modified 4-ary DFE With Soft Quantizer). In Fig. 8.5 we show the case of our modified DFE for $\mathbb{A} = \{\pm 1, \pm 3\}$ and a piecewise linear soft quantizer characteristic which employs three linear segments about the three decision thresholds at $\mathbb{D} = \{0, \pm 2\}$ (here we have chosen the slope to be all the same which is not strictly necessary but a sound choice). In this case, $\mathcal{P}_{\mathbf{A}}(\cdot)$ is the overlay of four $\mathcal{S}(\cdot)$ functions shown in Fig. 8.5.

Bringing the above results together we have the two equivalent systems in Fig. 8.6. We shall first explain the system on the left in Fig. 8.6. The upper block, which is linear and strictly proper (i.e., has no direct term), is a representation of (8.13), i.e., takes $\tilde{e}_k$ and generates the residual ISI r_k, and the bottom block is the non-linearity $\mathcal{P}_{a_k}(\cdot)$ which maps r_k, for a given a_k, to w_k.

We now consider the system on the right in Fig. 8.6 which differs cosmetically from the system on the left in three ways: i) there is an external input u_k, ii) the upper block has a zero initial state, and iii) we have indicated a virtual block for $\mathcal{P}_{\mathbf{A}}(\cdot)$.

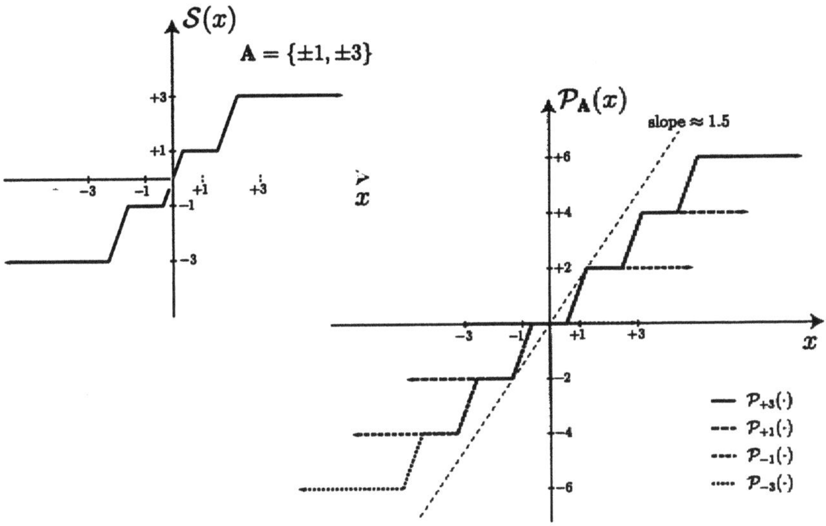

Fig. 8.5. Example of $\mathcal{S}(\cdot)$ and $\mathcal{P}_{\mathbf{A}}(\cdot)$ for the modified DFE where $M = 4$, $\mathbf{A} = \{\pm 1, \pm 3\}$ (4-ary) and $\mathcal{S}(\cdot)$ has a non-vertical linear segment around each of the three input decision thresholds $\mathbb{D} = \{0, \pm 2\}$. For the constituents of $\mathcal{P}_{\mathbf{A}}(\cdot)$: $\mathcal{P}_{+3}(\cdot)$ is shown as a solid line, $\mathcal{P}_{+1}(\cdot)$ is shown as a dashed line, $\mathcal{P}_{-1}(\cdot)$ is shown as a short dashed line and $\mathcal{P}_{-3}(\cdot)$ is shown as a dotted line. Note that $\mathcal{P}_{\mathbf{A}}(\cdot)$ is sector bounded in the first and third quadrants tightly bounded between a line of approximate slope 1.5 and a line of slope 0

In regard to explaining the nature of u_k, later we shall be considering all time signals as semi-infinite sequences – meaning that they are only defined on for $k \geq 0$ (see Section 8.3). This presents a minor problem for the upper block whose defining linear equation is given by the convolution (8.13) and whose input is, thereby, formally given by the doubly infinite sequence

$$\{\ldots, \widetilde{e}_{-2}, \widetilde{e}_{-1}, \widetilde{e}_0, \widetilde{e}_1, \widetilde{e}_2, \ldots\}. \tag{8.16}$$

However, by superposition we can rewrite convolution (8.13) as

$$r_k = \sum_{i=1}^{k} h_i \widetilde{e}_{k-i} + u_k, \tag{8.17}$$

where

$$u_k \triangleq \sum_{i=k+1}^{\infty} h_i \widetilde{e}_{k-i} \tag{8.18}$$

captures the effect of initial conditions at time $k = 0$. That is, the implicit semi-infinite sequence $\{\ldots, \widetilde{e}_{-2}, \widetilde{e}_{-1}\}$ used in (8.18) completely defines the initial conditions at time $k = 0$ for the upper block. The input to the linear upper block in the system on the right in Fig. 8.6 can now be given as the semi-infinite soft error sequence

$$\widetilde{e} \triangleq \{\widetilde{e}_0, \widetilde{e}_1, \widetilde{e}_2, \dots\} \tag{8.19}$$

and, from the first term in (8.17), effectively its initial state is zero.

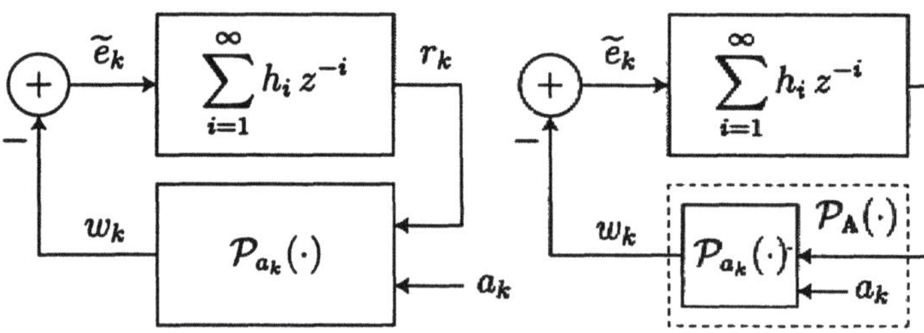

Fig. 8.6. Two equivalent representations for the error system for describing the propagation of the soft decision errors $\widetilde{e}_k$. The upper linear block in the system on the right has a zero initial state, at time k, with non-zero initial conditions modelled by an external signal u_k

For the conventional DFE, such as the binary case represented in Fig. 8.4, a special case of such an error system has been developed [3]. For the conventional DFE, the $\mathcal{P}_{\mathbf{A}}(\cdot)$ operator is a sector bounded memoryless non-linearity such as in Fig. 8.4, and this property enables the use of passivity methods to study the dynamics of the error system in Fig. 8.6. A key observation here is that $\mathcal{S}(\cdot)$ induces $\mathcal{P}_{\mathbf{A}}(\cdot)$, and there are a plethora of other useful $\mathcal{S}(\cdot)$ that preserve the sector boundedness of $\mathcal{P}_{\mathbf{A}}(\cdot)$. However, these $\mathcal{S}(\cdot)$ need to satisfy (8.5) to ensure the new structure acts like an equalizer. The non-linearity $\mathcal{S}(\cdot)$ represented in Fig. 8.5 is one such well-posed example. In summary, there is a considerable range of non-linearities $\mathcal{S}(\cdot)$ that are amenable to passivity techniques.

In the next subsection we develop enough of the general passivity machinery to enable analysis of the error system given in Fig. 8.6.

8.3 Passivity Concepts and Definitions

We begin with some definitions which are standard in input-output stability theory [1]. We focus on a Hilbert space structure composed of real valued sequences indexed by $k \in \mathbb{N}_+$ (non-negative integers). Then if we have two real sequences $x = \{x_0, x_1, \dots\}$ and $y = \{y_0, y_1, \dots\}$ their real inner product will be defined as

$$\langle x, y\rangle \triangleq \sum_{i=0}^{\infty} x_i y_i. \tag{8.20}$$

where it is clear that $\langle x, y\rangle = \langle y, x\rangle$. This inner product (8.20) induces a natural euclidean norm defined by

$$\|x\| \triangleq \langle x, x\rangle^{\frac{1}{2}} = \Big(\sum_{i=0}^{\infty} {x_i}^2\Big)^{\frac{1}{2}}. \tag{8.21}$$

We define the discrete function space l_2 which consists of all sequences satisfying

$$x \in l_2 \iff \|x\| < \infty. \tag{8.22}$$

The space l_2 is generally too restrictive an arena for deriving results, so we introduce the standard concept of an extended space l_2^e [1], defined by

$$x \in l_2^e \iff \|\mathcal{J}_T x\| < \infty, \quad \forall T \in \mathbb{N}_+ \tag{8.23}$$

where $\mathcal{J}_T$ is a truncation operator, parametrized by $T \in \mathbb{N}_+$, defined by

$$(\mathcal{J}_T x)(k) \triangleq \begin{cases} x_k & \text{if } k \le T, \\ 0 & \text{if } k > T. \end{cases} \tag{8.24}$$

Note (8.23) just says that $x \in l_2^e$ if and only if $|x_k| < \infty\ \forall k$, i.e., x does not have a finite escape time. So for example if $x \triangleq \{x_k = 2^k,\ \forall k \in \mathbb{N}_+\}$ then $x \in l_2^e$ but clearly $x \notin l_2$.

From definitions (8.22) and (8.23) it is apparent that $l_2 \subset l_2^e$. In our work all signals considered will lie in the extended space l_2^e (because we stipulate only that $h \in l_1$). However it is of great interest to show that particular signals also lie in the subset l_2. For example with the hard error signal (8.11), it is our aim to show $\widehat{e} \in l_2$. Then because $\widehat{e}_k \in \{0, \pm 2, \pm 4, \ldots, \pm 2(M-1)\}$ we have the following observation,

$$\widehat{e} \in l_2 \iff \widehat{e}_k = 0, \quad \forall k \ge K,\ K < \infty. \tag{8.25}$$

Later we will see that this is equivalent to saying that the DFE has recovered from error at some finite time K.

Now define $x_T \triangleq \mathcal{J}_T x$ and $\|x\|_T \triangleq \|\mathcal{J}_T x\|$. In relating l_2 and l_2^e we note the following properties of the inner product and its induced norms which we will use later without explicit reference:

Monotonicity Property: $\forall x \in l_2^e$, we have

$$T \mapsto \|x\|_T \text{ is monotonically non-decreasing.} \tag{8.26}$$

Convergence Property: $\forall x \in l_2$, we have

$$\lim_{T\to\infty} \|x\|_T = \|x\|. \tag{8.27}$$

Truncation Property: $\forall x, y \in l_2^e$, $\forall T \in \mathbb{N}_+$, we have

$$\langle x_T, y_T\rangle = \langle x_T, y\rangle = \langle x, y_T\rangle \triangleq \langle x, y\rangle_T. \tag{8.28}$$

This nomenclature leads to the crucial definitions of passivity.

Definition 1. An operator $\mathcal{H}\colon l_2^e \mapsto l_2^e$ is *passive* if $\exists$ constant η such that

$$\langle \mathcal{H}x, x\rangle_T \geq \eta, \quad \forall x \in l_2^e,\ \forall T \in \mathbb{N}_+. \tag{8.29}$$

If $\mathcal{H}$ were linear then η could be taken as zero [1].

Definition 2. An operator $\mathcal{H}\colon l_2^e \mapsto l_2^e$ is *strictly passive* if $\exists \delta > 0$ and $\exists \eta$ such that

$$\langle \mathcal{H}x, x\rangle_T \geq \delta \,\|x\|_T^2 + \eta, \quad \forall x \in l_2^e,\ \forall T \in \mathbb{N}_+. \tag{8.30}$$

Again if $\mathcal{H}$ were linear then η could be taken as zero. We label δ as the *degree of passivity.*

Example 1. As an example of passivity (but not strict passivity), which will be important later, we consider a special case of the sector boundedness property referred to in our error system model. Suppose $x \in l_2^e$ is the input to an operator $\widehat{\mathcal{H}}$ with output $y \triangleq \widehat{\mathcal{H}}x$, which satisfies $y_k x_k \geq 0$, $\forall k \in \mathbb{N}_+$ (a sign preserving operator). Then

$$\langle \widehat{\mathcal{H}}x, x\rangle_T = \sum_{k=0}^{T} y_k x_k \geq 0, \quad \forall x \in l_2^e,\ \forall T \in \mathbb{N}_+, \tag{8.31}$$

showing $\widehat{\mathcal{H}}$ is passive according to Definition 1 with $\eta = 0$. That is, if $\widehat{\mathcal{H}}$ is a non-linearity constrained to the first and third quadrants then it is passive (even if it is time-varying or has memory).

Example 2. Our second example which we state as a proposition will be important later and relates to the Definition 2 of strict passivity applied to *linear* operators.

Proposition 1 (Strictly Positive Real Linear Operator). *Suppose the linear operator $\mathcal{G}\colon l_2^e \mapsto l_2^e$ is defined by $\mathcal{G}u = g * u$, where $g \triangleq \{g_0, g_1, \ldots\} \in l_1$. Let $\delta > 0$. Then*

$$\langle \mathcal{G}u, u\rangle_T \geq \delta \,\|u\|_T^2, \quad \forall u \in l_2^e,\ \forall T \in \mathbb{N}_+ \tag{8.32}$$

if and only if

$$\Re(\widetilde{g}(e^{j\theta})) \geq \delta, \quad \forall \theta \in [0, 2\pi] \tag{8.33}$$

where $\widetilde{g}(z) \triangleq \sum_{i=0}^{\infty} g_i z^{-i}$ is the Z-transform of the impulse response g.

Proof. The proof follows from Parseval's Theorem, see [1]. □

Remark 1. This proposition says that a linear operator is strictly passive if and only if its Nyquist plot belongs to $\{z \in \mathbb{C}: \Re(z) \geq \delta > 0\}$.

We now come to the main passivity result. Figure 8.7 defines the signals and operators of interest. In it e and v are the input sequences to the operators $\mathcal{H}_1$ and $\mathcal{H}_2$ and $y = \mathcal{H}_1 e$ and $z = \mathcal{H}_2 v$ are the respective output sequences. There is a single external signal u. All signals shown are assumed to lie in l_2^e. The following proposition and proof are an adaptation of a more general result in [1].

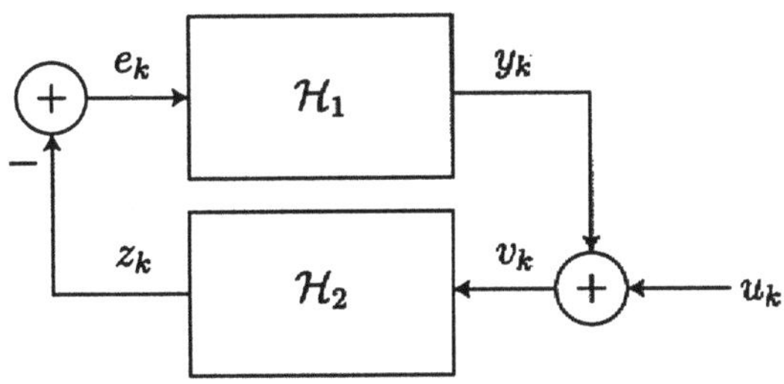

Fig. 8.7. Passivity Proposition setup

Proposition 2 (Passivity). *Suppose: i) operator $\mathcal{H}_1$ is strictly passive, i.e.,*

$$\langle \mathcal{H}_1 e, e \rangle_T \geq \delta_1 \|e\|^2 + \eta_1, \quad \forall e \in l_2^e, \ \forall T \in \mathbb{N}_+ \tag{8.34}$$

where $\delta_1 > 0$; and ii) operator $\mathcal{H}_2$ is a passive

$$\langle \mathcal{H}_2 v, v \rangle_T \geq \eta_2, \quad \forall v \in l_2^e, \ \forall T \in \mathbb{N}_+. \tag{8.35}$$

Then $u \in l_2 \implies e \in l_2$.

Proof. We show $e \in l_2$ by determining upper and lower bounds on $\langle \mathcal{H}_1 e, e \rangle_T + \langle \mathcal{H}_2 v, v \rangle_T$ which may be thought of as the total energy dissipated by the system in Fig. 8.7 from time 0 to time T. First we determine a lower bound. Using (8.34) and (8.35) we clearly have

$$\langle \mathcal{H}_1 e, e \rangle_T + \langle \mathcal{H}_2 v, v \rangle_T \geq \delta_1 \|e\|_T^2 + \eta \quad \forall T \in \mathbb{N}_+ \tag{8.36}$$

where, recall, $\delta_1 > 0$ is the constant associated with the degree of passivity of the $\mathcal{H}_1$ operator, and $\eta \triangleq \eta_1 + \eta_2$ (constants associated with the passivity of $\mathcal{H}_1$ and $\mathcal{H}_2$). An upper bound on (8.36) follows from the following simple calculation, using Fig. 8.7,

$$\begin{aligned}\langle \mathcal{H}_1 e, e\rangle_T + \langle \mathcal{H}_2 v, v\rangle_T &= \langle \mathcal{H}_1 e, e\rangle_T + \langle -e, v\rangle_T && (8.37a)\\ &= \langle -e, v - \mathcal{H}_1 e\rangle_T && (8.37b)\\ &= \langle -e, u\rangle_T \leq \|e\|_T \, \|u\|_T, \quad \forall T \in \mathbb{N}_+ && (8.37c)\end{aligned}$$

where the last line is an application of the Cauchy-Schwartz inequality. Then, whenever $\|e\|_T > 0$, combine (8.36) with (8.37c) to obtain

$$\|e\|_T \leq \delta_1^{-1}\Big(\|u\|_T - \frac{\eta}{\|e\|_T}\Big), \quad \forall T \in \mathbb{N}_+ \tag{8.38}$$

Consider two cases: i) $\|e\|_T < 1$, $\forall T \in \mathbb{N}_+$: which implies, as $T \to \infty$, that $\|e\| < 1$ and hence $e \in l_2$ (independently of u); ii) $\|e\|_T \geq 1$, $\forall T \geq T^*$ for some sufficiently large T^*: which implies from (8.38)

$$\|e\|_T \leq \delta_1^{-1}\big(\|u\|_T + |\eta|\big), \quad \forall T \geq T^* \tag{8.39}$$

Letting $T \to \infty$ we find $\|e\| \leq \delta_1^{-1}\big(\|u\| + |\eta|\big)$. In this case $u \in l_2 \implies e \in l_2$. This completes the proof. □

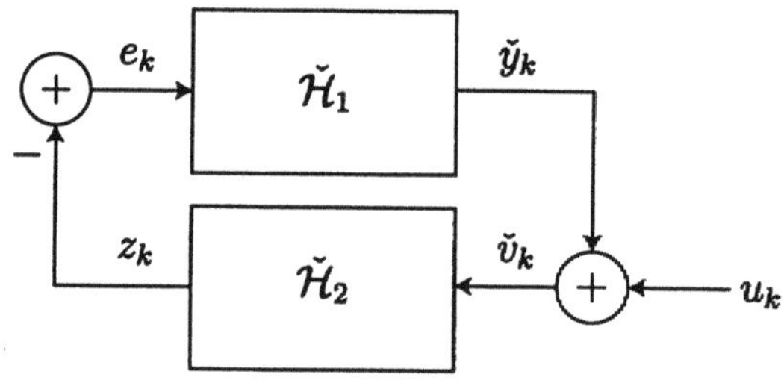

Fig. 8.8. Loop Transformed Passivity Proposition setup. In this case $\check{\mathcal{H}}_1$ is linear but need not be passive and $\check{\mathcal{H}}_2$ is a $[0, \alpha]$-sector bounded memoryless non-linearity

Remark 1. As it stands we cannot apply Proposition 2 directly to the error system in Fig. 8.6 whose upper block fails to be strictly passive (as it is strictly proper it lacks a direct term). However, since the upper block is linear it is always possible to add a sufficiently large positive direct term to its impulse response make it passive.

We can develop a more directly useful result by the use of a loop transformation [1,3]. The new system under consideration, analogous to that of Fig. 8.6, is shown in Fig. 8.8. By the addition of two mutually cancelling memoryless linear blocks of gain β we obtain the equivalent loop transformed version of Fig. 8.8 shown in Fig. 8.9 which will be used in the proof of the next proposition.

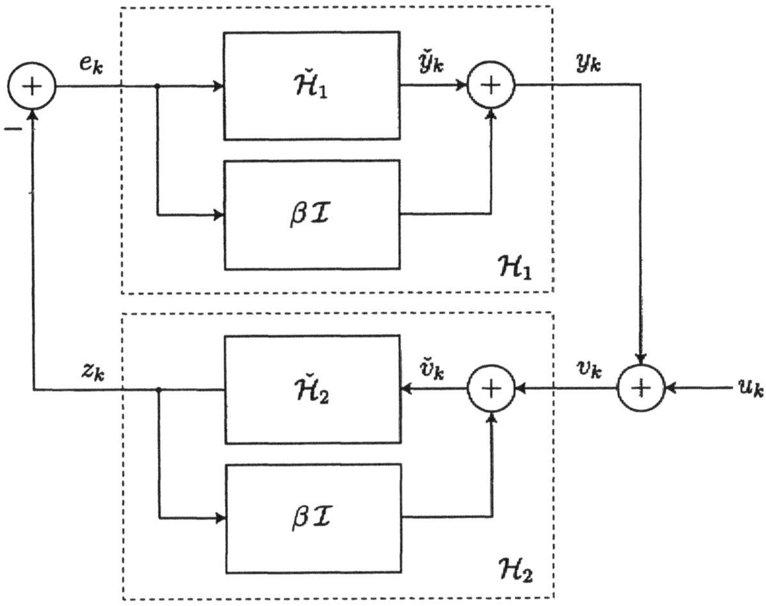

Fig. 8.9. Equivalent loop transformed system of the system considered in Fig. 8.8. By increasing the scalar $\beta \geq 0$ the $\mathcal{H}_1$ operator (defined as the operator mapping e to y) can be made more passive at the expense of the passivity of $\mathcal{H}_2$ operator (defined as the operator mapping v to w)

Proposition 3 (Loop Transformed Passivity). *Suppose: i) operator $\check{\mathcal{H}}_1$ is not necessarily passive; and ii) operator $\check{\mathcal{H}}_2$ is a $[0, \alpha]$-sector bounded memoryless non-linearity satisfying*

$$\alpha \|v\|_T \geq \langle \check{\mathcal{H}}_2 v, v \rangle_T \geq 0, \quad \forall v \in l_2^e, \ \forall T \in \mathbb{N}_+ \tag{8.40}$$

and is thus passive. Let $\mathcal{I}$ be the identity operator. If the operator given by $(\frac{1}{\alpha}\mathcal{I} + \check{\mathcal{H}}_1)$ is strictly passive then $u \in l_2 \implies e \in l_2$.

Proof. Let $\mathcal{H}_1$ and $\mathcal{H}_2$ be as in Fig. 8.7 and Proposition 2. Then by a loop transformation of Fig. 8.8, shown in Fig. 8.9, using a scalar $\beta \geq 0$, then we have the formal associations

$$\mathcal{H}_1 = \beta\mathcal{I} + \check{\mathcal{H}}_1 \tag{8.41a}$$
$$\mathcal{H}_2 = (\mathcal{I} - \beta\check{\mathcal{H}}_2)^{-1}\check{\mathcal{H}}_2 \tag{8.41b}$$

Then from (8.41b) since $\check{\mathcal{H}}_2$ is memoryless then $\mathcal{H}_2$ is memoryless and also acts like a variable positive gain with maximum $\alpha/(1 - \beta\alpha)$. Hence $\mathcal{H}_2$ is a sign preserving operator whenever $0 \leq \beta \leq 1/\alpha$. The transformed system is then in the form where Proposition 2 can be directly applied, provided

$0 \leq \beta \leq 1/\alpha$ (which ensures the transformed lower block remains passive) and $(\beta \mathcal{I} + \check{\mathcal{H}}_1)$ is strictly passive. This completes the proof. □

Remark 1. This result says that if there is sector bound on the lower memoryless block then we can bolster deficiencies in the upper linear block in an attempt to make it strictly passive. We are trading off degrees of relative passivity between the blocks.

Remark 2. The degree to which we can apply the passivity result depends on the size of α. A smaller α is better. This will be precisely our objective, that is, to find soft quantizers $\mathcal{S}(\cdot)$ that induce the least maximum gain, α, of their associated $\mathcal{P}_{\mathbf{A}}(\cdot)$ operator. This will then enlarge the class of linear systems, induced by the channel impulse response, for which passivity can be applied to show stability.

Remark 3. In our application of these results our $\mathcal{H}_1$ operator is linear, this was not required in the proof.

8.4 Stability and Error Propagation

8.4.1 Stability Concept

The idea of reformulating the error recovery problem as a stability problem originated with Cantoni et al. [5]. Our analysis here takes up this concept and it is natural to investigate the use of stability ideas in proving that under certain conditions a decision feedback structure has a finite recovery time (for all initial conditions and for all input sequences). The ideas led to a frequency domain condition that guarantees a finite error recovery time.

The problem of curtailing error propagation naturally leads to the study of error recovery defined as the time required for an initial error state to return to zero. This could mean that either the hard decision error becomes identically zero or the soft decision error becomes negligible for all time after some given time. Here, we choose to define error recovery it terms of the hard decision error:

Definition 3 (Error Recovery). Given an arbitrary initial error state the decision feedback structure has *recovered from error by time* K if

$$\widehat{e}_k = 0, \quad \forall k \geq K. \tag{8.42}$$

The relationship of this definition of error recovery to stability is given next.

Definition 4 (Stability). A decision feedback structure is *stable* if the error recovery time is finite for all possible initial error states and for all possible input sequences.

This is a stability concept in the sense that the persistent zero error condition can be regarded as the natural (desirable) condition and an initial error state acts like a transitory perturbation away from zero. Qualitatively the faster the structure can return to the zero error condition the better the performance will be in terms of lower transmission errors.

8.4.2 Conditions for Stability

In this subsection we determine technical conditions which are equivalent to the previous stability notion. These results form the basis of the key technical results which parallel those presented in [3].

The first result relates to the hard decision errors $\widehat{e}_k$:

Proposition 4. $\widehat{e} \in l_2 \Rightarrow \widehat{e}_k = 0, \quad \forall k \geq K, \; K < \infty.$

Proof. The result follows directly from the observation that $\widehat{e}_k$ takes on only discrete values in a finite alphabet $\mathbb{A}$ whose smallest nonzero values are bounded away from zero. □

Remark 1. Here l_2 denotes the space of square summable sequences[4]. Expressed in terms of Definition 3, this result says that if the hard error sequence is finite energy then the decision feedback structure has recovered from error at some finite time.

Our analysis makes use of the soft decision errors, $\widetilde{e}_k$, and, therefore, we need a second key result that relates (8.10) and (8.11):

Proposition 5. $\widetilde{e} \in l_2 \implies \widehat{e} \in l_2.$

Proof. Using the constraints on the soft decision quantizer, $\mathcal{S}(\cdot)$, i.e., that they leave the alphabet symbols unaffected (8.8) and locally continuous at those points (8.9):

$$\begin{aligned} \widetilde{e} \in l_2 &\Rightarrow \widetilde{e}_k \to 0 \quad \text{as } k \to \infty && (8.43a)\\ &\Rightarrow \widetilde{a}_k \to a_k \quad \text{as } k \to \infty && (8.43b)\\ &\Rightarrow \exists K(\epsilon) \quad \text{s.t. } |\widetilde{a}_k - a_k| < \epsilon \quad \forall k \geq K(\epsilon) && (8.43c)\\ &\Rightarrow \mathcal{Q}_{\mathbb{A}}(\widetilde{a}_k) = a_k \quad \forall k \geq K(\epsilon) && (8.43d) \end{aligned}$$

whenever

$$\epsilon < \min_{a^{(1)} \neq a^{(2)}} \left\{ \tfrac{1}{2} \left| a^{(1)} - a^{(2)} \right| \right\}, \quad a^{(1)}, a^{(2)} \in \mathbb{A} \tag{8.44}$$

4 The choice of l_2 for this argument is not crucial, indeed l_p can also be used for any $1 \geq p < \infty$. However, separate arguments relying on passivity ideas for which energy is a central concept heavily favor the use of l_2.

because $\mathcal{Q}_{\mathbf{A}}(\cdot)$ is a nearest neighbor rule. However, $\mathcal{Q}_{\mathbf{A}}(\widetilde{a}_k)-a_k = \widehat{a}_k-a_k = \widehat{e}_k$, whereby we infer

$$\widehat{e}_k = 0 \quad \forall k \geq K(\epsilon) \Rightarrow \widehat{e} \in l_2. \tag{8.45}$$

This completes the proof. □

Remark 1. This result implies it is sufficient that the soft error sequence belong to l_2 for there to be finite time error recovery associated with the hard errors.

8.4.3 Sector Boundedness of Memoryless Non-linearity

From (8.15) we showed how to determine the $\mathcal{P}_{\mathbf{A}}(\cdot)$ non-linearity from $\mathcal{S}(\cdot)$. The use of passivity (via Proposition 3) pivots around the degree of sector boundedness of the memoryless non-linearity $\mathcal{P}_{\mathbf{A}}(\cdot)$. It is of interest to ask, particularly for optimization, what are the conditions on $\mathcal{S}(\cdot)$ that guarantee a prescribed degree of sector boundedness on $\mathcal{P}_{\mathbf{A}}(\cdot)$?

Figure 8.10 depicts what it means for a non-linear mapping, such as $\mathcal{S}(\cdot)$, to be $[0, \alpha]$-sector bounded – the curve defining the input-output relationship is bounded above by a line of slope α, and below by a line of slope 0. This condition means that the instantaneous input to output gain is upper bounded by α. Note that it is the upper bound α and not the lower bound 0 that is crucial in using Proposition 3.

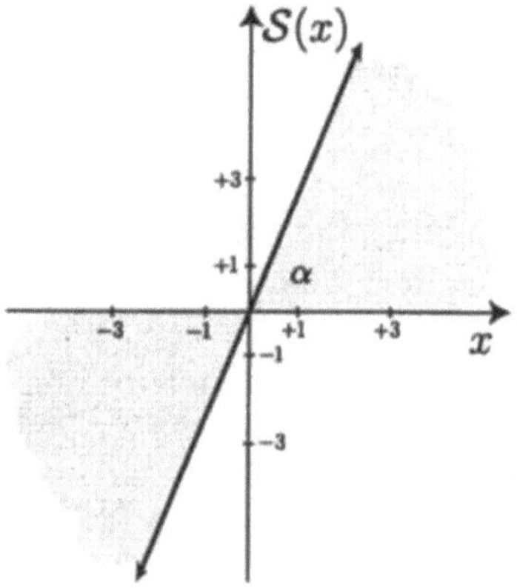

Fig. 8.10. The input output map of a $[0, \alpha]$-sector bounded non-linearity will have its input output characteristic constrained to the shaded region

We can use (8.14b) to determine conditions on $\mathcal{S}(\cdot)$ such that $\mathcal{P}_{\mathbf{A}}(\cdot)$ in (8.15) is $[0, \alpha]$-sector bounded. Just as we noted that $\mathcal{P}_{\mathbf{A}}(\cdot)$ could be determined from $\mathcal{S}(\cdot)$ via a simple translation (whose magnitude was determined from a_k); contrariwise, we can say that $\mathcal{S}(\cdot)$ can be determined via translation from $\mathcal{P}_{\mathbb{A}}(\cdot)$. The only interesting or nontrivial aspect of this reverse relationship is that there is ambiguity due to a_k which may range of all possible values in $\mathbb{A}$. The result is shown in Fig. 8.11 which shows the shaded region that results from the intersection of the translations of the region in Fig. 8.10. To be concrete, the illustration is for alphabet $\mathbb{A} = \{\pm 1, \pm 3\}$. Other sized alphabets are analogously determined. Note that the constraints given in (8.8) force the shaded region through the points (a, a) for all $a \in \mathbb{A}$.

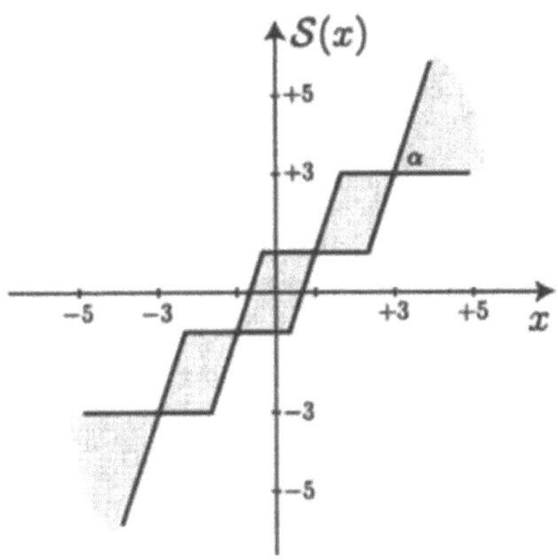

Fig. 8.11. Constraint on quantizer characteristic, $\mathcal{S}(\cdot)$, for the associated $\mathcal{P}_{\mathbb{A}}(\cdot)$ operator to be a $[0, \alpha]$-sector bounded non-linearity. Any function characteristic lying in the shaded region ensures the $[0, \alpha]$-sector boundedness property

The interpretation of Fig. 8.11 is the following[5]. Any soft quantizer function, $\mathcal{S}(\cdot)$, which lives in the shaded region is guaranteed to have a corresponding $\mathcal{P}_{\mathbb{A}}(\cdot)$ which is sector bounded to $[0, \alpha]$ as in Fig. 8.10. This is an interesting result which highlights that there is a broad range of $\mathcal{S}(\cdot)$ that achieve a given α specification. The objective will be to optimize the stability by minimizing the gain α as we shall see (Section 8.5.1).

8.5 Main Stability Results

For the error system in Fig. 8.6 we have the following key result:

Proposition 6 (Finite Error Recovery). *For the error system in Fig. 8.6, let $\mathcal{P}_{\mathbb{A}}(\cdot)$ be a $[0, \alpha]$-sector bounded operator for some $\alpha > 0$. If the channel*

[5] This region is easily generalized to the cases where $M \neq 4$.

coefficients $\{h_i\}$ *satisfy the positive real condition*

$$\frac{1}{\alpha} + \sum_{i=1}^{\infty} h_i \cos(i\theta) > \delta, \quad \forall \theta \in [0, 2\pi), \tag{8.46}$$

for some $\delta > 0$*, then the overall system is stable in the sense of error recovery in a finite time.*

Proof. The proof follows from a direct application of Proposition 3 to the system in Fig. 8.6. Condition (8.46) comes from the positive real condition of Proposition 1 which leads to the deduction that $\tilde{e}_k \in l_2$. In turn this implies, via Proposition 5, $\hat{e}_k \in l_2$, i.e., stability in the sense of error recovery in a finite time. This completes the proof. □

Example 1 (3 Tap Channel). As an illustration, if the ISI is limited in extent to two terms, i.e., the channel is given by $1 + h_1 z^{-1} + h_2 z^{-2}$, then the stability condition determines conditions on the admissible values of h_1 and h_2. Figure 8.12 shows the region in h_1 and h_2 space corresponding to (8.46) which in this case becomes the shaded rounded triangle with axis intercepts at magnitude α^{-1}. Clearly, the smaller is α the larger is the region. This inverse scaling property holds for arbitrary channel cases not just this special case.

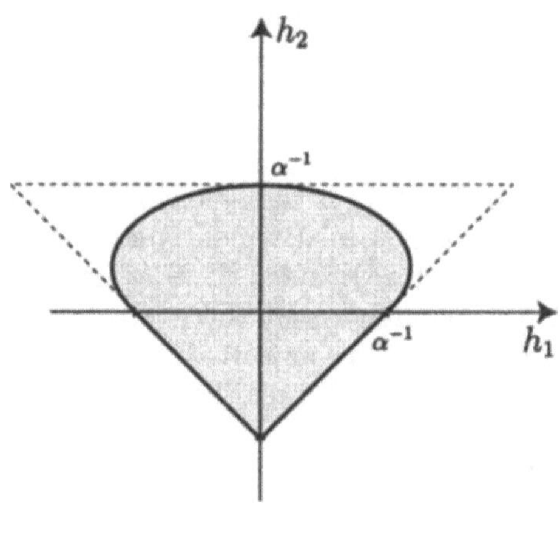

Fig. 8.12. Stability condition on the class of channels, $1 + h_1 z^{-1} + h_2 z^{-2}$, for error recovery in a finite time independent of the initial conditions and the input data sequence for any soft quantizer $\mathcal{S}(\cdot)$ whose associated operator $\mathcal{P}_{\mathbf{A}}(\cdot)$ is sector bounded in $[0, \alpha]$

8.5.1 Optimal Class of Memoryless Non-linearities

The objective in terms of designing a quantizer $\mathcal{S}(\cdot)$ is to maximize the stability region (8.46). As a designer we have control over $\mathcal{S}(\cdot)$ and this will influence α. We are now in a position to determine the optimal class of memoryless non-linearities:

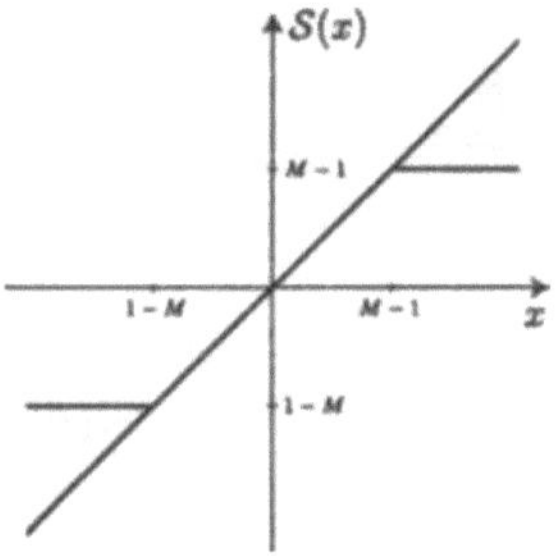

Fig. 8.13. Constraint on soft quantizer characteristic for the associated $\mathcal{P}_{\mathbb{A}}(\cdot)$ operator to be $[0,1]$-sector bounded non-linearity. Any quantizer satisfying this characteristic will be optimal in the sense of error recovery in a finite time

Proposition 7 (Optimal Class of Memoryless Non-linearities). *Soft quantizers $\mathcal{S}(\cdot)$ for which the associated operator $\mathcal{P}_{\mathbb{A}}(\cdot)$ is sector bounded in $[0,1]$ are optimum the sense of error recovery in a finite time. There are an infinite class of such soft quantizers including: i) the identity mapping, and ii) the classical saturation characteristic with saturation at $\pm(M-1)$.*

Proof. From Fig. 8.11 the equivalent objective is to minimize α. It is clear that $\alpha = 1$ is achievable leading to the nonempty region Fig. 8.13. Note that $\mathcal{S}(\cdot)$ necessarily passes through points (a, a) for all $a \in \mathbb{A}$ and, therefore, $\alpha = 1$ is the minimum. □

Remark 1. The constraint in this case reduces to a 45 degree line between $(1-M)$ and $(M-1)$ for the M-ary quantization levels, this is depicted in Fig. 8.13. Outside this range the quantization characteristic is not uniquely determined but still constrained. Curiously, having a simple identity mapping (no quantizer) for $\mathcal{S}(\cdot)$ whereby the modified decision feedback equalizer collapses to being a linear device also has the best stability.

Remark 2. In comparison with the soft saturation characteristic we can consider the hard nearest neighbor quantization function $\mathcal{Q}_{\mathbb{A}}(\cdot)$. With this selection it can be readily deduced that $\alpha = 2$ – indeed this is the result given in [3]. Thus, using Proposition 6 the stability region of the optimum soft quantizer is a two times scaling of the stability region for the hard quantizer.

8.5.2 Optimum Deadband Region

One means that has been suggested for suppressing error propagation is to map dodgy decisions, i.e., those near the nearest neighbor thresholds, to zero [6]. The idea is to avoid making a large mistake when one suspects an error is made and instead sacrifice a smaller mistake. For example, the minimum magnitude of a non-zero error is two. If we suspect that an error might be made we can just set $\widetilde{a}_k$ to the mid value of the two adjacent nearest neighbor symbols and, thereby, force an error whose minimum magnitude is one, i.e., the non-zero error is now one. The question is how large should we make the region about the nearest neighbor thresholds within which we average two adjacent nearest neighbor symbols[6]?

In the context of our passivity theory we can determine the best deadband region (or erasure region [6]). Firstly let $M = 2$, then a deadband soft quantizer is given by

$$\mathcal{D}_\Delta(x) \triangleq \tfrac{1}{2}\big(\operatorname{sgn}(x-\Delta)+\operatorname{sgn}(x+\Delta)\big), \quad x \in \mathbb{R},\ 0<\Delta<1 \tag{8.47}$$

which reduces to $\operatorname{sgn}(\cdot)$ when $\Delta \to 0$. Note that the width of the deadband region is $2\,\Delta$. This is the building block for the M-ary deadband soft quantizer, which takes an analogous form to (8.3) with $\mathcal{D}_\Delta(\cdot)$ replacing $\operatorname{sgn}(\cdot)$, given in the following proposition:

Proposition 8 (Optimum Deadband Size). *Given the deadband quantization characteristic*

$$\mathcal{S}_\Delta(x) \triangleq \sum_{j=1}^{M-1} \mathcal{D}_\Delta(x-M+2j), \quad x \in \mathbb{R},\ 0<\Delta<1 \tag{8.48}$$

the optimum deadband size Δ in the sense of error recovery in a finite time is $\Delta = 1/3$.

Proof. The gain of the induced non-linearity $\mathcal{P}_{a_k}(\cdot)$ is determined by the maximum slope, call this α_Δ, of the shifted versions of the (8.48) characteristic according to (8.14b) given by

$$\alpha_\Delta = \max_x \max_{a\in\mathbf{A}} \big\{\big(\mathcal{S}_\Delta(x+a)-a\big)/x\big\}\,. \tag{8.49}$$

The piecewise linear nature of (8.48) means the maximization over x can be restricted to the set of $4(M-1)$ discrete points

$$\mathcal{X} \triangleq \big\{3-2M\pm\Delta,\dots,2M-5\pm\Delta, 2M-3\pm\Delta\big\}. \tag{8.50}$$

[6] A small technical issue here is that if the soft decisions map to the threshold value, as they do more regularly with the deadbands defined here, then this will cause ambiguity with the hard decisions (8.5). In this case it will be better to feed the hard quantizer $\mathcal{Q}_{\mathbf{A}}(\cdot)$ in Fig. 8.3 with the input signal to the soft quantizer.

corresponding to knee points in the monotonically non-decreasing function $\mathcal{P}_{a_k}(\cdot)$. This means from (8.49)

$$\alpha_\Delta = \max_{x\in\mathcal{X}} \max_{a\in\mathbf{A}} \left\{ \left(\mathcal{S}_\Delta(x+a) - a\right)/x \right\} \tag{8.51a}$$

$$= \max\left\{ \tfrac{1}{1-\Delta}, \tfrac{2}{1+\Delta}, \tfrac{3}{3-\Delta}, \tfrac{4}{3+\Delta}, \tfrac{5}{5-\Delta}, \tfrac{6}{5+\Delta}, \dots \right\} \tag{8.51b}$$

$$= \max\left\{ \tfrac{1}{1-\Delta}, \tfrac{2}{1+\Delta} \right\} \tag{8.51c}$$

by tedious verification. This maximum gain, α_Δ, can be minimized by equating the gradients at these two dominating gradient knee points implying the optimum deadband region Δ satisfies $1 - \Delta = (1 + \Delta)/2$. The solution is $\Delta = 1/3$. □

Remark 1. At this optimum deadband setting, $\Delta = 1/3$, the gain satisfies $\alpha_\Delta|_{\Delta=1/3} = 1.5$ which lies at the midpoint of the conventional DFE gain $\alpha = 2$ and the optimal gain, derived in the previous subsection, which equals $\alpha = 1$.

Remark 2. Deadband soft quantizers are suboptimal in the sense of error recovery in a finite time.

8.5.3 Simulations

The model used in the simulations was geared towards demonstrating the extent of error propagation. A simple impulsive noise model was adopted consisting of a gaussian noise spike (variance 10) every 100 data samples. Depending on the amplitude of the noise at each of these instants a hard decision error (primary error) may or may not have been made. The data record length was 10^6 samples and the symbols were binary. Table 8.1 gives results for 3 indicative channels. The conventional (hard) decision feedback equalizer (DFE) and the soft decision feedback equalizer that uses a saturation type soft quantizer are compared. We call this the Saturation DFE.

The first channel in Table 8.1 represents a channel such that both types of equalizer are stable in the error propagation sense. For this channel there is no discernible difference in the error propagation. The ISI for this channel is quite mild but sufficiently large to close the eye pattern. The second channel is one for which only the Saturation DFE is stable and the improvement in bit error rate is evident. The third channel is one for which both channels are stable but the stability margin of the Saturation DFE is much greater which explains the performance advantage. This indicates, qualitatively, that the degree of stability relates inversely with the extent of error propagation.

8.6 Conclusions

We have used passivity theory to examine the stability of a generalized decision feedback structure which incorporates a general memoryless nonlinearity in lieu of the traditional nearest neighbor quantizer used for feeding

Channel Model	DFE	Saturation DFE
$1.0 + 0.3z^{-1} - 0.3z^{-2}$	0.00374	0.00374
$1.0 + 0.6z^{-1} - 0.6z^{-2}$	0.01474	0.00874
$1.0 + 0.6z^{-1} + 0.6z^{-2}$	0.01128	0.00626

Table 8.1. Bit error rate performance comparison

back data estimates. In the sense of error recovery in a finite time, which was our defined stability notion, we determined the optimal class of memoryless non-linearities. This class includes the saturation characteristic. However, somewhat surprisingly, this class contains the case of direct linear feedback. Among constrained quantization rules, we also determine the optimal size of a quantizer deadband region – a problem that has been previously been raised in the literature on erasures in decision feedback equalization.

These new general structures were shown theoretically to have superior error propagation suppression characteristics to the conventional decision feedback equalizer. Limited simulations corroborated these findings.

Overall, this study demonstrates that the theoretical tools of input output stability, and in particular Passivity Theory, can have wide applicability even to problems in telecommunications which do not normally easily lend themselves to the use of stability concepts.

References

1. C.A. Desoer and M. Vidyasagar, *Feedback Systems: Input-Output Properties*, Academic Press, New York, NY, 1975.
2. B.D.O. Anderson, R.R. Bitmead, C.R. Johnson Jr., P.V. Kokotovic, R.L. Kosut, I.M.Y. Mareels, L. Praly, and B.D. Riedle, *Stability of Adaptive Systems: Passivity and Averaging Analysis*, The MIT Press, Cambridge, Massachusetts, 1986.
3. R.A. Kennedy, B.D.O. Anderson, and R.R. Bitmead, "Channels leading to rapid error recovery for decision feedback equalizers," *IEEE Trans. Commun.*, vol. 37, no. 11, pp. 1146–1155, November 1989.
4. R.A. Kennedy and C.R. Johnson Jr, "Encoder stability study for the 32kbits/s CCITT G.721 ADPCM standard based on a simplified error model," in *Proc. The Second Int. Sym. on Signal Processing and its Applic., ISSPA'90*, Gold Coast Australia, August 1990, pp. 744–747.
5. A. Cantoni and P. Butler, "Stability of Decision Feedback Inverses," *IEEE Trans. Commun.*, vol. 24, no. 9, pp. 1064–1075, September 1976.
6. M. Chiani, "Introducing Erasures in Decision-Feedback Equalization to Reduce Error Propagation," *IEEE Trans. Commun.*, vol. 45, no. 7, pp. 757–760, July 1997.
7. S.U.H. Qureshi, "Adaptive equalization," *Proc. of the IEEE*, vol. 73, pp. 1349–1387, September 1985.

8. R.A. Kennedy and B.D.O. Anderson, "Recovery times of decision feedback equalizers on noiseless channels," *IEEE Trans. Commun.*, vol. 35, no. 10, pp. 1012–1021, October 1987.
9. M. Austin, *Decision Feedback Equalization for digital communication over dispersive channels*, MIT, MIT Res. Lab. Electron., Tech. Rep.461, August 1967.
10. M.V. Eyuboglu and S.U.H. Qureshi, "Reduced-State Sequence Estimation with Set Partitioning and Decision Feedback," *IEEE Trans. Commun.*, vol. COM-36, no. 1, pp. 13–20, January 1988.
11. A. Duel-Hallen and C. Heegard, "Delayed Decision Feedback Sequence Estimation," *IEEE Trans. Commun.*, vol. COM-37, no. 5, pp. 428–436, May 1989.
12. D.L. Duttweiler, J.E. Mazo, and D.G. Messerschmitt, "An upper bound on the error probability in decision feedback equalizers," *IEEE Trans. Inform. Theory.*, vol. 20, no. 4, pp. 490–497, July 1974.
13. J.J. O'Reilly and A.M. de Oliveria Duarte, "Error Propagation in Decision Feedback Receivers," *IEE Proc. F. Comm., Radar and Sig. Proc.*, vol. 132, no. 7, pp. 561–566, December 1985.
14. A.M. de Oliveria Duarte and J.J. O'Reilly, "Simplified Technique for Bounding Error Statistics for DFB Receivers," *IEE Proc. F. Comm., Radar and Sig. Proc.*, vol. 132, no. 7, pp. 567–575, December 1985.
15. R.A. Kennedy, B.D.O. Anderson, and R.R. Bitmead, "Tight bounds on the error probabilities of decision feedback equalizers," *IEEE Trans. Commun.*, vol. 35, no. 10, pp. 1022–1029, October 1987.

9 Connecting Steiglitz-McBride Identification, Active Noise Control, and Coefficient Filtering to a Common Framework

C. Richard Johnson Jr.[1] and William A. Sethares[2]

[1] School of Electrical Engineering, Cornell University, Ithaca, NY 14853
[2] Department of Electrical and Computer Engineering, University of Wisconsin, Madison, WI 53706

Abstract. There are many adaptive algorithms that incoporate filtered versions of the regressor vector, the error signal, the parameter vector, or the update term. These modifications can often be viewed as attempts to recapture or improve the known stability properties of linear LMS algorithms, despite operating with IIR parameterizations or within a feedback loop. This chapter reviews some basic results on the (exponential) stability of adaptive algorithms, and then applies them in three situations: to the Steiglitz-McBride family of algorithms, to an analysis of algorithm performance in acoustic noise cancellation systems that usually incorporate the 'filtered-u' approach, and to a consolidation of coefficient filtering and update smoothing schemes such as momentum LMS and leakage. The principle of trading-off linear filters on various signals within the adaptive schemes is a powerful unifying principle that can be used to create 'new' algorithms, and it allows translation of stability results between seemingly unrelated algorithm forms.

9.1 Introduction

The basic adaptive algorithm has the form:

$$\begin{pmatrix} new \\ coefficients \end{pmatrix} = \begin{pmatrix} old \\ coefficients \end{pmatrix} + (stepsize) \begin{pmatrix} regressor \\ vector \end{pmatrix} \begin{pmatrix} prediction \\ error \end{pmatrix}$$

and the incorporation of 'additional' filters into this basic structure has a long history. Sometimes, the additional filters are intrinsic to the application. Sometimes, they are added to help speed convergence or to improve robustness of the adaptive mechanism. Sometimes, it seems, they are added without a complete appreciation for the possible repercussions. In an earlier survey article [24], we were able to identify three basic approaches to the understanding of the behavior of such adaptive algorithms. These were the 'expected value' approach typified by Widrow [29], the ODE (or stochastic approximations) approach of Ljung and coworkers [16], and the deterministic averaging approach, for which Anderson and coworkers [1] were substantively responsible. Our sense is that this is a powerful method of analyzing the behavior of algorithms. The next section reviews this methodology, which is exploited throughout the chapter both as an analytical tool and as a way to create new algorithm forms specifically tailored for particular application environments.

Section 9.3 shows how the Steiglitz-McBride family of algorithms [27] fits into this framework, and demonstrates (local) stability of the associated error system. A key ingredient in the presentation is that certain signals within the adaptive mechanism are not used directly, rather, they are filtered before use. It is often possible to trade-off a filtering of the regressor vector with a corresponding filtering of the prediction error, and this idea is exploited to derive several "new" Steiglitz-McBride-like algorithms in which the filters are applied to different signals.

Section 9.4 investigates the problem of active noise cancellation. While normally conceived of as an IIR adaptive filtering problem, the 'filtered-u' algorithm [7] is shown to have more in common with the linear equation error algorithms familiar from adaptive control than with the nonlinear IIR output error schemes. In particular, it is possible to use the "key technical lemma" of [11] to demonstrate global stability of the error system in the ideal case. The various filters that are present arise from the problem setting itself, and not as a part of the adaptive algorithm - they are quite distinct from (and more benign than) the kinds of filtering that arise in [1] or [25].

Continuing the theme of trading-off filters in the adaptive algorithms, section 9.5 shows that algorithms which incorporate filtering of the coefficient vector can be rewritten using a corresponding filtering of the update. Conversely, algorithms which filter the update can be rewritten using a corresponding filtering of the coefficient vector. Moreover, these algorithms (such as LMS with leakage, momentum LMS, etc.) can be analyzed using the deterministic averaging approach.

9.2 Technical Background

Consider a linear plant parameterized by an unknown constant vector θ^* which maps a bounded scalar input $u(k)$ to a scalar output $y(k) = X^T(k)\theta^*$, where $X(k)$ is a regressor vector of past inputs and/or outputs. Let $\hat{\theta}(k)$ be an estimate of θ^*, and form the estimated output $\hat{y}(k) = X^T(k)\hat{\theta}(k)$. The error $e(k) = y(k) - \hat{y}(k)$ between the measured output and the estimated output can be used to improve the estimates of the unknown parameters. Using gradient descent techniques or L_2 minimization ideas leads to the parameter update scheme

$$\hat{\theta}(k+1) = \hat{\theta}(k) + \mu X(k)e(k) \tag{9.1}$$

where μ is a small positive stepsize. This is called the LMS algorithm or the "equation error" algorithm, depending on the exact definition of the signals within the regressor vector. For purposes of analysis, it is often more convenient to work with the error system

$$\theta(k+1) = \theta(k) - \mu X(k)\{X^T(k)\theta(k)\} \tag{9.2}$$

where $\theta(k) = \theta^* - \hat{\theta}(k)$. This error system is Lyapunov stable [19], and the equilibrium at $\theta = 0$ is exponentially asymptotically stable (EAS) if $X(k)$ is persistently spanning [4], that is, if there is a finite time window m such that for every j,

$$\sum_{k=j}^{j+m} X(k)X^T(k) \tag{9.3}$$

is uniformly positive definite. The exponential asymptotic stability of the error system (9.2) implies that $\hat{\theta}(k)$ of (9.1) is exponentially convergent to θ^*.

In certain applications, the measured prediction error $e(k)$ is not precisely of the form $X^T(k)\theta(k)$, a number of which are listed in [25]. A simple case would be when $e(k)$ contains a known filtering by a stable rational transfer function $F(q^{-1})$, that is, $e(k) = F(q^{-1})\{X^T(k)\theta(k)\}$. If $F(q^{-1})$ is stably invertible, then it is natural to consider an algorithmic form which filters the prediction error

$$\theta(k+1) = \theta(k) - \mu X(k)F^{-1}(q^{-1})\{e(k)\} = \theta(k) - \mu X(k)\{X^T(k)\theta(k)\} \tag{9.4}$$

to regain the known stability properties of (9.2). An alternative approach is to recognize that when the stepsize is small, the dynamics of $\theta(k)$ are much slower than the dynamics of $X(k)$. The prediction error $e(k)$ is then approximately equal to $\mathbf{F}(q^{-1})\{X^T(k)\}\theta(k)$ where $\mathbf{F}(q^{-1})$ operates on each component of the vector $X(k)$ in exactly the same way that $F(q^{-1})$ operates on $e(k)$. Thus

$$e(k) = F(q^{-1})\{X^T(k)\theta(k)\} = \mathbf{F}(q^{-1})\{X^T(k)\}\theta(k) + O(\mu), \tag{9.5}$$

The $O(\mu)$ term can be considered a small "disturbance" that is dominated by the exponential stability and so it is reasonable to consider an alternative algorithmic form that filters the regressor $X(k)$ to recapture the stability properties of (9.2),

$$\begin{aligned}\theta(k+1) &= \theta(k) - \mu\mathbf{F}(q^{-1})\{X(k)\}e(k), \\ &\approx \theta(k) - \mu\mathbf{F}(q^{-1})\{X(k)\}\mathbf{F}(q^{-1})\{X^T(k)\}\theta(k).\end{aligned} \tag{9.6}$$

Thus $\mathbf{F}(q^{-1})\{X(k)\}$ plays the same role in (9.6) that $X(k)$ plays in (9.2), and both algorithm forms (9.4) and (9.6) can be expected to retain the desirable behaviors of the LMS form (9.1) in a modified problem setting.

In ([25]), the authors along with B. D. O. Anderson investigated such expectations precisely and concretely. Motivated by the reasoning above, we considered the general algorithm form

$$\theta(k+1) = \theta(k) - \mu\mathbf{L}(q^{-1},k)\{X(k)\}M(q^{-1},k)\{X^T(k)\theta(k)\}. \tag{9.7}$$

where $\mathbf{L}(q^{-1},k)$ is a possibly time varying filtering of the regressor and $M(q^{-1},k)$ is a filtering of the prediction error. To be specific, we detailed conditions on μ, $\mathbf{L}$, M, and $X(k)$ under which (9.7) is EAS. The three steps in this investigation were:

i. Find conditions on $M(q^{-1},k)$ and μ so that exponential stability of

$$\theta(k+1) = \theta(k) - \mu\mathbf{L}(q^{-1},k)\{X(k)\}\mathbf{M}(q^{-1},k)\{X^T(k)\}\theta(k) \qquad (9.8)$$

implies exponential stability of (9.7). Observe that the regressor in (9.8) is filtered by $\mathbf{M}$ whereas in (9.7) the error is filtered by M (again, the bold letter indicates an element by element vector filtering).

ii. Use results from averaging theory to give conditions on $X(k)$ and the time varying filters $\mathbf{L}$ and $\mathbf{M}$ under which (9.8) is EAS. Basically, this requires that for some integer m, all eigenvalues of the sliding average

$$\bar{A}_k(m) = \frac{1}{m}\sum_{i=1}^{m}\mathbf{L}(q^{-1},k+i-1)\{X(k+i-1)\}$$
$$\mathbf{M}(q^{-1},k+i-1)\{X^T(k+i-1)\} \qquad (9.9)$$

have positive real parts for each k.

iii. Translate the conditions on the time varying operators $\mathbf{L}(q^{-1},k)$ and $\mathbf{M}(q^{-1},k)$ to conditions on related time invariant operators. The simplest of these assume that the related "frozen" operators $\mathbf{L}(q^{-1},p)$ and $\mathbf{M}(q^{-1},p)$ are passive (or SPR) for each fixed p, and that the time variation in the filters is suitably slow.

The bulk of this chapter exploits and extends these results, examining the character and behavior of a number of useful adaptive algorithms.

9.3 Steiglitz-McBride Algorithms

While the "equation error" method of identifying the parameters of a linear IIR filter is straightforward to derive (it is simply an instantaneous gradient descent applied to an appropriate cost function), and is correspondingly easy to analyze (because it is linear in the unknown parameters), it is also well known to return biased estimates of the parameters when the output is distorted by measurement noise, even when the noise is white. The "output error' methods are capable of removing this bias, but they require a nonlinear update rule which contains a derivative that can only be approximated recursively; analysis of such methods is correspondingly limited to local results near the equilibrium points. Steiglitz and McBride [27] discovered another approach which attempts to maintain the simplicity of the equation error formulation while also removing the bias.

The essential idea is to "distort" the cost function in the equation error formulation using a sequence of prefilters, and the resulting block oriented algorithm was shown by Stoica and Soderstrom [28] to have a single stable stationary point when the correct order of the system is known, and when the disturbance is white. Since the prefilters are determined from the solution to a linear equation error problem, the algorithm can be rewritten in an on-line form as in Fan and Jenkins [8], and this version also returns unbiased estimates of the parameters when the model order is known and the noise is white.

Unfortunately, this distortion of the cost function means that the algorithm is not really following a path of steepest descent, that is, the algorithm is not necessarily converging to minimize some cost. As Regalia mentions [22], this is one reason why analysis of the scheme is tricky. He then proceeds to develop off line block forms, on line adaptive forms, as well as lattice forms.

In this section, we show how the Steiglitz and McBride algorithms can be viewed as a kind of filtered-regressor equation error algorithm, in direct analogy with the filtered-regressor output error algorithms of Stearns [9] and Ljung and Soderstrom [16]. Using the idea of recapturing the LMS-like properties as suggested in the previous section, we then propose "new" algorithms which filter the error signal in the equation error format, in direct analogy with the filtered-error algorithms of Landau [15] and Ljung [17].

9.3.1 Two "Derivations" of the Algorithm

Regalia's [22] exposition of the (direct form) Steiglitz-McBride algorithm begins with inputs $u(k)$, outputs $y(k)$, and "prefilter states" $u_f(k)$ and $y_f(k)$. The error signal on which the algorithm is based is $e(k) = y(k) - \hat{y}(k)$, where

$$\hat{y}(k) = \sum_{j=0}^{M} \hat{b}_j(k) u_f(k-j). \tag{9.10}$$

Let $\hat{a}_i(k)$ and $\hat{b}_j(k)$ be the coefficients of the filters $\hat{A}$ and $\hat{B}$ at time k, the coefficient updates are

$$\hat{a}_i(k+1) = \hat{a}_i(k) + \mu y_f(k-i)e(k)$$

and

$$\hat{b}_j(k+1) = \hat{b}_j(k) + \mu u_f(k-j)e(k)$$

while the "output prefilter" is

$$y_f(k) = y(k) + \sum_{i=1}^{N} \hat{a}_i(k) y_f(k-i)$$

and the "prefilter state recursion" is

$$u_f(k) = u(k) + \sum_{i=1}^{N} \hat{a}_i(k) u_f(k-i). \tag{9.11}$$

To understand the Steiglitz-McBride algorithm, observe that it can be viewed as attempting to recapture a simple linear combiner form. In figure 9.1, for instance, the path from u_f to $\hat{y}$ has "transfer function" $\hat{B}$ (this is in quotes because the actual system is time varying; as long as the time variation in $\hat{B}$ is slow, such notational abuse may be useful), as required by (9.10) above. Similarly, the transfer function from u to u_f is $\frac{1}{1-\hat{A}}$ (from (9.11)), and so the transfer function from u to $\hat{y}$ is $\frac{B}{1-\hat{A}}$.

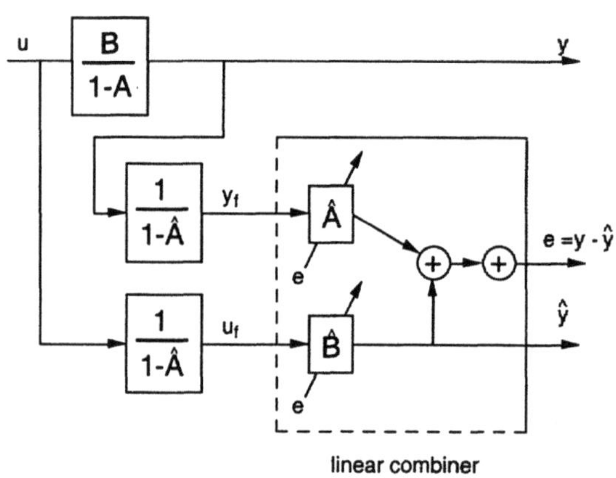

Fig. 9.1. Linear combiner view of the Steiglitz-McBride algorithm.

Thus, it is reasonable to consider the IIR filter structure

$$\hat{y} = \hat{A}\{\hat{y}\} + \hat{B}\{u\}.$$

as a model for the "identification" of the linear system

$$y = A\{y\} + B\{u\}$$

Referring again to figure 9.1, the path from y to e contains $(\frac{1}{1-\hat{A}})(1-\hat{A})$, which is a close approximation to unity. In other words, by prefiltering both the inputs and outputs of the system with the "transfer function" $\frac{1}{1-\hat{A}}$, the signals u_f and y_f have been synthesized. These can then be considered as the "inputs" to the dashed box, which is precisely a linear combiner structure. The associated LMS-style (instantaneous gradient) algorithm can be readily written down for this structure, and it is identical to the equations above.

Next, the algorithm is rederived in a way that formalizes the above "notational abuse" as a standard $O(\mu)$ approximation. The procedure is to expand $e = y - \hat{y}$ by adding "zero" in the form $\hat{A}\{y\} - \hat{A}\{y\}$, and to then regroup.

$$\begin{aligned} e = y - \hat{y} &= A\{y\} + B\{u\} - \hat{A}\{\hat{y}\} - \hat{B}\{u\} - \hat{A}\{y\} + \hat{A}\{y\} \\ &= (A - \hat{A})\{y\} + (B - \hat{B})\{u\} + \hat{A}\{y - \hat{y}\} \\ &= (1 - \hat{A})^{-1}\{X^T\theta\} \end{aligned} \tag{9.12}$$

where $X^T = [y \quad u]$ and $\theta^T = [(A - \hat{A}) \quad (B - \hat{B})]$. This can be rewritten for small μ as

$$e = (1 - \hat{A})^{-1}\{X^T\}\theta + O(\mu)$$

and hence

$$e \approx (1 - \hat{A})^{-1}\{X^T\}\theta \tag{9.13}$$

which can be formalized using Lemmas 1 and 2 of [25] to show that, to within order μ, the appropriate algorithm would be

$$\hat{\theta}(k+1) = \hat{\theta}(k) + \mu \mathbf{L}(q^{-1}, k)\{X(k)\}e(k) \tag{9.14}$$

where the regressor filter is

$$\mathbf{L}(q^{-1}, k) = (1 - \hat{A})^{-1} \tag{9.15}$$

and the parameter estimates are $\hat{\theta}^T = [\hat{A} \quad \hat{B}]$. Clearly, this is just a rewriting of the direct form of the Steiglitz-McBride algorithm of (9.10) to (9.11).

In contrast with the 'output error' algorithms of [25], the regressor X in (9.14) includes y and u, i.e. the output and input of the system being identified, rather than $\hat{y}$ and u, i.e. the output and input of the adapted model. Hence the Steiglitz-McBride algorithm is a form of 'equation error' in which the regressor is filtered, and it falls within the general algorithm form (9.7) with $\mathbf{L}$ defined in (9.15) and $M(q^{-1}, k) = (1 - \hat{A})^{-1}$. Regalia has commented [22] "The most confusing aspect of the algorithm is the use of time varying prefilters." From the current point of view, we see that these prefilters are used in an attempt to 'recapture' the known and desirable behavior of simpler algorithms.

9.3.2 Analysis of the Algorithm

Since the Steiglitz-McBride algorithm is of the general form (9.7), the analysis in [25] can be directly applied. To establish the EAS of the algorithm, (9.9) must hold, that is, the sliding average

$$\bar{A}_k(m) = \frac{1}{m}\sum_{i=1}^{m}(1 - \hat{A})^{-1}\{X(k+i-1)\}(1 - \hat{A})^{-1}\{X^T(k+i-1)\} \tag{9.16}$$

must be positive definite for every k, $(1 - \hat{A})^{-1}$ must be stable, and μ must be sufficiently small.

i. If the regressor $X(k)$ fails to be persistently spanning (9.3), then (9.16) cannot hold. Hence persistence of excitation condition is necessary condition for local exponential stability of the Steiglitz-McBride algorithm.
ii. The operators $\mathbf{L}$ and M (which are both copies of $(1 - \hat{A})^{-1}$) cannot be guaranteed to be exponentially stable for all initial conditions, since the evolution of $\hat{A}$ is governed by the iteration itself (recall that $\hat{\theta}$ contains $\hat{A}$). However, if the $\hat{A}$ is initialized near the correct answer (which must be stable since y is assumed bounded), then the exponential stability will hold. Accordingly, the stability results are local (not global) in the initial conditions.
iii. Theorem 2 of [25] requires that $\mathbf{M}(q^{-1}, k)\{\mathbf{L}^{-1}(q^{-1}, k)\}$ be strictly passive. For the Steiglitz-McBride algorithm, the corollary to Lemma 5 shows that this always holds (The frozen systems $\mathbf{M}(q^{-1}, p)\mathbf{L}^{-1}(q^{-1}, p)$ are exactly equal to the identify for all fixed p, so it retains passivity for slow enough time variations.)
iv. Since the time evolution of $(1 - \hat{A})$ is governed by the stepsize μ, both $\mathbf{L}$ and M can be made to vary as slowly as needed by requiring that μ be small. Hence, assessing the local exponential stability of the frozen filters is enough to guarantee the exponential stablity of the time varying filters (from Lemma 3 in [25]).
v. A major reason for focusing on the exponential stability of the adaptive error system is that exponenial convergence gurantees a certain robustness in the presence of nonidealities such as unmodelled dynamics or measurement noise, and allows consideration of the situation in which θ^* is itself slowly varying. For instance, if θ^* varies on a time scale of $\frac{1}{\mu^2}$ or slower, then the analysis needed to demonstrate that $\hat{\theta}$ tracks θ^* is substantially unchanged from the case when θ^* is fixed.
vi. Regalia [22] conducts an ODE analysis of the Steiglitz-McBride algorithm and demonstrates robust convergence. As in the analysis of this section, the unknown filter $(1 - \hat{A})^{-1}$ must be stable , minimum phase, and convergence can only be shown locally. The 'unbiased' nature of the convergence holds only for white measurement noise. In the deterministic averaging results here, the only assumption on the disturbance is 'smallness'. Hence the disturbances can arise from noise in the system, from undermodelling, from nonlinearities, or from time variations. The EAS results guarantee that such disturbances do not change the equilibrium, though the convergent points may be biased away from the equilibrium in such a way that the bias vanishes as the disturbance vanishes.
vii. A number of researchers ([5] and [22]) have observed that the global convergence properties of the Steiglitz-McBride algorithm are appealing in the undermodelled case, and several case studies have shown that the Steiglitz-McBride algorithm does not become trapped in local minima to

which the output error schemes are vulnerable. As Regalia comments, this is good evidence that the algorithm deserves further study, although it is reasonable to wonder if there may be other problem settings in which the Steiglitz-McBride algorithm becomes trapped at local minima to which other algorithms are immune.

9.3.3 "New" Steiglitz-McBride-like Algorithms

The basic philosophy of section 9.2 is that there is always more than one way to 'undo' a filtering operation. Since the Steiglitz-McBride algorithm can be viewed as incorporating the regressor filtering as a way of undoing the intrinsic error prefiltering, an alternative strategy is to post-filter the error. This can be implemented simply as

$$\hat{\theta}(k+1) = \hat{\theta}(k) + \mu X(k)(1 - \hat{A})\{e(k)\}.$$

In this case, $\mathbf{L}(q^{-1}, k) = I$ and $M(q^{-1}, k) = (1 - \hat{A})(1 - \hat{A})^{-1}$. As before, the averaging analysis of [25] leads to a local EAS error system whenever $(1 - \hat{A})$ and $(1 - \hat{A})^{-1}$ are stable, and when μ is small. Observe that (9.16) is full rank whenever (9.3) is full rank, and comments analogous to those of the previous section are germane. This algorithm may be considered a kind of 'filtered-error equation error' algorithm, by analogy with the filtered error output algorithms.

Alternatively, if some *a priori* information about the $\hat{A}$ system is available, then this may also be exploited. Suppose, for example, that $\hat{A}$ is unknown but that a fixed stable transfer function $C(q^{-1})$ is known for which $\frac{C}{1-\hat{A}}$ is passive. Then C can be used as either a regressor filter

$$\hat{\theta}(k+1) = \hat{\theta}(k) + \mu C(q^{-1})\{X(k)\}e(k)$$

or as an error filter

$$\hat{\theta}(k+1) = \hat{\theta}(k) + \mu X(k)C(q^{-1})\{e(k)\}.$$

The passivity of $\frac{C}{1-\hat{A}}$ guarantees the passivity of $\mathbf{M}\{\mathbf{L}^{-1}\}$, and local EAS results follow for both algorithms under analogous excitation and stepsize assumptions. It is an interesting and open question as to whether such modified Steiglitz-McBride algorithms will retain the immunity to local minima shown in [5].

9.4 Active Noise Control and the Filtered-u Algorithms

This section provides a stability analysis of a class of acoustic noise control algorithms such as the filtered-u algorithm, by showing that the adapted models have more in common with (nonlinear, FIR) equation error models than with the (IIR) output error models they superficially resemble. Stability results from the adaptive control literature are applied to show global stability in the noise free case, and to show exponential stability when the input is persistently excited. The latter demonstrates a robustness to mismodelling errors, disturbances such as noises, and allow results on the tracking time varying systems. Application areas include HVAC noise control and enclosure noise control for automobiles, aircraft, and elevators.

The preferred compensator structure for parameter adaptive, feedforward, active noise control in ducts is a direct form infinite impulse response (IIR) model. Thus, adaptation algorithms, and their analytical justification and interpretation, have been sought among those intended for adaptive IIR filters in a parallel system identification configuration. The point of this section is to show that the IIR structure may be misleading and that the stability analysis can proceed by rewriting the problem in a nonlinear equation error formulation. This allows direct application of theoretical results from the adaptive control literature. In particular, the filtering used in the 'filtered-u' algorithm plays a very different role from the filtered regressor and filtered error terms of the previous sections.

9.4.1 A Stability Perspective on Algorithm Development

Active noise cancellation schemes such as those from [6] and [14] can be portrayed as in Figure 9.2. The simplified case considered here for analysis, which illustrates the main points with less technical complexity, is when $C = 1$. The following assumptions specify the problem setting in terms of the variables defined in Figure 9.2.

(A1) Disturbance. The source signal s is uniformly bounded

$$|s(k)| \leq \alpha \quad \forall k. \tag{9.17}$$

(A2) Feedback. The detection microphone signal u is described by

$$u(k) = s(k) + \sum_{i=1}^{Q_2} f_i \hat{y}(k-i) \tag{9.18}$$

where the coefficients f_i are bounded

$$|f_i| \leq \beta \ \forall i. \tag{9.19}$$

(A3) Error Signal. The source propagated to the error microphone plus $\hat{y}$ yields the error signal

$$\epsilon(k) = \hat{y}(k) + \sum_{i=0}^{Q_1} p_i s(k-i) \tag{9.20}$$

where the coefficients are bounded

$$|p_i| \leq \gamma. \tag{9.21}$$

(A4) Control Output. The speaker output at the error microphone (with $C = 1$) is

$$\hat{y}(k) = \sum_{i=1}^{Q_1+Q_2} \hat{a}_i(k)\hat{y}(k-i) + \sum_{i=0}^{Q_1} \hat{b}_i(k)u(k-i). \tag{9.22}$$

Thus the order of the adaptive filter is assumed to match the order of the error signal (for the moving average portion) and the order of the feedback path plus the error signal (for the autoregressive portion).

(A5) Adaptive Algorithm. The algorithm for updating the filter coefficients in (9.22) is a modified version of the filtered-u algorithm [20]. The time-varying parameters in (9.22) are the adapted coefficients of the IIR feedforward controller. The parameter adaptation algorithm is

$$\hat{a}_i(k+1) = \hat{a}_i(k) - \mu(k)\epsilon(k)\hat{y}(k-i) \tag{9.23}$$

$$\hat{b}_i(k+1) = \hat{b}_i(k) - \mu(k)\epsilon(k)u(k-i) \tag{9.24}$$

where the time-varying, normalized stepsize is

$$\mu(k) = \frac{\mu}{1 + \sum_{i=1}^{Q_1+Q_2} \hat{y}^2(k-i) + \sum_{i=0}^{Q_1} u^2(k-i)} \tag{9.25}$$

for small fixed positive values of δ and μ.

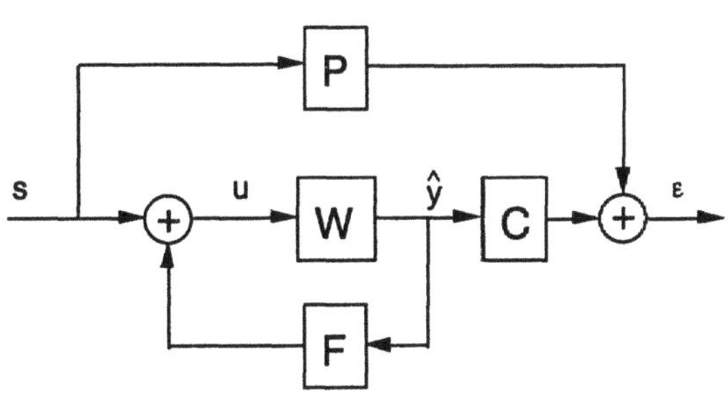

Fig. 9.2. Block diagram of a simplified active noise cancellation problem.

The problem set up (A1)-(A5) captures a fairly typical, if somewhat simplified adaptive noise cancellation setting. As shown below, this can be rewritten as an 'adaptive control' problem in which the adaptation of the $\hat{a}_i$'s and $\hat{b}_i$'s are viewed as an implicit identification of an unknown system $\frac{B}{A}$. Observe that when $W = \frac{P}{1-FP}$, the error signal ϵ is identically zero, and hence $B \leftrightarrow P$ and $A \leftrightarrow 1 - FP$ define the system to which the $\hat{a}_i$ and $\hat{b}_i$ must converge in order to minimize the squared error.

Accordingly, define the 'unknown' parameter vector

$$\theta^* = [-a_1, ..., -a_{Q_1+Q_2}, -b_1, ..., -b_{Q_1}]^T$$

where $b_i = p_i$ and $a_\ell = -\sum_{i+j=\ell} p_i f_j$, the 'parameter estimates'

$$\hat{\theta}(k) = [\hat{a}_1(k), ..., \hat{a}_{Q_1+Q_2}(k), \hat{b}_0(k), ..., \hat{b}_{Q_1}(k)]^T,$$

and the parameter error vector $\theta(k) = \hat{\theta}(k) - \theta^*$. Also, define the 'regressor vector'

$$X(k-1) = [\hat{y}(k-1), ..., \hat{y}(k - Q_1 - Q_2), u(k), ..., u(k - Q_1)]^T$$

and the prediction error

$$e(k) = X(k-1)^T \theta(k-1). \tag{9.26}$$

The 'projection algorithm' from Goodwin and Sin [11] is then

$$\hat{\theta}(k) = \hat{\theta}(k-1) + \frac{\mu X(k-1)}{\delta + X(k-1)^T X(k-1)} e(k). \tag{9.27}$$

Theorem 1. *Consider the acoustic noise control problem as shown in Figure 9.2 under assumptions (A1)-(A5). Then*

i. The error signal converges to zero for all initial conditions, that is,

$$\epsilon(k) \to 0 \quad k \to \infty.$$

ii. The regressor $X(k)$ is uniformly bounded.

Proof: In order to apply Lemma 3.3.2 of [11], it is only necessary to show that $\epsilon(k)$ of (9.20), (9.23) and (9.24) is equal to $e(k)$ of (9.26) and (9.27). Combining (9.22) and (9.20) gives

$$\epsilon(k) = \sum_{i=0}^{Q_1} p_i s(k-i) + \sum_{i=1}^{Q_1+Q_2} \hat{a}_i(k)\hat{y}(k-i) + \sum_{i=0}^{Q_1} \hat{b}_i(k) u(k-i). \tag{9.28}$$

Equation (9.18) can be rewritten as

$$s(k-i) = u(k-i) - \sum_{j=1}^{Q_2} f_j \hat{y}(k-j-i), \tag{9.29}$$

which can be used in (9.28) to give

$$\epsilon(k) = \sum_{i=0}^{Q_1} p_i u(k-i) - \sum_{i=0}^{Q_1} p_i \sum_{j=1}^{Q_2} f_j \hat{y}(k-i-j) + \sum_{i=1}^{Q_1+Q_2} \hat{a}_i(k)\hat{y}(k-i) + \sum_{i=0}^{Q_1} \hat{b}_i(k)u(k-i). \quad (9.30)$$

Since $a_\ell = -\sum_{i+j=\ell} p_i f_j$ and $b_i = p_i$, b_i can be added to each side of (9.24) and a_i to each side of (9.23). Hence (9.30) can be rewritten

$$\begin{aligned}\epsilon(k) &= \sum_{i=1}^{Q_1+Q_2} [a_i + \hat{a}_i(k)]\,\hat{y}(k-i) + \sum_{i=0}^{Q_1} \left[b_i + \hat{b}_i(k)\right] u(k-i) \\ &= X(k-1)^T \theta(k-1) \\ &= e(k).\end{aligned} \quad (9.31)$$

To apply the 'key technical lemma' 6.2.1 of [11], observe that the bounds in (9.17) and (9.21) can be substituted into (9.20) to give

$$\hat{y}(k) \le |\hat{y}(k)| \le Q_1 \gamma \alpha + \max_{0 \le \tau \le k} |\epsilon(\tau)| . \quad (9.32)$$

Similarly, the bounds (9.19) and (9.21) are substituted into (9.18) to give

$$|u(k)| \le \gamma + Q_2 \beta \max_{0 \le \tau \le k} |\hat{y}(\tau)| . \quad (9.33)$$

Merging (9.32) and (9.33) shows that $X(k)^T X(k) \le C_1 + C_2 \max_{0 \le \tau \le k} |\epsilon(\tau)|$ and hence Lemma 6.2.1 provides the desired conclusions.

Remarks:

i. The case where $C \neq 1$ but is minimum phase follows from [11,12], but the issue of nonminimum phase C remains. Moreover, efficient and stable algorithms for the adaptation of C are not obvious.
ii. This stability result applies to "recursive LMS" algorithms based on a (nonlinear) equation error, rather than an output error, problem formulation. The nominal behavior is that of a globally stable algorithm.
iii. The uniform boundedness of the regressor is with respect to any norm in $\Re^{2Q_1+Q_2+1}$ over time.
iv. Theorem 1 also shows that $\epsilon(k) \in \ell_2$, that is, the prediction error sequence is square summable.

The conclusions of Theorem 1 imply that $||\hat{\theta}(k) - \theta^*|| \le ||\hat{\theta}(k-1) - \theta^*||$ for all k. Hence once an estimate of a given quality is made, the algorithm never makes a worse estimate. However, the theorem does not imply that the parameter estimates $\hat{\theta}$ converge to their desired values θ^*. But if an assumption of richness or complexity of the regressor is made, then this convergence can be guaranteed. The conditions (9.3) can be rewritten

(A6) Persistence of Excitation. Assume that the unknown system θ^* represents a stable linear filter and that the regressor vector satisfies

$$\frac{1}{K}\sum_{k=\ell}^{\ell+K-1}\frac{X(k)X(k)^T}{1+X(k)^TX(k)} \geq \alpha I > 0 \tag{9.34}$$

for some $\alpha > 0$, some $K > 0$, and all ℓ.

Theorem 2. *Consider the acoustic noise control problem as shown in Figure 9.2 under assumptions (A1)-(A6). Then the parameter estimates $\hat{\theta}(k)$ converge exponentially to θ^* from any initial condition.*

Proof: This is a straightforward application of Theorem 3.5.19 of [18].
Remarks:

i. The persistence of excitation condition on the regressor vector X can be transferred to a similar condition on the signals u (and hence ultimately onto s) following the development in [2].
ii. Exponential stability is a strong property. It guarantees that the algorithm (9.27) remains stable even under persistent perturbations, that is, it guarantees robustness to small noises or disturbances in the inputs, outputs, measurements, and modelling parameters that inevitably arise in implementation.
iii. Theorem 2 also guarantees that if the underlying system (the P and F transfer functions) change slowly, then the algorithm will track the changes. (See for instance Theorem 3.5.29 in [18] for a detailed statement.) Hence robustness to time variations is also guaranteed.

The filtered-u style algorithm has stability properties closer to those of a nonlinear equation error (the projection algorithm) than to the output error form often associated with IIR adaptive filtering [12]. In particular, the stability is global (rather than local) as would be expected in an IIR setting. While the algorithm superficially appears to be an output error form (and hence would require an appropriate filtering of the error and/or regressor as in the previous sections) the structure of the problem actually reveals a linear setting.

9.5 Coefficient Filtering and Update Smoothing

A number of proposed modifications to the basic adaptive algorithm (9.1) can be viewed as a linearly filtering the old coefficients, i.e. "coefficient filtering". Others can be viewed as linear filtering of the product of the the regressor and the prediction error, i.e. "update smoothing". This section establishes a simple condition on such filters that is vital to the local stability of these "improved" algorithms, using the technical apparatus assembled in [25] and reviewed in section 9.2.

A variety of modifications have been proposed in [3], [10], [13], [21], and [23], to improve the performance of the adaptive mechanism in particular circumstances. This section shows how these may be viewed as either update smoothing

$$\hat{\theta}(k+1) = \hat{\theta}(k) + \mu F(q^{-1})\{X(k)e(k)\} \tag{9.35}$$

or coefficient filtering

$$\hat{\theta}(k+1) = G(q^{-1})\{\hat{\theta}(k)\} + \mu X(k)e(k) \tag{9.36}$$

where F and G are linear filters. Section 9.5.1 also shows how these two seemingly distinct modifications are interchangeable with appropriate (re) definitions.

Then section 9.5.2 studies the local stability of the error system underlying algorithms that incorporate update smoothing. The results also apply to algorithms that use coefficient filtering, and to those that may combine update smoothing with coefficient filtering. The result we establish is that, if the added filtering is interpreted as an update smoother, local stability in the idealized identification-style setting is guaranteed if the regressor is persistently spanning, the update smoother is stable, and the sum of its impulse response coefficients is positive (which corresponds to a positive DC gain).

9.5.1 Coefficient Filtering and Update Smoothing are the Same

In [10] and [21] the case is made for the benefits (and detriments) of certain types of update smoothing as in (9.35). In [10] one of the update smoothers proposed has the form of (9.35) with

$$F(q^{-1}) = \frac{\beta}{1-(1-\beta)q^{-1}}$$

which, if $0 < \beta < 1$, is stable and has a DC gain that is unity, and therefore positive.

In [23] and [26], the Momentum LMS (MLMS) variant of LMS is analyzed, which uses a coefficient filter of the form of (9.36) with

$$G(q^{-1}) = (1+\alpha) - \alpha q^{-1} \tag{9.37}$$

where $|\alpha| < 1$. Restriction of α to $0 < \alpha < 1$ is said to improve the convergence rate of MLMS relative to LMS, at a cost of increased misadjustment. Plus, when $|\alpha| = 1$, instability is shown to result.

Another coefficient filtering scheme arises with a generalization of leaky LMS

$$\hat{\theta}(k+1) = (1-\lambda)\hat{\theta}(k) + \mu X(k)e(k) \tag{9.38}$$

where the leakage factor λ is a scalar constant between 0 and 1. However, if λ is extended to be a linear filter, (9.38) can be seen to become a variant of the coefficient filter of (9.36) with $G(q^{-1}) = 1 - \lambda(q^{-1})$. This extension of leaky LMS is termed generalized leakage in [13], which shows that quite a variety of modifications to the LMS algorithm, motivated by various concepts for improved tracking capabilities can all be collected under the coefficient filtering rubric. Generalized leakage and MLMS are just two of the numerous algorithm modifications that are collected in [13] as falling in the class of coefficient filtering.

The connection of update smoothing and coefficient filtering is effectively made in [13] with the discussion of the multi-step algorithms of [3], which [13] writes initially in the form of update smoothing in (9.35) and then converts to (9.36). In our notation here, (9.35) can be converted to (9.36) via

$$G(q^{-1}) = \frac{(F(q^{-1}) - (1 - q^{-1}))}{q^{-1}F(q^{-1})}$$

or conversely (9.36) can be converted to (9.35) with

$$F(q^{-1}) = \frac{(1 - q^{-1})}{(1 - q^{-1}G(q^{-1}))}. \tag{9.39}$$

Thus, a large group of modifications to the basic algorithm form (9.1) can be viewed as either update smoothing or coefficient filtering. The next section analyzes them all simultaneously.

9.5.2 Local Stability

Consider the update smoothing modification of the LMS algorithm in (9.35), with both sides subtracted from θ^* to yield the error system

$$\theta(k+1) = \theta(k) + \mu F(q^{-1})\{X(k)e(k)\}. \tag{9.40}$$

Given the equation error form of $e = y - \hat{y}$, and an assumption of the smallness of the stepsize μ, θ in (9.40) can be closely approximated by $\bar{\theta}$ with

$$\bar{\theta}(k+1) = (I - \mu F(q^{-1})\{X(k)X^T(k)\})\bar{\theta}(k). \tag{9.41}$$

In fact, as formalized in [25], the exponential asymptotic stability of (9.41) to $\bar{\theta} = 0$ implies the local stability of (9.40) about $\theta = 0$.

For simplicity, we will assume that the input $u(k)$ is periodic with period p_1, which implies that X is also periodic with period p_1. Further, suppose that F is an exponentially stable, time-invariant, linear filter with impulse response coefficients f_i, i.e.

$$F(q^{-1}) = \sum_{i=0}^{p_2} f_i q^{-i} \tag{9.42}$$

Note that a further simplification has been made with the assumption that F has a finite impulse response, i.e. $p_2 < \infty$. Define p as the least common multiple of p_1 and p_2.

From Theorem 1 of [25], (9.41) is exponentially asymptotically stable if

$$\gamma_1 I \geq \sum_{i=j}^{j+p} F(q^{-1})\{X(i)X^T(i)\} \geq \gamma_2 I \tag{9.43}$$

for some positive γ_1 and γ_2 and all j. Since F can be unity, as in ordinary LMS, to guarantee (9.43), we will assume that X is persistently spanning, i.e., that (9.3) holds. Thus, the local stability of (9.40) now rests on what conditions on $F(q^{-1})$ satisfy (9.43) given (9.3). To foreshadow the result of the following theorem, imagine the case where the regressor and parameter vectors of (1.3) and (1.4) are scalars, i.e. $n = 0$, and u is a nonzero constant. Then (9.3) is trivially satisfied and (9.43) is satisfied only if F is stable and has a positive DC gain.

Theorem 3. : *Consider the system (9.40) where $F(q^{-1})$ is an FIR filter of order p_2 with $\sum f_i > 0$ and suppose that X is persistently spanning as described in (9.3). Then, there exists a μ^* such that (9.40) is uniformly exponentially stable for all μ within $0 < \mu < \mu^*$.*

Proof. As noted earlier, given Theorem 1 in [25], we need only show that (9.43) is satisfied. Given (9.42)

$$\sum_{i=j}^{j+p} F(q^{-1})\{X(i)X^T(i)\} = \sum_{i=j}^{j+p} \sum_{l=0}^{p_2-1} f_l q^{-l}\{X(i)X^T(i)\}.$$

Due to the periodicity of X, this becomes

$$\sum_{i=j}^{j+p} F(q^{-1})\{X(i)X^T(i)\} = \bar{f} \sum_{i=j}^{j+p} X(i)X^T(i)$$

where

$$\bar{f} = \sum_{l=0}^{p_2-1} f_l$$

Thus, (9.43) holds given (9.3) with $\gamma_i = \bar{f}\delta_i$ for $i = 1$ and 2.

Remarks:

i. Given a suitably small stepsize, the persistently spanning condition of (9.3) is required for parameter convergence and exponential asymptotic stability of θ to zero by unmodified LMS. So, the sole added requirement for similar parameter convergence by small-stepsize LMS with update smoothing is that the stable added filter have a positive DC gain.

ii. The result can be extended to include IIR F with $p_2 = \infty$. This can be done by truncating the impulse response sequence of the stable F for a large enough p_2 so that

$$||F_T(q^{-1})\{X(i)X^T(i)\} - F(q^{-1})\{X(i)X^T(i)\}|| < \epsilon$$

where F_T represents the truncated version of F. The exponential stability is then retained for suitably small ϵ. Essentially, the requirement of stability and positive DC gain for F then applies whether F is FIR or IIR.

iii. Due to the periodicity of X, $\sum_{i=j}^{j+p} F(q^{-1})\{X(i)X^T(i)\}$ is independent of j, which is a helpful simplification. However, the result can be readily extended to include non-periodic inputs via analysis similar to that in [25].

iv. As discussed in [25], the exponential stability demonstrates robustness of the algorithm in (9.35) (and the algorithm in (9.36) with satisfaction of the condition $\sum f_i > 0$ on the stable F arising via (9.39) due to the particular G in use) to small disturbances, modest undermodelling or nonlinearities, and slow time-variations of the true system.

v. If the filter F (or G) is slowly time-varying, lemma 3 of [25] can be applied. Label the "frozen" versions of the time-varying F as $F^m(q^{-1})$ with impulse responses $\{f_i^m\}$. Then, if $\sum f_i^m > \rho > 0$ for every m, the results of the Theorem still hold.

vi. The condition commonly associated with LMS algorithm modifications of regressor and/or error filtering, as examined in [25], is a strictly positive real condition. Note that the positive DC gain condition on the update smoothing F is quite different.

vii. From (9.37) and (9.39), for MLMS the associated F reduces to $1/(1 - \alpha q^{-1})$. Thus, the local stability condition on F from the Theorem in this section agrees with the stability condition indicated in section 2 above from [23], i.e. $|\alpha| < 1$.

A large variety of modifications to the LMS algorithm can be subsumed into the format of update smoothing (or coefficient filtering). These algorithms have been proposed for their various performance enhancements that may be desired in particular circumstances. The stable convergence of some, but by no means all, of these algorithms has been established separately in the literature. This section succinctly handles the small stepsize versions of all of them at once.

9.6 Conclusion

This chapter has applied the deterministic averaging approach to a collection of algorithms of recent interest. Stating the adaptive algorithms in this way does more than allow a unified analysis - it also provides a powerful way to conceptualize the design of new algorithms. Examples such as the new Steiglitz-McBride algorithms, and possible combinations of the coefficient filtering/update smoothing methods show how it is possible to "play" the filters against each other, effectively creating alternative implementations of the same underlying error system.

An issue of future interest is the combination of regressor and error filtering from [25] with the coefficient filtering and update smoothing of section 9.5. Their combination, due to the regressor-spectrum-dependent conditions associated with regressor and error filtering is likely to place requirements on more than the stability and DC gain of the update smoothing filter. The most general linear form would involve three filters $\mathbf{N}$, $\mathbf{L}$, and M

$$\theta(k+1) = \theta(k) - \mu \mathbf{N}(q^{-1},k)\{\mathbf{L}(q^{-1},k)\{X(k)\}M(q^{-1},k)\{X^T(k)\theta(k)\}\}$$

and it is interesting to contemplate the exact form appropriate averaging results might assume. This general form includes many algorithm combinations, for instance, the Steiglitz-McBride algorithm with leakage, the filtered output error algorithm with momentum, and many others that have no obvious names.

References

1. B. D. O. Anderson, R. R. Bitmead, C. R. Johnson, Jr., P. V. Kokotovic, R. L. Kosut, I. M. Y. Mareels, L. Praly, and B. D. Reidle, *Stability of Adaptive Systems, Passivity and Averaging Analysis*, MIT Press 1986.
2. B. D. O. Anderson and C. R. Johnson, Jr., "Exponential convergence of adaaptive identification and control algorithms," *Automatica* 18:1 Jan. 1982.
3. A. Benveniste, "Design of adaptive algorithms for the tracking of time-varying systems," *Int. Journal of Adaptive Control and Signal Processing*, vol. 1, pp. 3-29, September 1987.
4. R. R. Bitmead, "Persistence of excitation conditions and the convergence of adaptive schemes," *IEEE Transactions on Information Theory* IT-30:183-191, (1984).
5. J. E. Cousseau, "Adaptive IIR filtering: available results," *Circuits and Systems Society Newsletter*, 10:3, 1999.
6. L. J. Eriksson, "Development of filtered-u algorithms for active noise cancellation," *Journal of the Acoustical Society of America* 89:257-265, Jan. 1991.
7. L. J. Eriksson, M. C. Allie and R. A. Greiner, "The selection and application of an IIR adaptive filter for use in active sound attenuation," *IEEE Transaction on Acoustics, Speech and Signal Processing*, ASSP-35:433-437, April 1987.
8. H. Fan and W. K. Jenkins, "A new adaptive IIR filter," *IEEE Transaction on Circuits and Systems*, CAS-33:939-947, 1986.
9. B. Friedlander, "System identification techniques for adaptive signal processing," *IEEE Transactions on Acoustics, Speech and Signal Processing*, ASSP-30, 240-246, 1982.
10. J. R. Glover, Jr., "High order algorithms for adaptive filters," *IEEE Transactions on Communications*, COM-27, pp. 216-221, January 1979.

11. G. C. Goodwin and K. S. Sin, *Adaptive Filtering, Prediction and Control,* Prentice-Hall, 1984.
12. C. R. Johnson, Jr., *Lectures on Adaptive Parameter Estimation,* Prentice-Hall, 1987.
13. G. Kubin, "Coefficient filtering - a common framework for the adaptation in time-varying environments," *Proc. 1st COST Project #229 WG.2 Workshop,* Bayona la Real, Spain, March 1991.
14. S. M. Kuo and D. R. Morgan, *Active Noise Control Systems,* Wiley, NY 1996.
15. I. D. Landau, "Unbiased recursive identification using model reference adaptive techniques," *IEEE Transactions on Automatic Control,* AC-21:194-202, 1976.
16. L. Ljung and T. Söderström, *Theory and Practice of Recursive Identification,* MIT Press, Cambridge, MA, 1983.
17. L. Ljung, "On positive real transfer functionsand the convergence of soem recursive schemes," *IEEE Transactions on Automatic Control,* AC-22:539-551, 1977.
18. I. M. Y. Mareels and J. W. Polderman, *Adaptive Systems: An Introduction,* Birkhäuser, 1996.
19. J. M. Mendel, *Discrete Techniques of Parameter Estimation: The Equation Error Formulation,* Marcel Dekker, NY 1973.
20. P. A. Nelson and S. J. Elliott, *Active Control of Sound,* Academic Press, London, 1992.
21. J. G. Proakis, "Channel identification for high speed digital communicatins," *IEEE Transactions on Automatic Control,* AC-19, pp. 916-922, December 1974.
22. P. A. Regalia, *Adaptive IIR Filtering in Signal Processing and Control,* Marcel Dekker, NY, 1995.
23. S. Roy and J. J. Shynk, "Analysis of the momentum LMS algorithm," *IEEE Transactions on Acoustics, Speech, and Signal Processing,* ASSP-38: 2088-2098, December 1990.
24. W. A. Sethares, "The LMS Family," in *Efficient System Identification and Signal Processing Algorithms,* Ed. N. Kalouptsidis and S. Theodoridis Springer-Verlag, 1993.
25. W. A. Sethares, B. D. O. Anderson, and C. R. Johnson, Jr., "Adaptive algorithms with filtered regresor and filtered error," *Mathematics of Control, Signals, and Systems,* 2:381-403, 1989.
26. R. Sharma, W. A. Sethares, and J. A. Bucklew, "Analysis of momentum adaptive filtering algorithms," *IEEE Transactions on Signal Processing,* 46:5:1430-1434, May 1998.
27. K. E. Steiglitz and L. E. McBride, "A technique for the identification of linear systems," *IEEE Transactions on Automatic Control,* AC-10:464, 1965.
28. P. Stoica and T. Söderström, "The Steiglitz-McBride identification algorithm revisited - convergence analysis and accuracy aspects," *IEEE Transactions on Automatic Control,* AC-26:712-717, 1981.
29. B. Widrow, J. M. McCool, M. G. Larimore, and C. R. Johnson, Jr., "Stationary and nonstationary learning characteristics of the LMS adaptive filter," *Proc. IEEE,* vol. 64, pp. 1151-1162, August 1976.

Part IV

Fundamental Design Issues in Adaptive Control

10 Iterative Adaptive Control: Windsurfing with Confidence

Robert L. Kosut[1,2]

[1] SC Solutions, Inc., 3211 Scott Blvd., Santa Clara, CA 95054, USA
kosut@scsolutions.com
[2] Research supported by DARPA, Applied Computation & Mathematics Program, under NASA Grant No. NAG-1-1964.

10.1 Introduction

SEEKER: How do I gain Good Judgement?
TEACHER: You must first acquire Wisdom.
SEEKER: How do I gain Wisdom?
TEACHER: From Bad Judgement.

Windsurfing, Learning, and Unfalsification

Learning, apparently, is the result of making mistakes. A good example is learning to windsurf. The approach I took is simple. Fall Often. This is the approach I seem to use regularly when learning anything. It helped a great deal that I did this in Tahiti where both the air and water were 80 degrees and there was little wind. I was in Tahiti following my first visit with Brian Anderson in Canberra in 1984. A few years later when I was describing my learning experience at windsurfing to Brian, who by the way is an excellent windsurfer, we realized that how a human learns windsurfing might be a procedure that could be used for adaptive control. We speculated about human learning of physical activities as follows. The human first learns to control over a limited bandwidth and learning pushes out the bandwidth over which an accurate knowledge of the "plant" is known. The human first implements a low-gain controller and learning allows the loop to be tightened. The human tends to "over-react" when the environment changes, often resorting to high-gain or no-gain. Then falls. The few instants just before falling may provide some new information about the next higher increments in the bandwidth of the system. So of course mistakes are fundamental to learning. (This is why stabilizing feedback is *negative*; the popular concept of *positive* feedback may be useful, in fact, when it is actually negative!) Our speculations on this "windsurfing" approach to adaptive control was first expressed in [2] and subsequently filled out in a series of papers [24,25].

In general, the learning process has the following steps: make observations, construct a model, and validate the model against new data. The last step means the model is used for some purpose and tested against reality in some way. If the intended purpose of the model is control design, then the penultimate validation of the model is to *implement* a controller based on the model. Of course the implementation step generates new data, and this is usually where new mistakes are uncovered. For example, just prior to "falling."

Validation is perhaps a misnomer. One can never prove that a model will be able to accurately predict the future. More precisely, the data can *falsify* a model, *i.e.*, the model may prove to be incapable of fully explaining the data. This step, that is, discarding bad models, could be done before implementation. To emphasize this kind of in-validation, we use the more precise, but awkward term: *unfalsification*. The essence of unfalsification is this: we use data from one experiment and ask if it could have come from another experiment. This second experiment need not be performed because if it is shown that the data is not consistent with the second "thought experiment," then there is no need to perform it. If it cannot be proven inconsistent, then the only recourse is to do the second experiment, *e.g.*, implement the controller under consideration. No guarantees are offered other than it fails to be proven inconsistent.

In order to relate this to control design it is natural to think in terms of a model based design, *i.e.*, design the control using a model of the system being controlled. However, no mathematical model of a physical system is ever exact. Hence, a complete model for purposes of control design should include both the nominal model as well as an explicit description of the uncertainty between the model and the physical system. Such a model is referred to as an *uncertainty model* and is the starting point for robust control design [40]. Typically this uncertainty characterization is a guess, although an informed one. Inferring the uncertainty from data alone is not provided by a straight forward application of the classical method of indirect adaptive control *i.e.*, system identification followed by control design. Classical system identification alone does not offer this feature, *e.g.*, [27]. Recent efforts, however, which modify the classical approach to system identification, specifically for control design, do address this problem, *e.g.*, [28].

Classical Adaptive Control

Classical adaptive control methods, typified by the "indirect" scheme shown in figure 10.1, can work well in many instances, *e.g.*, when the plant variations are slow and the input excitation is concentrated at those frequencies which are critical. But even in this almost ideal case, initial plant-model mismatch can induce unacceptably large transients. In more volatile circumstances the adaptive system can exhibit instabilities, limit cycles, and chaotic behavior, even when initialized at a desired controller parameter setting. Theoretical

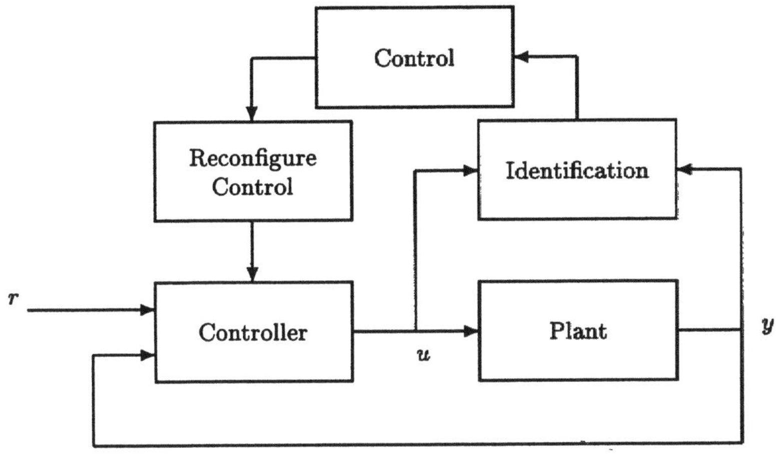

Fig. 10.1. Classical indirect adaptive control: the control parameters are adjusted "indirectly" via identification and control re-design based on the (model) estimate.

studies have exposed some of the underlying mechanisms for stability and instability of these classical adaptive control approaches, *e.g.*, [3], [1]. A rough explanation is that these classical adaptive schemes require gradients of the error, which depend of course on the unknown plant, and hence, have to be approximated. (There are some recently developed methods which provide "unbiased" estimates of the gradients from data, *e.g.*, [12], but they require specialized experiments and thus are not al all suited to handling rapid or unexpected system variations.) The same problems plague the "direct" adaptive schemes where controller parameters are directly adjusted to minimize the error between the adaptively controlled plant and a reference system.

Unfasified Adaptive Control

Over the past several years a new approach to adaptive control, based on the paradigm of "unfalsification," has been under investigation. The approach is based on a new perspective to the adaptive control problem and is important in two ways: (1) as an "indirect" adaptive robust control method it provides a model error bound from the experimental data which can be used to assess the model quality for robust and/or fault tolerant control design, and (2) as a "direct" adaptive control method it can be used to eliminate, *without further implementation*, those controllers that would not satisfy the performance goals. These approaches appear to provide the theoretical foundation for synthesizing reliable adaptive control systems which are not prone to the difficulties encountered by classical approaches.

Unfalsification provide a means to estimate an uncertainty model from finite data which includes both disturbance and dynamic uncertainty as well as the usual parameters in the standard prediction error models. In this paper we will show how unfalsification can be used for *direct iterative adaptive control*, *i.e.*, the control parameters are adjusted directly. In our earlier work on iterative (windsurfing) adaptive control [24,25], tests for unfalsification were not available.

Uncertainty model unfalsification using finite time domain data, which is the underlying basis for the work presented in this paper, was first described by Poolla *et al.*[33]. Earlier work using frequency domain data was presented in [35], and some precursors to unfalsification can be found in [16–18]. Further extensions and applications to plant uncertainty model unfalsification can be found in [19], [20], [21], and [26]. A method employing a probabilistic description of dynamic uncertainty is given in [10]. The origin of the ideas for direct controller unfalsification are presented by Safonov and Tsao in [34] and the references therein. The mathematical basis for unfalsification of linear-time-invariant systems can be found in [11] and [8]. Computations using convex programming is discussed in [37,38].

10.2 Iterative Adaptive Control

The roots of iterative adaptive control can be traced to the dual control concept (see, *e.g.*, [3, Ch. 7]), which typically involves *indirect adaptation, i.e.*, identification followed by control parameter adjustment. A survey of iterative identification and control schemes is given in [9]. Of particular relevance to the work presented here – for purposes of comparison – is [39], [5], and [4] which describe how data filters can be selected to make the identification and control criteria merge; the windsurfer approach to adaptation and learning, as described in [24,25], where the closed-loop bandwidth is gradually increased every iteration; and [12], which describes a direct iterative controller design method.

A generic iterative adaptive control system is depicted in figure 10.2. The adaptive part of the controller consists of a parameter estimator and a control design algorithm connected in series through a sample and hold. The latter is what makes the system "iterative." That is, the next controller design is based on data collected while the previous controller was in place.

The system consists of two feedback "loops" each operating at different sampling rates. The inner loop, operating at the fast rate, consists of the plant and controller, where u is the control input to the plant, y is the sensed output from the plant, and r is the reference command to the controller. The outer loop, operating at the slow rate, consists of the plant parameter estimator and control parameter design, The sequence of parameter estimates, $\bar{\theta}$, are produced at the end of every data collection interval of ℓ-samples, and hence, depend on the prior applied sequence of controller parameters, $\hat{\alpha}$, which are based on $\hat{\theta}$, the prior plant parameter sequence, and so on. Thus, $\hat{\theta}$ and $\hat{\alpha}$, are piece-wise constant vector sequences, *i.e.*, constant over every ℓ-samples.

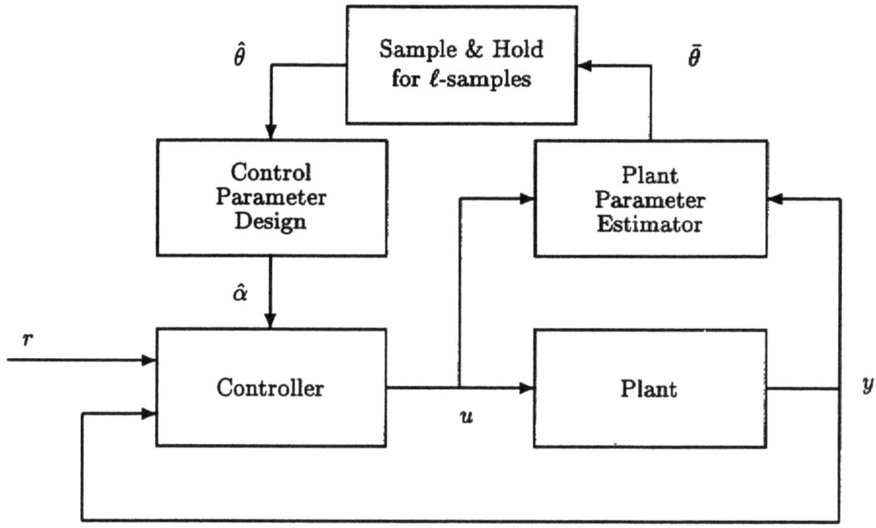

Fig. 10.2. Iterative adaptive control system.

Specifically, during the i-th iteration (ℓ-data collection interval), that is, for $t = 1 + (i-1)\ell, \ldots, i\ell$, let $\theta_\ell^i \in \mathbf{R}^p$ denote the plant parameters and let $\alpha_\ell^i \in \mathbf{R}^q$ denote the corresponding control parameters. The relation between θ_ℓ^i and α_ℓ^i is typically algebraic and depends on the design procedure, *i.e.*,

$$\alpha_\ell^i = \kappa(\theta_\ell^i) \tag{10.1}$$

Let $(y^i, u^i) \in \mathbf{S}^\ell \times \mathbf{S}^\ell$ denote[1] the plant output and input data recorded during the i-th data collection period, *i.e.*,

$$\begin{aligned} y^i &= \{\ y_{1+(i-1)}, \ldots, y_{i\ell}\ \} \\ u^i &= \{\ u_{1+(i-1)}, \ldots, u_{i\ell}\ \} \end{aligned} \tag{10.2}$$

After every data collection period of ℓ samples, a new parameter estimate, denoted by θ_ℓ^{i+1}, is determined by solving an optimization problem of the form[2],

$$\theta_\ell^{i+1} = \arg \min_{\theta \in \Theta} V_\ell^i(\theta) \tag{10.3}$$

[1] $\mathbf{S}^\ell$ denotes the set of scalar sequences of length ℓ, *i.e.*, $x \in \mathbf{S}^\ell \leftrightarrow x = \{\ x_t \mid t \in [1, \ell]\ \}$ where t are the uniformly spaced integer samples.

[2] The optimization operation "arg min" as used here and throughout the paper is to be understood to mean the minimizing argument, or if there is no unique minimum, then "arg min" refers to the set of (local) minima, *i.e.*, $\arg\min f(x) = \{\ x \mid f_x(x) = 0,\ f_{xx}(x) > 0\ \}$.

where $\Theta \subset \mathbf{R}^p$ is the set of possible parameters, which in most cases is simply $\mathbf{R}^p$. The objective function, $V_\ell^i(\theta)$, depends on the data $(y^i, u^i) \in \mathbf{S}^\ell \times \mathbf{S}^\ell$, as denoted by the superscript i and the subscript ℓ, and on θ, where the precise dependence is determined by the estimation and/or control design criterion. For "indirect adaptive" schemes, the criterion is related to providing a good fit to the data based on an assumed uncertainty model for the plant system. Hence, $\alpha = \kappa(\theta)$ simply denotes how the plant uncertainty model parameters relate to the corresponding control design parameters. For the "direct adaptive" schemes, as the name implies, the control parameter is itself adjusted, *i.e.*, $\alpha \equiv \theta$, and the criterion is directly related to closed-loop performance. Actually even in the indirect case, the criterion is constructed to be as close as possible to a closed-loop performance criterion or at least useful for control design.

10.2.1 Convergence

Convergence analysis is very difficult because the data $(y^i, u^i) \in \mathbf{S}^\ell \times \mathbf{S}^\ell$ depends on *all* past control parameter switching $\{\ \alpha_\ell^1, \ldots, \alpha_\ell^i\ \}$, which in turn depend on all past plant parameter estimates $\{\ \theta_\ell^1, \ldots, \theta_\ell^i\ \}$. However, for long data collection periods, provided that all past controllers are stabilizing, the system memory of past controllers fades. Hence, in the limit, with infinite data collected during every iteration, the data collected during the i-th interval only depends on the last parameter values (α^i, θ^i), *i.e.*,

$$\lim_{\ell \to \infty} y^i = y(\alpha^i), \quad \lim_{\ell \to \infty} u^i = u(\alpha^i), \quad \alpha^i = \kappa(\theta^i) \tag{10.4}$$

Similarly, the objective function becomes,

$$\lim_{\ell \to \infty} V_\ell^i(\theta) = V(\theta, \alpha^i), \quad \alpha^i = \kappa(\theta^i) \tag{10.5}$$

So in the limiting case of infinite data, the parameter estimation step at iteration i is,

$$\theta^{i+1} = \arg \min_{\theta \in \Theta} V(\theta, \alpha^i), \quad \alpha^i = \kappa(\theta^i) \tag{10.6}$$

As observed in [13], for infinite data, convergent parameter values are equivalently fixed-points of the mapping $\Gamma : \mathbf{R}^p \mapsto \mathbf{R}^p$, where

$$\Gamma(\theta) = \arg \min_{\psi \in \Theta} V(\psi, \alpha) \text{ subject to } \alpha = \kappa(\theta) \tag{10.7}$$

Hence, if $\hat{\theta}$ is a fixed-point, then

$$\hat{\theta} = \arg \min_{\psi \in \Theta} V(\psi, \hat{\alpha}), \quad \hat{\alpha} = \kappa(\hat{\theta}) \tag{10.8}$$

and must satisfy the necessary condition for optimality, namely,

$$\frac{\partial}{\partial\theta}V(\theta,\kappa(\hat{\theta}))\Big|_{\theta=\hat{\theta}} = 0 \tag{10.9}$$

However, the minimizer, θ_{opt}, of $V(\theta,\kappa(\theta))$ satisfies,

$$\frac{\partial}{\partial\theta}V(\theta,\kappa(\theta))\Big|_{\theta=\theta_{opt}} = 0 \tag{10.10}$$

As pointed out in [13], the fixed-point $\hat{\theta}$ is not likely to be the same as θ_{opt}. So it appears that iterative schemes have a built-in flaw. Even if the infinite-data estimation criterion, $V(\theta,\kappa(\theta))$, is constructed to be a sensible control criterion, no iterative algorithm can be guaranteed to reach the minimum, or at least a local minimum. This has led some researchers to seek another path, *e.g.*, in [12] the authors show how to obtain an unbiased estimate of the gradient (and Hessian) of error signals with respect to control parameters by performing a series of specialized experiments.

10.2.2 Relation to Slow Adaptation

This "flawed" property of iterative adaptation is a recrudescence of the identical property of all slowly varying parameter adaptive algorithms, *e.g.*, see [1], [32], or the chapter on averaging analysis in [3]. As an illustration, consider the adaptive algorithm,

$$\begin{aligned} \theta_{t+1} &= \theta_t + \mu\phi_t(\alpha_t)\varepsilon_t(\theta_t,\alpha_t) \\ \varepsilon_t(\theta,\alpha) &= y_t(\alpha) - \phi_t(\alpha)^T\theta_t \\ \alpha &= \kappa(\theta) \end{aligned} \tag{10.11}$$

where $\mu > 0$ adjusts the speed of adaptation and where $\varepsilon_t(\theta_t,\alpha_t)$ is the prediction error associated with an ARX plant model being identified in closed-loop with the control parameters set to α_t. If the adaptation is slow, *i.e.*, μ is sufficiently small, then as shown in [1], there exists $(\theta_t^\mu,\alpha_t^\mu)$ satisfying (10.11) in an order-μ neighborhood of $(\hat{\theta},\hat{\alpha})$ obtained from[3] ,

$$\Big\langle \phi(\hat{\alpha})\varepsilon(\hat{\theta},\hat{\alpha})\Big\rangle = 0, \quad \hat{\alpha} = \kappa(\hat{\theta}) \tag{10.12}$$

Equivalently,

$$\begin{aligned} \hat{\theta} &= \Big\langle \phi(\hat{\alpha})\phi(\hat{\alpha})^T\Big\rangle^{-1}\Big\langle \phi(\hat{\alpha})y(\hat{\alpha})\Big\rangle \\ \hat{\alpha} &= \kappa(\hat{\theta}) \end{aligned} \tag{10.13}$$

---

[3] $\langle x\rangle$ denotes time average, *i.e.*, $\langle x\rangle = \lim_{\ell\to\infty}\frac{1}{\ell}\sum_{t=1}^{\ell} x_t$

This is precisely the solution to (10.9) if $V(\theta, \alpha)$ is the standard RMS (least-squares) identification criterion,

$$V(\theta, \alpha) = \left\langle \varepsilon(\theta, \alpha)^2 \right\rangle \tag{10.14}$$

In this case the reason that $(\hat{\theta}, \hat{\alpha})$ is not the optimum solution given by (10.10) is that

$$\phi(\kappa(\theta)) \neq \frac{\partial}{\partial \theta} \varepsilon(\theta, \kappa(\theta)) \tag{10.15}$$

The conditions which insure that $(\theta_t^\mu, \alpha_t^\mu)$ satisfies (10.11) for small μ do not insure that the solution is stable. Further conditions are required [1], [32]. Similar results hold for "direct" adaptive algorithms, *e.g.*,

$$\begin{aligned} \alpha_{t+1} &= \alpha_t + \mu \psi_t(\alpha_t) \varepsilon_t(\alpha_t) \\ \varepsilon_t(\alpha) &= y_t(\alpha) - y_t^{\text{ref}} \\ u_t &= \psi_t(\alpha_t)^T \alpha_t \end{aligned} \tag{10.16}$$

In this case for sufficiently small μ, there is a solution α_t^μ in an order-μ neighborhood of $\hat{\alpha}$ satisfying

$$\left\langle \psi(\hat{\alpha}) \varepsilon(\hat{\alpha}) \right\rangle = 0 \tag{10.17}$$

As in the previous case, $\hat{\alpha}$ is not the solution to (10.9) with the logical choice,

$$V(\alpha) = \left\langle \varepsilon(\alpha)^2 \right\rangle \tag{10.18}$$

because

$$\psi(\alpha) \neq \frac{\partial}{\partial \alpha} \varepsilon(\alpha) \tag{10.19}$$

But all is not lost for iterative schemes. By introducing an unfalsification step we can develop iterative schemes which avoid the above "convergence" problems. Perhaps convergence is not an appropriate concept when conclusions are being drawn, in this case eliminated, strictly from finite data. Let's leave this to the philosophers.

10.3 Parametrization and Performance

10.3.1 Parametrization

We will assume throughout that the control is given by,

$$u = C(\alpha)(r - y) \tag{10.20}$$

where for fixed $\alpha \in \mathbf{R}^q$, $C(\alpha^i) \in \mathbf{LTI}$[4] The controller parameters, α, are to be adaptively adjusted. There are of course many possible ways to parametrize the controller. For example, the control parameters, α, can consist of all the numerator and denominator transfer function coefficients up to a specified degree, thereby restricting the controller order. For example, all order-n controllers can be parametrized as follows:[5]

$$\begin{aligned} C(\alpha) &= N(\alpha)D(\alpha)^{-1} \\ N(\alpha) &= b_0 + b_1 z^{-1} + \cdots + b_n z^{-n} \\ D(\alpha) &= 1 + a_1 z^{-1} + \cdots + a_n z^{-n} \\ \alpha &= [a_1 \cdots a_n \; b_0 \cdots b_n]^T \in \mathbf{R}^{2n+1} \end{aligned} \tag{10.21}$$

Another parameterization is all the PI controllers, *i.e.*,

$$C(\alpha) = \alpha_P + \alpha_I \frac{z^{-1}}{1 - z^{-1}}, \qquad \alpha = [\alpha_P \; \alpha_I \;]^T \in \mathbf{R}^2 \tag{10.22}$$

Obviously many parameterizations are possible, *e.g.*, co-prime factorizations where the factors are affine in the control parameters, or direct affine parameterizations with orthogonal basis functions [36].

There are a number of other interesting possibilities. For example, a list of controllers each corresponding to some measurable condition of the system. This approach is somewhat like "multiple model control" or "gain-scheduling." Control structures could have free parameters (α) corresponding to a prescribed set of anticipated faults, *e.g.*, changes in sensor bias and scale factors, changes in actuator gains and bandwidth, and changes in a mechanical structure resulting in spillover of elastic modes into the closed-loop bandwidth. A similar approach was investigated in [7] for a switching missile autopilot.

10.3.2 Performance

We would like the closed-loop system to behave like the reference system,

$$y_{\text{ref}} = T_{\text{ref}} \; r \tag{10.23}$$

for a specified system $T_{\text{ref}} \in \mathbf{LTI}$. "Behave like" can have a variety of meanings. For example, it could mean that the output error, $y - T_{\text{ref}}\, r$, should be small relative to the size of the command r. An example of such a specification is that[6] ,

$$\|y - T_{\text{ref}}\, r\|_{rms} \leq \rho \|r\|_{rms}, \qquad \forall \|r\|_{rms} < \infty \tag{10.24}$$

---

[4] **LTI** is the set of linear-time-invariant systems with rational transfer functions.

[5] z^{-k} denotes the k-delay operator.

[6] The RMS-norm of a sequence (technically a semi-norm) is defined as $\|x\|_{rms} = \left(\lim_{\ell \to \infty} \frac{1}{\ell} \sum_{t=1}^{\ell} x_t^2 \right)^{1/2}$

In this case, we are not looking for a response to a *specific* r, such as a sinusoid at one frequency or a step, rather, for every possible r such that $\|r\|_{rms} < \infty$. For example, if the plant system is given by,

$$y = Pu \tag{10.25}$$

where $P \in \mathbf{LTI}$, and the controller is given by,

$$u = C(r - y) \tag{10.26}$$

with $C \in \mathbf{LTI}$, then (10.24) is equivalent to,

$$\|T(P, C) - T_{\text{ref}}\|_{\mathbf{H}_\infty} \le \rho \tag{10.27}$$

with $T(P, C)$ given by,

$$T(P, C) = (1 + PC)^{-1} PC \tag{10.28}$$

This is clearly a measure of the error between the transfer functions of the closed-loop and reference systems. Hence, for a given controller parametrization, $C(\alpha) \in \mathbf{LTI}$, and a given plant parametric model, $P(\theta) \in \mathbf{LTI}$, the optimal control design (10.1) is,

$$\kappa(\theta) = \arg\min_{\alpha} \|T(P(\theta), C(\alpha)) - T_{\text{ref}}\|_{\mathbf{H}_\infty} \tag{10.29}$$

Hence, the *designed* closed-loop system $T(P(\theta), C(\kappa(\theta))$ is the closest $\mathbf{H}_\infty$ approximation to T_{ref}, the reference system. Since it is unrealistic to expect that the plant is in the model set, *i.e.*, $P \neq P(\theta)$, it follows that the actual, or *achieved* closed-loop system, $T(P, C(\kappa(\theta))$, may be quite different. Of course the most desirable goal is,

$$\alpha_{opt} = \arg\min_{\alpha} \|T(P, C(\alpha)) - T_{\text{ref}}\|_{\mathbf{H}_\infty} \tag{10.30}$$

The performance measure can be modified to penalize control activity, *e.g.*,

$$\|y - T_{\text{ref}}\, r\|_{rms}^2 + \lambda \|u\|_{rms}^2 \le \rho^2 \|r\|_{rms}^2, \quad \forall \|r\|_{rms} < \infty \tag{10.31}$$

Again, if the plant and controller are LTI, then this measure is equivalent to,

$$\left\| \begin{bmatrix} T(P, C) - T_{\text{ref}} \\ \lambda Q(P, C) \end{bmatrix} \right\|_{\mathbf{H}_\infty} \le \rho \tag{10.32}$$

where

$$Q(P, C) = (1 + PC)^{-1} C \tag{10.33}$$

10.4 Uncertainty Model Unfalsification

10.4.1 Unfalsification

The generic uncertainty model unfalsification problem is as follows:

Given scalar data sequences $e, v \in \mathbf{S}^\ell$, establish necessary and sufficient conditions for the existence of a disturbance sequence $w \in \mathbf{S}^\ell$ and a causal system Δ such that

$$w \in \mathbf{W}(\sigma), \quad \Delta \in \Delta(\delta) \tag{10.34}$$

and which are consistent with the model

$$e_t = w_t + (\Delta v)_t, \quad t \in [1, \ell] \tag{10.35}$$

The sets $\mathbf{W}(\sigma)$ and $\Delta(\delta)$ denote, respectively, a set of sequences with norm bounded by σ and a set of systems with gain bounded by δ.

10.4.2 Uncertainty Model Forms

The data sequence e is often obtained as the *prediction error* associated with an assumed model of the system, and v is a function of other sensed signals, the choice reflecting the type of dynamic uncertainty, or model error. For example, consider the standard prediction error form in [27],

$$e = H^{-1}(y - Pu), \qquad P, H \in \mathbf{LTI} \tag{10.36}$$

If $v = u$, then Δ represents *additive* model error, *i.e.*, the uncertainty model set is given by,

$$\mathbf{J}(\sigma, \delta) = \{\ y, u \mid y = Pu + H\Delta u + Hw, \quad w \in \mathbf{W}(\sigma), \quad \Delta \in \Delta(\delta)\ \}$$

If $v = Gu$, then Δ represents *multiplicative* model error, *i.e.*, the uncertainty model set becomes,

$$\mathbf{J}(\sigma, \delta) = \{\ y, u \mid y = Pu + H\Delta Gu + Hw, \quad w \in \mathbf{W}(\sigma), \quad \Delta \in \Delta(\delta)\ \}$$

There are clearly many variations one could include, *e.g.*, combinations of additive and multiplicative model errors, co-prime factor uncertainty, and so on, ultimately leading to the uncertainty structures described by the more inclusive linear fractional representation familiar in robust control design, *e.g.*, [30]. In addition, the error could be obtained from a *parametric* prediction error model with parameters associated with transfer function coefficients which characterize the input/output and disturbance dynamics, *i.e.*,

$$e(\theta) = H(\theta)^{-1}\,(y - P(\theta)u) \tag{10.37}$$

with $\theta \in \Theta$, the set of parameters for which the predictor is stable [27].

10.4.3 Disturbance Uncertainty

There are many ways to characterize the disturbance set $\mathbf{W}(\sigma)$. For example, consider the following sets of finite sequences:

- **Rms-bounded noise**

$$\mathbf{W}_{\text{rms}}(\sigma) = \left\{ w \in \mathbf{S}^\ell \;\middle|\; \frac{1}{\ell}\|w\|^2 \leq \sigma^2 \right\} \tag{10.38}$$

- **Time-domain white noise** [31]

$$\mathbf{W}_{\text{wht\_time}}(\gamma, m) = \{ \; w \in \mathbf{S}^\ell \mid |r_w(\tau)| \leq \gamma r_w(0) \; \} \tag{10.39}$$

where $r_w(\tau)$ is the auto-correlation of w,

$$r_w(\tau) = \frac{1}{\ell}\sum_{t=1}^{\ell-\tau} w_t w_{t+\tau}, \quad \tau \in [0, m-1] \leq \ell \tag{10.40}$$

Observe that $r_w(0) = \|w\|^2/\ell$.

- **Frequency-domain white noise** [29]

$$\mathbf{W}_{\text{wht\_freq}}(\sigma, \epsilon, m) = \{ \; w \in \mathbf{S}^\ell \mid |eig\{R_m(w)\}/\sigma^2 - 1| \leq \epsilon \; \} \tag{10.41}$$

where

$$R_m(w) = \begin{bmatrix} r_w(0) & \cdots & r_w(m-1) \\ \vdots & \ddots & \vdots \\ r_w(m-1) & \cdots & r_w(0) \end{bmatrix} \tag{10.42}$$

The disturbance set $\mathbf{W}_{\text{rms}}(\sigma)$ is the simplest of choices for deterministically characterizing "noise." The main advantage is that it is a convex set and therefor easy to handle in optimization. However, there are no restrictions preventing correlation with inputs and so the "worst-case" can occur. As shown above, characterizations of deterministic sets which resemble white noise have been examined in [29] in the frequency domain with application to system identification and in [31] for both time and frequency domains with application to robust control. The set $\mathbf{W}_{\text{wht\_time}}(\gamma, m)$ is essentially one of the standard white noise test where γ is chosen from χ^2 distribution tables; m is the *lag window* used to smooth the correlation function. The set $\mathbf{W}_{\text{wht\_freq}}(\sigma, \epsilon, m)$ is shown in [29] to also be useful for white noise testing; m again is the lag window, σ^2 is the rms-level of w and hence, the average level of the spectrum of w, and $\epsilon \in (0, 1)$ determines the "flatness" of the spectrum. Clearly these latter sets do preserve the character of white noise, but they are not convex. However, they are no worse than quadratic and so may be quite amenable to conjugate-gradient methods of optimization. The work reported in [23] shows a two-step procedure involving a Kalman filter for unfalsifying stochastic disturbance signals.

10.4.4 Gain-Bounded Dynamic Uncertainty

Uncertain dynamics can also be characterized in a number of ways. Consider the following gain-bounded, time-invariant (**TI**) dynamic uncertainty sets:

- **Linear (LTI)**

$$\Delta_{LTI}(\delta) = \{ \Delta \in \mathbf{LTI} \mid \|\Delta v\|_{rms} \leq \delta \|v\|_{rms}, \ \forall \|v\|_{rms} < \infty \} \tag{10.43}$$

Since $\Delta \in \mathbf{LTI}$, the gain bound condition is equivalent to the frequency domain bound:

$$|\Delta(e^{j\omega})| \leq \delta, \quad \omega \in [-\pi, \pi] \tag{10.44}$$

- **Incrementally nonlinear (INTI)**

$$\Delta_{\mathrm{INTI}} = \{\Delta \in \mathbf{TI} \| \Delta v_1 - \Delta v_2\|_{rms} \leq \delta \|v_1 - v_2\|_{rms}, \ \forall \|v_1\|_{rms}, \|v_2\|_{rms} < \infty\} \tag{10.45}$$

- **Nonlinear (NTI)**

$$\Delta_{\mathrm{NTI}} = \{ \Delta \in \mathbf{TI} \mid \|\Delta v\|_{rms} \leq \delta \|v\|_{rms}, \ \forall \|v\|_{rms} < \infty \} \tag{10.46}$$

10.4.5 Unfalsification

Consequences of unfalsification are summarized in the following.

(i) **Finite-Data Test**
Given data sequences $e, v \in \mathbf{S}^\ell$, there exists a sequence $w \in \mathbf{S}^\ell$ and a causal system Δ such that,

$$e_t = w_t + (\Delta v)_t, \quad t \in [1, \ell] \tag{10.47}$$

with $w \in \mathbf{W}_{\mathrm{rms}}(\sigma)$ if and only if

$$\frac{1}{\ell}\|w\|^2 \leq \sigma^2 \tag{10.48}$$

and such that:

- *$\Delta \in \Delta_{\mathrm{LTI}}(\delta)$ if and only if,*

$$\mathcal{T}\{e - w\}^T \mathcal{T}\{e - w\} - \delta^2 \mathcal{T}\{v\}^T \mathcal{T}\{v\} \leq 0 \tag{10.49}$$

with $(\mathcal{T}\{e\}, \mathcal{T}\{v\}, \mathcal{T}\{w\})$ the $\ell \times \ell$ Toeplitz matrices formed from the sequences (e, v, w), respectively, e.g.,

$$\mathcal{T}\{e\} = \begin{bmatrix} e_1 & 0 & \cdots & 0 \\ e_2 & e_1 & \cdots & 0 \\ \vdots & \vdots & \ddots & \vdots \\ e_\ell & e_{\ell-1} & \cdots & e_1 \end{bmatrix}$$

- $\Delta \in \Delta_{\mathbf{INTI}}(\delta)$ *if and only if,* $\forall m - n = 0 : \ell$ *and* $\forall t \in [1, \ell]$,

$$\left\|(z^n - z^m)(e - w)\right\|_{\mathbf{L}_2[1,t]} \leq \delta \left\|(z^n - z^m)v\right\|_{\mathbf{L}_2[1,t]} \tag{10.50}$$

where z^k *is the k-forward shift operator, i.e., if* $x = \{ x_1, x_2, \dots \}$ *then* $z^k x = \{ 0, \dots, 0, x_1, x_2, \dots \}$ *with k-zeros.*

- $\Delta \in \Delta_{\mathbf{NTI}}(\delta)$ *if and only if* $\forall t \in [1, \ell]$,

$$\left\|e - w\right\|_{\mathbf{L}_2[1,t]} \leq \delta \left\|v\right\|_{\mathbf{L}_2[1,t]} \tag{10.51}$$

(ii) **Uncertainty Tradeoff**
The (σ, δ) *boundary between falsified and unfalsified uncertainty models for a given finite data set with* $w \in \mathbf{W}_{\mathrm{rms}}$ *and* $\Delta \in \Delta_\mu(\delta)$, $\mu =$ **LTI, INTI, NTI** *is determined by solving:*

$$\sigma_\mu(\delta) = \min\left\{ \ell^{-1}\left\|e - \Delta v\right\|_{\mathbf{L}_2[1,\ell]} \mid \Delta \in \Delta_\mu(\delta) \right\}, \quad 0 \leq \delta \leq \delta_\mu \tag{10.52}$$

with

$$\delta_\mu = \min\{ \delta \mid e_t = (\Delta v)_t,\ t \in [1, \ell],\ \Delta \in \Delta_\mu(\delta) \} \tag{10.53}$$

(iii) **Nesting**
For all $\delta \geq 0$,

$$\sigma_{\mathbf{NTI}}(\delta) \leq \sigma_{\mathbf{INTI}}(\delta) \leq \sigma_{\mathbf{LTI}}(\delta) \tag{10.54}$$

and

$$\delta_{\mathbf{NTI}} < \delta_{\mathbf{INTI}} < \delta_{\mathbf{LTI}} \tag{10.55}$$

Comments

(1) The results in part (i) for $\Delta \in \Delta_{\mathbf{LTI}}(\delta)$ and the necessity for $\Delta \in \Delta_{\mathbf{NTI}}(\delta)$ (which is the same as the necessary and sufficient conditions for gain-bounded linear-time-varying (LTV) systems) is found in [33]. Proof of the remaining results in (i) can be found in [21].

(2) The tradeoff and nesting results follow from convexity of the uncertainty sets. All the results can be extended when the error is formed from the ARX parametric prediction error model with efficient computations using LMIs, [6]. Output error and other linear fractional parameter forms are not convex sets, and the nesting and tradeoff results are thus not guaranteed. A further discussion of the relation of NTI and LTI uncertainty sets, can be found in [19–21].

10.5 Controller Unfalsification

In a series of papers by Safonov *et al.*(see [34] and the references therein), it is shown how to *directly* falsify a candidate controller before it is implemented. The procedure for controller unfalsification is essentially the same as that for uncertainty model unfalsification, but applied to the closed-loop specification. Specifically, the closed-loop specification (10.24) can be viewed as the uncertainty model set,

$$\mathbf{J}(\rho) = \{\ y, r \ \mid \|y - T_{\text{ref}}\ r\|_{rms} \leq \rho \|r\|_{rms}\ ,\ \ \forall \|r\|_{rms} < \infty\ \} \tag{10.56}$$

The goal is to adjust the parameters $\alpha \in \mathbf{R}^q$ such that the controller,

$$u = C(\alpha)(r - y) \tag{10.57}$$

makes ρ as small as possible. From the previous discussion on unfalsification of uncertainty models, without any further assumptions about the plant system, the specification set is equivalently expressed as,

$$\mathbf{J}(\rho) = \{\ y, r \ \mid y - T_{\text{ref}}\ r = \Delta r,\ \ \Delta \in \Delta_{\mathbf{NTI}}(\rho)\ \} \tag{10.58}$$

If we make the additional assumption that the closed-loop system is LTI, then the specification becomes the LTI uncertainty set,

$$\mathbf{J}(\rho) = \{\ y, r \ \mid y - T_{\text{ref}}\ r = \Delta r,\ \ \|\Delta\|_{\mathbf{H}_\infty} \leq \rho\ \} \tag{10.59}$$

The finite data tests provided in the previous section can now be used to compute the performance ρ. Let $(y, u, r) \in \mathbf{S}^\ell \times \mathbf{S}^\ell \times \mathbf{S}^\ell$ denote the measured data for $t \in [1, \ell]$, where ℓ is essentially the current time. If $\Delta \in \mathbf{NTI}$, then it follows from (10.51) that the *measured* performance is,

$$\rho^\ell_{meas} = \min \left\{ \rho \ \middle|\ \|\varepsilon\|_{\mathbf{L}_2[1,t]} \leq \rho \|r\|_{\mathbf{L}_2[1,t]}\ ,\quad t \in [1, \ell]\ \right\} \\ \varepsilon = y - T_{\text{ref}}\ r \tag{10.60}$$

If $\Delta \in \mathbf{LTI}$, then it follows from (10.49) that the *measured* performance is,

$$\rho^\ell_{meas} = \min \left\{ \rho \ \middle|\ \mathcal{T}\{\varepsilon\}^T \mathcal{T}\{\varepsilon\} - \rho^2 \mathcal{T}\{r\}^T \mathcal{T}\{r\} \leq 0\ \right\} \tag{10.61}$$

Both of these measure the performance of the already implemented controller, which may have already switched (adapted) several times. The question arises: could the existing data record be informative about an untried *candidate controller*, say $C(\alpha)$, whose predicted performance might be better than the measured performance? To answer the question, consider, the following "thought experiment:"

If a candidate controller $C(\alpha)$ had produced the measured plant output/input data $(y, u) \in \mathbf{S}^\ell \times \mathbf{S}^\ell$, then the reference input **would have been** *the sequence $r(\alpha)$ satisfying*

$$u = C(\alpha)\,(r(\alpha) - y) \tag{10.62}$$

Assuming the indicated inverse exists,

$$r(\alpha) = y + C(\alpha)^{-1} u \tag{10.63}$$

Thus, the error, $\varepsilon(\alpha)$, **would have been,**

$$\begin{aligned} \varepsilon(\alpha) &= y - T_{\text{ref}}\; r(\alpha) \\ &= S_{\text{ref}}\; y - T_{\text{ref}} C(\alpha)^{-1} u \end{aligned} \tag{10.64}$$

with

$$S_{\text{ref}} = 1 - T_{\text{ref}}$$

Hence, the set of all controller parameters that achieve a performance level, ρ, **would have been,**

$$\mathbf{A}_\ell(\rho) = \{\ \alpha \in \mathbf{R}^q \mid \varepsilon(\alpha) = \Delta r(\alpha), \quad \Delta \in \Delta(\rho)\ \} \tag{10.65}$$

By this argument, $\mathbf{A}_\ell(\rho)$ is the the set of all controller parameters which are unfalsified by the available data, with respect to performance level ρ.

If $\Delta \in \mathbf{NTI}$, then

$$\begin{aligned} \mathbf{A}_\ell(\rho) &= \left\{\ \alpha \in \mathbf{R}^q \;\middle|\; \|\varepsilon(\alpha)\|_{\mathbf{L}_2[1,t]} \le \rho \|r(\alpha)\|_{\mathbf{L}_2[1,t]}\,, \quad t \in [1, \ell]\ \right\} \\ &= \bigcap_{t \in [1,\ell]} \left\{\ \alpha \in \mathbf{R}^q \;\middle|\; \|\varepsilon(\alpha)\|_{\mathbf{L}_2[1,t]} \le \rho \|r(\alpha)\|_{\mathbf{L}_2[1,t]}\ \right\} \end{aligned} \tag{10.66}$$

This last expression shows that since $\mathbf{A}_\ell(\rho)$ is an intersection of inequalities, one for each data point, it follows that as more data is recorded, *i.e.*, as ℓ increases, the unfalsified parameter set, $\mathbf{A}_\ell(\rho)$, can only get smaller. The *falsified* parameter set, its complement, therefor, can only increase. Similarly, if $\Delta \in \mathbf{LTI}$, then

$$\mathbf{A}_\ell(\rho) = \left\{\ \alpha \in \mathbf{R}^q \;\middle|\; \mathcal{T}\{\varepsilon(\alpha)\}^T \mathcal{T}\{\varepsilon(\alpha)\} - \rho^2 \mathcal{T}\{r(\alpha)\}^T \mathcal{T}\{r(\alpha)\} \le 0\ \right\} \tag{10.67}$$

In this case as ℓ increases, the Toeplitz matrices get larger thereby incorporating more constraints imposed by the data.

Having obtained the unfalsified set of controller parameters, essentially a set of candidate controllers, it remains to choose a controller parameter to implement in the next iteration, $\ell+1$. An aggressive choice is one which produces the smallest ρ, *i.e.*,

$$(\alpha^{\ell}_{unf}, \rho^{\ell}_{unf}) = \arg\min \{ \rho \mid \alpha \in \mathbf{A}_{\ell}(\rho) \} \tag{10.68}$$

A cautious choice, reflecting the distribution of elements in $\mathbf{A}_{\ell}$, is the average, or the geometric center of the set, *i.e.*,

$$(\alpha^{\ell}_{unf}, \rho^{\ell}_{unf}) = \arg avg \{ \rho \mid \alpha \in \mathbf{A}_{\ell}(\rho) \} \tag{10.69}$$

No matter the choice, the control parameters are considered likely to be updated whenever the unfalsified performance level, ρ^{ℓ}_{unf}, is smaller than the measured performance, ρ^{ℓ}_{meas}. Thus, the controller parameter update rule is,

$$\alpha^{\ell+1} = \begin{cases} \alpha^{\ell}_{unf}, & \rho^{\ell}_{unf} < \rho^{\ell}_{meas} \\ \alpha^{\ell}, & \rho^{\ell}_{unf} \geq \rho^{\ell}_{meas} \end{cases} \tag{10.70}$$

Clearly if the exogenous inputs are not sufficiently rich from iteration to iteration, then it is likely that the control will not switch.

As discussed in [34], there are several advantages to this data-based control design approach:

i. The approach is nonconservative; *i.e.*, it gives "if and only if" conditions on the candidate controller $C(\alpha)$ to be unfalsified.
ii. The unfalsified set of candidate controllers is determined from past data only – *no candidate controller is implemented if it is falsified*, and the test is applied *without actually implementing the candidate controller* $C(\alpha)$. Equally important, if the test fails, those candidate controllers, $C(\alpha)$, which have been falsified, again without implementation, *can all be discarded from any future consideration.*
iii. The test for controller unfalsification is "plant-model free." No plant model is needed to test its conditions. It depends only on the data, the controller, and the specification.
iv. The data which falsifies a controller may be open loop data or data generated by some other control law which may or may not be in the parametric set.
v. Controller falsification implies falsification of any underlying uncertainty model for the plant model, based on the same data, which would have resulted in the same controller. The converse, however, is not true: a falsified uncertainty model of the plant does not imply falsification of a controller based on this falsified uncertainty model. As a result, using the same data set, direct controller unfalsification can produce less conservative control than plant unfalsification followed by robust control design.

10.5.1 Relation to Windsurfing

A key element in the windsurfing approach to adaptation is to vary the reference system bandwidth. This is one of many possible performance parameters, *e.g.*, overshoot, settling time, and so on. To illustrate the idea let β denote the reference system bandwidth and let $T_{\text{ref}}(\beta)$ denote the reference system. Now the set of unfalsified control parameters depends not only on ρ, the performance error, but also on β,

$$\begin{aligned} \mathbf{A}_\ell(\rho,\beta) &= \{\ \alpha \in \mathbf{R}^q \mid \varepsilon(\alpha,\beta) = \Delta r(\alpha), \quad \Delta \in \Delta(\rho)\ \} \\ \varepsilon(\alpha,\beta) &= y - T_{\text{ref}}(\beta)\ r(\alpha) \end{aligned} \tag{10.71}$$

We can now generate a *family* of unfalsified parameter updates which depend on β. For example,

$$(\alpha_{\text{un}f}^\ell(\beta), \rho_{\text{un}f}^\ell(\beta)) = \arg\min \{\ \rho \mid \alpha \in \mathbf{A}_\ell(\rho,\beta)\ \} \tag{10.72}$$

This will be illustrated in the example in section 10.6.1.

Computational Issues

Measured performance, ρ_{meas}, can be obtained directly from either (10.60) or (10.61). However, solving the the optimization problem for $(\rho_{\text{un}f}, \alpha_{\text{un}f})$ has two difficulties. First, it is not in general convex, hence, there is no guaranty of finding the optimum. For PID and/or lead-lag type controllers, which have a small number of parameters, a combinatorial search is very effective as has been shown in [34] and will be demonstrated in the example in section 10.6.1.

Difficulties with the optimization are to be expected, because in essence, we are trying to solve the fixed-order control design problem, which is generically hard even when the plant is known, *e.g.*, [15]. In the case here, the plant is not known, and the problem is compounded further by using data! However, as in the output error identification problem, which is related, there are some instances where there are no local minima for parameters restricted to a region where a certain transfer function is passive [27, Ch.10,p.301]. Even if this could be applied here, obviously more assumptions about the plant are required.

The second issue is that the problem size increases as time goes on because more data is recorded, effectively adding more constraints. We offer an approximate solution to this problem in the next section on iterative unfalsification. Essentially the data is only recorded over a fixed length window which slides along with current time. This makes sense if the data record length, ℓ, is long compared to the assumed system memory.

Dealing with the ever increasing problem size can be addressed by developing recursive methods which provide outer and inner ellipsoid bounds on earlier data, hence compressing the earlier data into matrices on the order of the parameters. This is akin to least-squares estimation which compresses prior information into parameter sized covariance matrices. (This is still an idea.)

For LTI closed-loop uncertainty sets, the unfalsification tests can be recast as convex optimizations. In [37,38] it is shown that a reformulation of the performance specification as a receding horizon criterion allows the problem to be cast as a convex optimization. Another approach is described next.

10.5.2 LTI Unfalsification

Consider the closed-loop specification represented by the uncertainty model set (10.59),

$$\mathbf{J}(\rho) = \left\{ \; y, r \; \middle| \; y - T_{\text{ref}} \; r = \Delta r, \quad \|\Delta\|_{\mathbf{H}_\infty} \leq \rho \; \right\} \tag{10.73}$$

To test the candidate controller,

$$u = C(\alpha)(r - y)$$

form

$$r(\alpha) = y + C(\alpha)^{-1} u$$

where (y, u) are given data of length ℓ. The closed-loop uncertainty model is then

$$(1 - T_{\text{ref}})y - T_{\text{ref}} C(\alpha)^{-1} u = \Delta(y + C(\alpha)^{-1} u), \quad \|\Delta\|_{\mathbf{H}_\infty} \leq \rho$$

Suppose that $C(\alpha)$ is parametrized as given by (10.21). If ℓ is sufficiently large, then both sides of the the above LTI uncertainty model can be operated on by $N(\alpha)$ to give,

$$\begin{aligned} e(\alpha) &= \Delta f(\alpha), \quad \|\Delta\|_{\mathbf{H}_\infty} \leq \rho \\ e(\alpha) &= N(\alpha)(1 - T_{\text{ref}})y - D(\alpha) T_{\text{ref}} u \\ f(\alpha) &= N(\alpha) y + D(\alpha) u \end{aligned} \tag{10.74}$$

The set of unfalsified controller parameters is then,

$$\mathbf{A}_\ell(\rho) = \left\{ \; \alpha \in \mathbf{R}^q \; \middle| \; \mathcal{T}\{e(\alpha)\}^T \mathcal{T}\{e(\alpha)\} - \rho^2 \mathcal{T}\{f(\alpha)\}^T \mathcal{T}\{f(\alpha)\} \leq 0 \; \right\} \tag{10.75}$$

Since $N(\alpha)$ and $D(\alpha)$ are affine in α, so are $e(\alpha)$ and $f(\alpha)$. Hence, for a fixed value of ρ, the above set is convex in α. Solving for α which produces the smallest value of ρ is in the form of a "generalized eigenvalue problem" and can be solved using convex programming [6].

10.6 Iterative Controller Unfalsification

In this section the unfalsification paradigm is used to develop an iterative direct adaptive controller.

During the i-th iteration (data collection period) the controller is held fixed at $C(\alpha^i) \in \mathbf{LTI}$, *i.e.*,

$$u = C(\alpha^i)(r - y) \tag{10.76}$$

Suppose each data collection period contains ℓ samples, where $(y^i, u^i, r^i) \in \mathbf{S}^\ell \times \mathbf{S}^\ell \times \mathbf{S}^\ell$ is the data measured. Define the corresponding *measured* performance error by,

$$\rho^i_{\text{meas}} = \min\{\ \rho\ \mid \varepsilon^i = \Delta r^i,\quad \Delta \in \Delta_{\mathbf{NTI}}(\rho)\ \} \tag{10.77}$$

where

$$\varepsilon^i = y^i - T_{\text{ref}}\ r^i \tag{10.78}$$

> **Caveat emptor** - *This definition of measured performance for the i-th data collection period is reasonable only if the period data length, ℓ, is sufficiently large so as to make negligible any effects due to controller adjustments or exogenous disturbances in previous periods.* **Assume from now on that this is the case.**

From (10.51) it follows that the measured performance is given by,

$$\rho^i_{\text{meas}} = \min\left\{\ \rho\ \middle|\ \|\varepsilon^i\|_{\mathbf{L}_2[1,t]} \le \rho \|r^i\|_{\mathbf{L}_2[1,t]},\quad t \in [1, \ell]\ \right\} \tag{10.79}$$

Based solely on the data collected in the i-th period, the set of unfalsified controller parameters that achieve a performance level, ρ, is,

$$\mathbf{A}^i_\ell(\rho) = \left\{\ \alpha \in \mathbf{R}^q\ \middle|\ \|\varepsilon^i(\alpha)\|_{\mathbf{L}_2[1,t]} \le \rho \|r^i(\alpha)\|_{\mathbf{L}_2[1,t]},\quad t \in [1, \ell]\ \right\} \tag{10.80}$$

It follows that the set of all controller parameters which are unfalsified, with respect to performance level ρ, up to and including the i-th interval, is the intersection of these sets, *i.e.*,

$$\mathbf{A}^i(\rho) = \bigcap_{k \in [1,i]} \mathbf{A}^k_\ell(\rho) \tag{10.81}$$

As before, it remains to choose a controller to implement in the next iteration. The aggressive choice produces the smallest ρ, *i.e.*,

$$(\alpha^i_{\text{unf}}, \rho^i_{\text{unf}}) = \arg\min\{\ \rho\ \mid \alpha \in \mathbf{A}^i(\rho)\ \} \tag{10.82}$$

whereas the cautious choice, reflecting the distribution of elements in $\mathbf{A}^i$, is the average, or the geometric center of the set, *i.e.*,

$$(\alpha^i_{unf}, \rho^i_{unf}) = \arg avg \{ \rho \mid \alpha \in \mathbf{A}^i(\rho) \} \tag{10.83}$$

We then propose to update the control parameters whenever the unfalsified performance level, ρ^i_{unf}, is smaller than the best measured performance,

$$\rho^k_{meas} = \min_{j \in [1,i]} \rho^j_{meas} \tag{10.84}$$

If not, then control is returned to $C(\alpha^k)$, the controller which produced the best measured performance. Thus, the controller parameter update rule is,

$$\alpha^{i+1} = \begin{cases} \alpha^i_{unf}, \ \rho^i_{unf} < \rho^k_{meas} \\ \alpha^k, \quad \rho^i_{unf} \geq \rho^k_{meas} \end{cases} \tag{10.85}$$

This is a slightly different procedure than in the previous "one-step-at-a-time" case. Here, because the control is held fixed at $C(\alpha^i)$ for a long time, we have (we assume) a good reading of the performance with this control. In the previous formulation, the control can switch at every instant when new data is acquired.

10.6.1 Simulation Example: PI Control of a Nonlinear System

The iterative procedure is simulated with the following nonlinear plant system:

$$\begin{aligned} y &= G\Big(v + N(u)\Big) \\ G &= \frac{.1z^{-1}}{1 - .4z^{-1}} \\ N(u) &= \begin{cases} 0, & |u| \leq d \\ u - d, \ u > 0 \\ u + d, \ u < 0 \end{cases} \\ \|v\|_{rms} &\leq \sigma \end{aligned} \tag{10.86}$$

The plant system is thus a linear system, G, driven by an RMS-bounded disturbance, v, and controlled through a deadband nonlinearity, $N(\cdot)$, with deadband of size d. The control is given by the PI control,

$$u = C(\alpha)(r - y)$$

where

$$C(\alpha) = \alpha_P + \frac{\alpha_I z^{-1}}{1 - z^{-1}} \tag{10.87}$$

The reference system is

$$T_{\text{ref}} \doteq \frac{(1-a)z^{-1}}{1 - az^{-1}}, \quad a = \exp(-2\pi f_{\text{ref}})$$

Figures 10.3-10.6 show the results of the simulations. Each figure has two rows and four columns. Each row corresponds to a different bandwidth (f_{ref}) of the reference system. The rows are as follows:

- row 1: the initial output response, before adaptation, compared to the reference system output.
- row 2: the final output response, after adaptation, compared to the reference system output.
- row 3: the per iteration values of ρ^i_{meas}, ρ^i_{unf}, and the $\mathbf{H}_\infty$-norm of the error between the linearized system and the reference system.
- row 4: the PI gains per iteration.

The simulations were performed under the following conditions:

- The control was initialized as the low gain integrator:

$$C = \frac{.01z^{-1}}{1 - z^{-1}}$$

- A single repeating cycle of the reference input is given by:

$$r = \begin{cases} 1 & t = 1:200 \\ -1 & t = 201:400 \\ 0 & t = 401:600 \end{cases}$$

- There are two cycles of 4 iterations each of this reference.
  - During cycle 1, the reference system bandwidth

    $$f_{\text{ref}} = .005 \text{ hz}$$

    The results are shown in column 1 of all the figures.
  - During cycle 2, the reference system bandwidth

    $$f_{\text{ref}} = .05 \text{ hz}$$

    The results are shown in column 2 of all the figures.
- The deadband width (d) and RMS-disturbance level (σ) were set as follows:

  figure 10.3 $d = 0 \quad \sigma = 0$
  figure 10.4 $d = 0 \quad \sigma = .1$
  figure 10.5 $d = 1.5 \; \sigma = 0$
  figure 10.6 $d = 1.5 \; \sigma = .1$

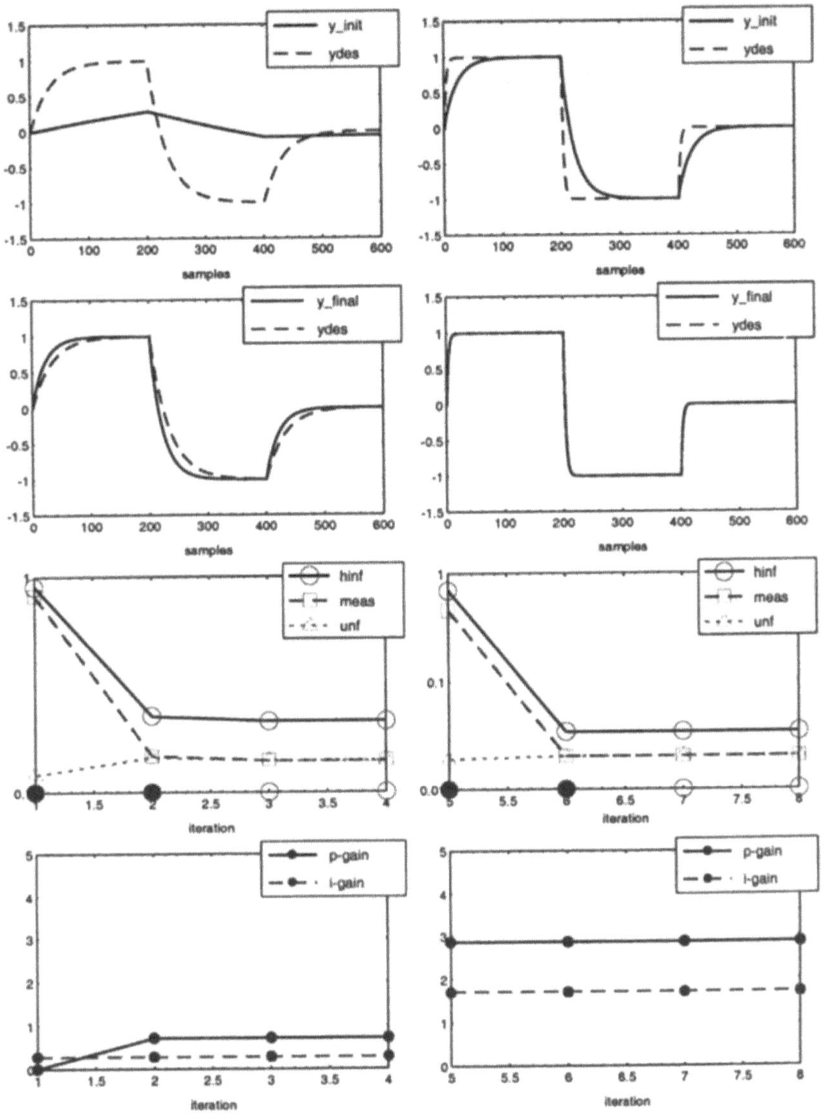

Fig. 10.3. Deadband width $d = 0$; RMS-disturbance $\sigma = 0$; reference bandwidth $f_{\text{ref}} = .005$ hz (col 1), $f_{\text{ref}} = .05$ hz (col 2).

We see in all cases that the iterative unfalsified adaptation works very well despite some extreme variations and no prior knowledge about the plant system. Although not shown, the intermediate time responses are not very much different than the final responses (after 4 iterations).

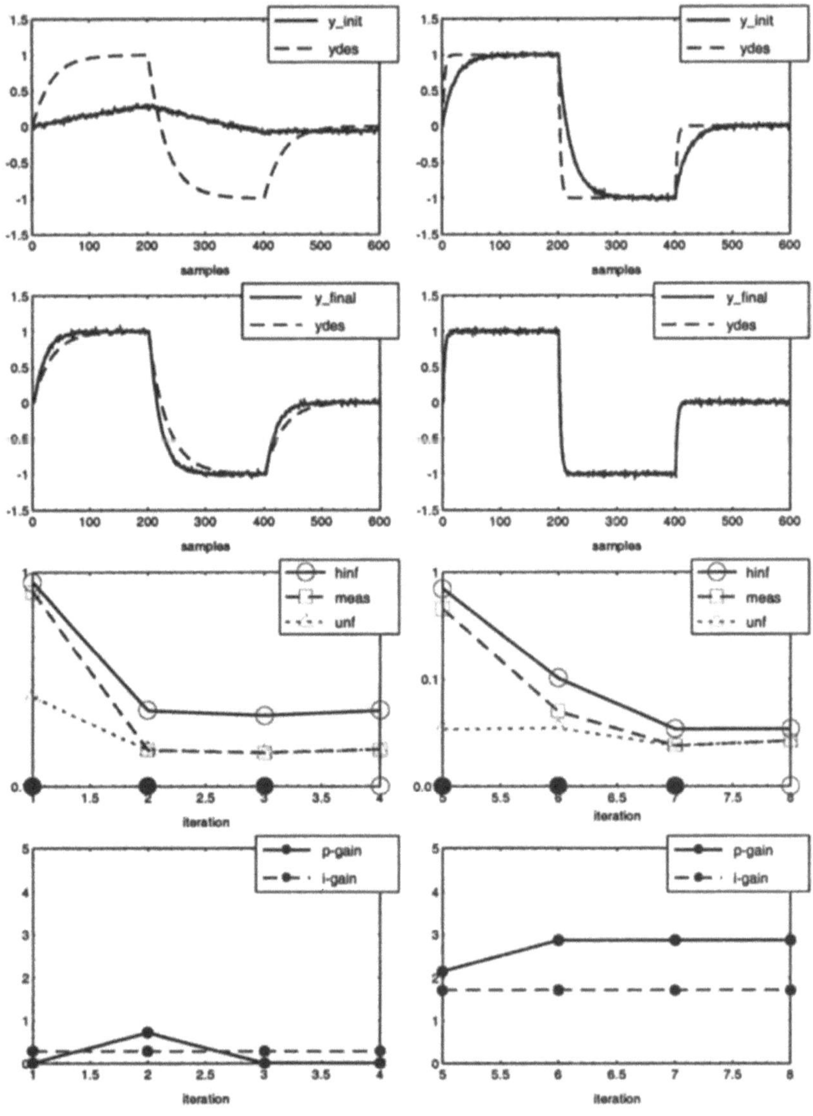

Fig. 10.4. Deadband width $d = 0$; RMS-disturbance $\sigma = .1$; reference bandwidth $f_{\text{ref}} = .005$ hz (col 1), $f_{\text{ref}} = .05$ hz (col 2).

10.7 Concluding Remarks

There are several intriguing aspects of unfalsification as applied to direct adaptive control. First, existing data can be used to falsify an experiment you would like to perform, but cannot. Secondly, controllers can be proven to be unable to meet the closed-loop performance specification without being

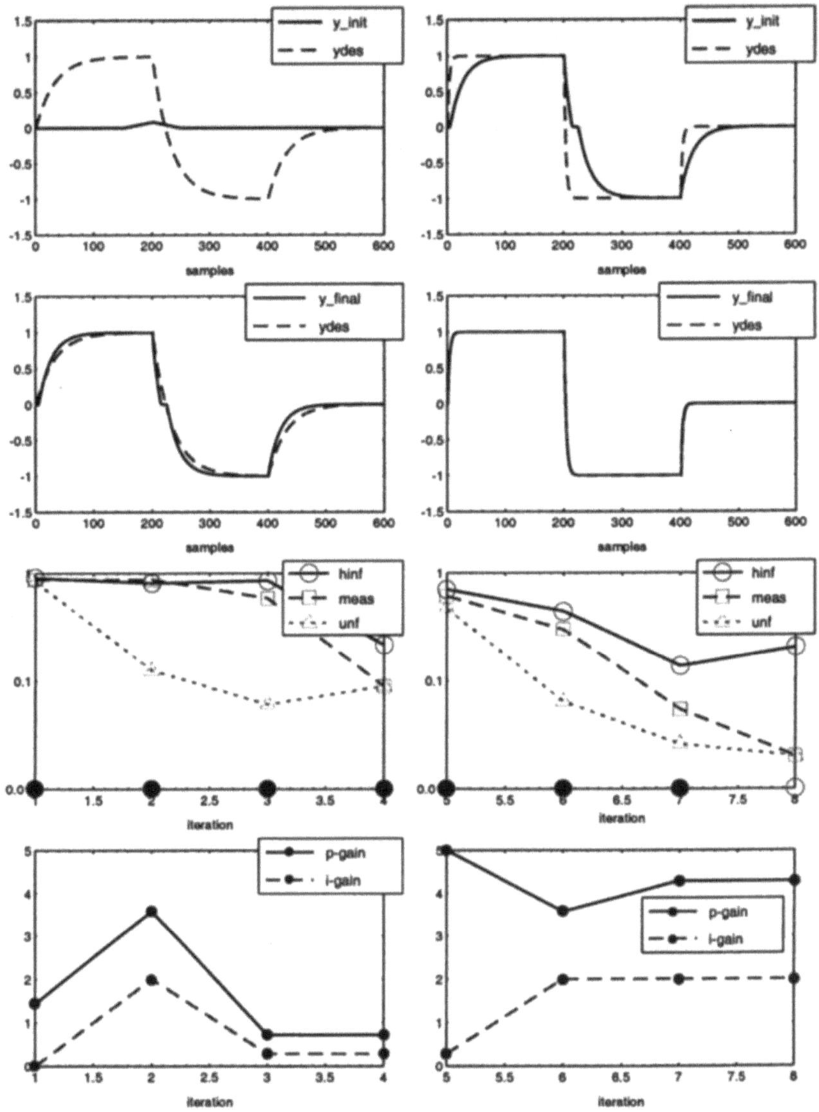

Fig. 10.5. Deadband width $d = 1.5$; RMS-disturbance $\sigma = 0$; reference bandwidth $f_{\mathrm{ref}} = .005$ hz (col 1), $f_{\mathrm{ref}} = .05$ hz (col 2).

implemented. This reduces the set of unfalsified controllers, and this reduction is non-conservative. But what about convergence? The answer to this could be: *why convergence?* If adaptation is meant to be used in the face of highly uncertain systems, which may exhibit large variations over time and operating conditions, there is no convergence. We just keep throwing away bad controllers. A well respected American football player, when asked why

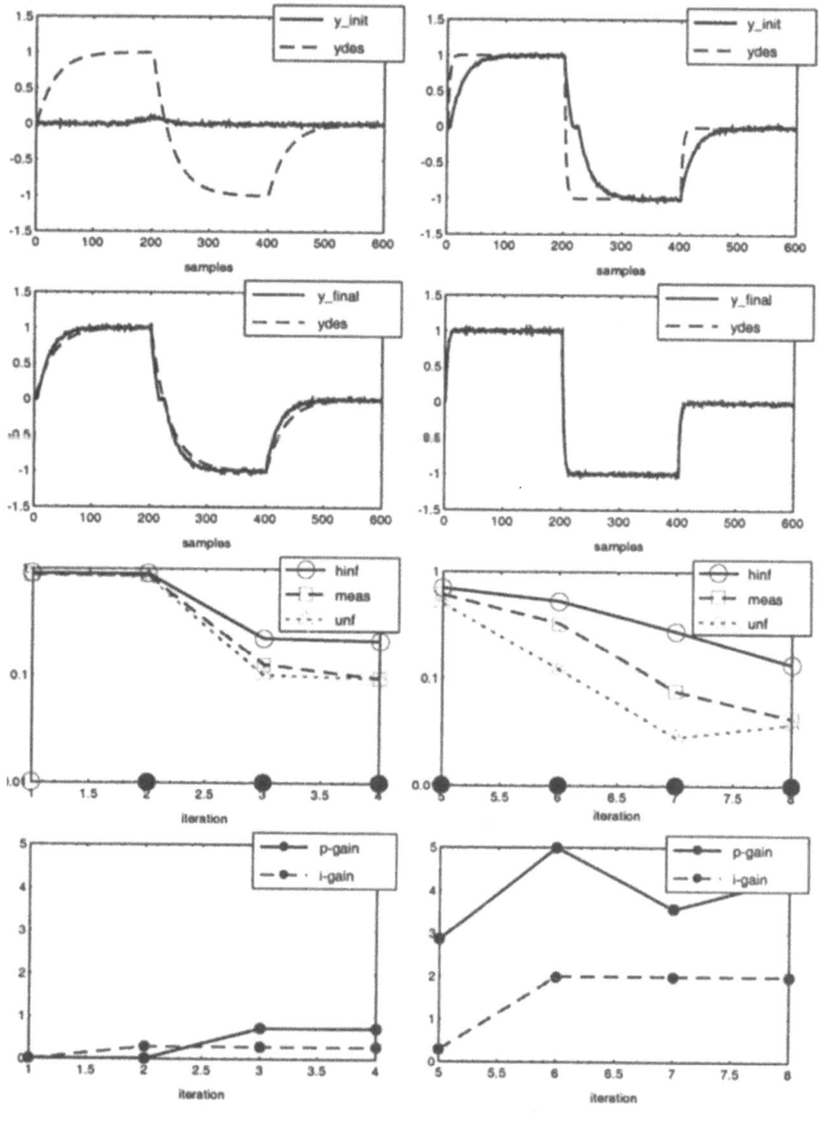

Fig. 10.6. Deadband width $d = 1.5$; RMS-disturbance $\sigma = .1$; reference bandwidth $f_{\text{ref}} = .005$ hz (col 1), $f_{\text{ref}} = .05$ hz (col 2).

he was such a good defender against the run, replied, "I just keep knock'n 'em down 'till I get to the one with the ball."

An issue not discussed here is the general topic of estimating an uncertainty model from data. As shown in [22], unfalsification of an uncertainty model consisting of a Gaussian white disturbance and an $\mathbf{H}_\infty$ bounded model error leads to conclusions, in the control design context, that are not all that different from other methods, *e.g.*, ***stochastic embedding*** [10] and ***model-error-modeling*** [28]. All these methods are utilizing "least-squares" in some form, and perhaps that is the connecting piece, despite the fact that the interpretations seem different. Although often extremely effective, I view all these methods as somewhat piece-meal, and not satisfactorily elegant. Perhaps this is due to fundamental limitations, *e.g.*, it is not possible to uniquely separate dynamic and disturbance uncertainty from finite data. For control design, however, this limitation may have no consequence. No model is needed in direct controller falsification [34] or direct tuning [12,14] – the data "speaks" to the closed-loop specification.

The answer will come, I believe, by resolving what I consider a fundamental challenge: *to develop a self consistent theory of identification and validation from a single set of assumptions.*

Acknowledgements

Brian D. O. Anderson.

References

1. B.D.O. Anderson, R.R. Bitmead, C.R. Johnson, Jr., P.V. Kokotovic, R.L. Kosut, I.M.Y. Mareels, L. Praly, and B.D. Riedle, (1986), *Stability of Adaptive Systems: Passivity and Averaging Analysis*, MIT Press, 1986.
2. B.D.O.Anderson, and R.L.Kosut, Adaptive Robust Control: On-Line Learning, Proc CDC'91, Brighton, England, Dec 1991.
3. K. J. Åström and B. Wittenmark, *Adaptive Control*, Addison-Wesley, 1995.
4. K. J. Åström and J. Nilsson, "Analysis of a scheme for iterated identification and control," *Proc. IFAC Symp. on Identification*, pp. 171-176, Copenhagen, Denmark, 1994.
5. K. J. Åström , "Matching criteria for control and identification," *Proc. 2nd European Control Conference*, pp.248-251, Groningen, Holland, June 1993.
6. S. Boyd, L. El Ghaoui, E. Feron, and V. Balakrishnan, *Linear Matrix Inequalities in System and Control Theory*, SIAM Studies in Applied Mathematics, 1994.
7. P. B. Brugarolas, V. Fromion, and M. G. Safonov, "Robust switching missile autopilot", *Proc. 1998 ACC*, pp.3665-3669, Philadelphia, PA, June 24-26,1998.
8. C. Foias and A. E. Frazho, *The Commutant Lifting Approach to Interpolation Problems*, Birkhauser-Verlag, 1990.
9. M. Gevers, "Towards a joint design of identification and control," *Essays on Control: Perspectives in the Theory and its Applications*, editors: H.L. Trentelman and J.C. Willems, pp. 111-151, Birkhauser, Boston, MA, 1993.
10. G.C. Goodwin, M. Gevers, and B. Ninness, "Quantifying the error in estimated transfer functions with applications to model order selection," *IEEE Trans. on Automatic Control*, vol. 37, no.7, pp. 913-928, July 1992.
11. U. Grenander and G. Szebo, *Toeplitz Forms and Their Applications*, Chelsea Publishing Co., New York, 1958.

12. H. Hjalmarsson, S. Gunnarsson, and M. Gevers, "A convergent iterative restricted complexity control design scheme," *Proc. 1994 CDC*,pp. 1735-1740, Orlando, Florida, Dec. 1994.
13. H. Hjalmarsson, S. Gunnarsson, and M. Gevers, "Optimality and suboptimality of iterative identification and control design schemes," *Proc. 1995 ACC*, pp.2559-2563, Seattle, Washington, June 1995.
14. L. Kammer, R. Bitmead, P. Bartlett, "Direct iterative tuning via spectral analysis," *Proc. 1998 CDC*, pp.2874-2879, Tampa, FL, Dec. 1998.
15. R. L. Kosut, "Sub-optimal control of linear time-invariant systems subject to control structure constraints," *IEEE Trans. on Auto. Contr.*, AC-15(5):557-563, June 1970.
16. R. L. Kosut, "Adaptive calibration: An approach to uncertainty modeling and on-line robust control design," *Proc. 25th IEEE CDC*, Athens, Greece, Dec. 1986.
17. R.L. Kosut, G.C. Goodwin, and M.P. Polis, *editors*, "Special Issue on System Identification for Robust Control Design," *IEEE Trans. on Automatic Control*, vol. 37, no.7, pp. 899-1008, July 1992.
18. R. L. Kosut, M. Lau, and S. Boyd, "Set-membership identification of systems with parametric and nonparametric uncertainty," "Special Issue on Identification for Robust Control Design," *IEEE Trans. Aut. Contr.*, vol. 37, no. 7, July, 1992.
19. R.L. Kosut, "Uncertainty model unfalsification: a system identification paradigm compatible with robust control design," *Proc. 1995 CDC*, New Orleans, LA, Dec. 1995.
20. R. L. Kosut, "Iterative adaptive robust control via uncertainty model unfalsification," *Proc. 1996 IFAC World Congress*, June, 1996, San Francisco, CA
21. R. L. Kosut and B. D. O. Anderson, "Uncertainty model unfalsification," *Proc. 1997 CDC*, San Diego, CA, Dec, 1997.
22. R. L. Kosut, "Uncertainty model unfalsification," CD-ROM SYSID 2000, Santa Barbara, CA, 21-23 June 2000.
23. M. V. P. Kruger and K. Poolla, "Validation of uncertainty models in the presence of stochastic disturbance signals," *Selected Topics in Modelling and Control*, Vol. 11,pp.1-8, Delft University Press, Dec. 1998.
24. W.S. Lee, B.D.O. Anderson, R.L. Kosut, and I.M.Y. Mareels, "A new approach to adaptive robust control," *Int. J. of Adaptive Control and Signal Processing*, vol. 7, pp. 183-211, 1993.
25. W.S. Lee, B.D.O. Anderson, I.M.Y. Mareels, and R.L. Kosut, "On some key issues in the windsurfer approach to robust adaptive control", *Automatica*, Vol. 31, No. 11, pp. 1619-1636, 1995.
26. M.M. Livestone, M.A. Dahleh, and J.A. Farrell, "A framework for robust control based model invalidation", *Proc. 1994 ACC*, pp. 3017-3020, Baltimore, MD, June 1994.
27. L. Ljung, *System Identification: Theory for the User*, First Edition, Prentice-Hall, 1987.
28. L. Ljung, "Model validation and model error modeling," *The Åström Symposium on Control*, pp.15-42, Lund, Sweden, August 27-28, 1999.
29. M. Massoumnia and R. L. Kosut, "A family of norms for system identification problems," *Proc. 1993 ACC*, San Francisco, CA, June 1993, and *IEEE Trans. Aut. Control*, vol. 39, no. 5, pp. 1027-1031, May 1994.

30. M. P. Newlin and R. S. Smith, "A generalization of the structured singular value and its application to model validation," *IEEE Trans. Aut. Contr.*, vol. 43, no. 7, pp. 901-907, July 1998.
31. F. Paganini, "A set-based approach for white noise modeling," *IEEE Trans. Aut. Contr.*, vol. 41, no. 10, pp. 1453-1465, Oct. 1996.
32. S. M. Phillips, R. L. Kosut, G. F. Franklin, "An averaging analysis of discrete-time indirect adaptive control", *Proc. 1988 American Control Conference*, Atlanta, GA, 1988.
33. K. Poolla, P. Khargonnekar, A. Tikku, J. Krause, and K. Nagpal, "A time-domain approach to model validation", *Proc. 1992 ACC*, Chicago, IL, June 1992, and *IEEE Trans. Aut. Contr.*, vol. 39, no. 5, pp. 951-959, May 1994.
34. M. G. Safonov and T. C. Tsao, "The unfalsified control concept and learning", *IEEE Trans. Aut. Contr.*, vol. 42, no. 6, pp. 843-847, June 1997.
35. R. S. Smith and J. C. Doyle, "Model validation – a connection between robust control and identification," *Proc. 1989 ACC*, pp. 1435-1440, June 1989, and *IEEE Trans. Aut. Contr.*, vol. 37, pp. 942-952, July 1992.
36. B. Wahlberg and P.M. Makila (1996), "On approximation of stable linear dynamical systmes using Laguerre and Kautz functions," *Automatica*, 32(5), 693-708.
37. B. R. Woodley, R. L. Kosut, and J. P. How, "Uncertainty model unfalsification with simulation," *Proc. 1998 ACC*, pp. 2754-2755, Philadelphia, PA, June 24-26, 1998.
38. B. R. Woodley, J. P. How, and R. L. Kosut, "Direct unfalsified controller design - solution via convex optimization," submitted 1999 ACC, San Diego, CA, June 2-4, 1999.
39. Z. Zang, R. R. Bitmead, and M. Gevers, "$\mathbf{H}_2$ iterative model refinement and control robustness enhancement," *Proc. 30th IEEE CDC*, pp. 279-284, Brighton, UK, Dec. 1991.
40. K. Zhou and J. C. Doyle, *Essentials of Robust Control*, Prentice Hall, 1998.

11 Model Quality for the Windsurfer Scheme of Adaptive Control

Robert R. Bitmead and Wayne J. Dunstan

University of California San Diego, La Jolla CA 92093-0411, USA.

Abstract. The Windsurfer Approach to adaptive control combines Internal Model Control (IMC) design parametrized by the desired closed-loop bandwidth with event-driven system identification using Hansen's scheme on closed-loop data. Our aim in this work is to investigate the nature of model quality applicable to the windsurfer approach. We do this by expanding on recent activities dealing with the complete class of all models suited to model reference control design. In particular, we interpret the conditions for simultaneous stabilization of the plant and the model by an IMC controller. This extends the model reference results to a different domain in which the controller is characterized algebraically by the model but also inherently includes a specification of design performance objective.

11.1 Introduction

The Windsurfer Approach to Adaptive Control [1,2] is based on an anthropomorphic learning model in which the difficulty of the control task is progressively increased until the closed-loop performance exposes modeling inadequacies, when a new model is fitted. The comparison is with the way a human (in this case Brian Anderson) learns to ride a windsurfer. The bandwidth of the control is progressively increased as experience improves, usually with several (wet) restarts after poor achieved closed-loop performance. The model adjustment phase occurs following the detection of the controller's inability to provide performance at the requested closed-loop bandwidth. Coupled with a cautious model-based control design, using the Internal Model Control (IMC) approach of Morari and Zafiriou [3], is control-relevant system identification using closed-loop data and Hansen's scheme [4].

As an approach to adaptive control, the windsurfer method sits among iterative identification and control, where mutually cogniscent criteria are selected for the identification and control design phases [5–9]. The pace of adaptation is slow, with a single fixed linear controller remaining in the closed loop sufficiently long to determine reliably the achieved performance. The control redesign continues at each step with an ever-increasing demanded closed-loop bandwidth. By contrast, the identification is event-driven and is invoked only after poor achieved performance is detected with a controller designed using the current model. That is, model refinement is based on assessed need from closed-loop data. Thus models are not adjusted unless required by the ultimate control specification. When the model is adjusted, typically, the complexity of the model increases.

Our aim in this study is to begin to examine some of the questions of complexity for models to be used in windsurfer methods. We do this by considering the class of models for which the real plant and the model are simultaneously stabilized by the IMC controller computed using the model. Further, we consider how the demands of the control specification — the closed-loop bandwidth here — impinge upon the requisite model quality. We appeal to recent work by Blondel *et al.* [10,11] which treats the complete class of models, $\mathcal{P}$, for which a model-based stabilizing control design is assured of stabilizing the actual system. The control design in these works is based on model reference control. This permits the construction of a simple bijective map between plant model and controller which, in turn, allows investigation of the internal stability properties of the actual plant-controller closed loop. The results reduce to a simply parametrized set of admissible models which must satisfy interpolation conditions at points in the closed right-half complex plane. This is extensible to guarantees of performance between the model-controller and plant-controller loops.

These studies of stabilizing model sets are couched in the framework of answering the question "What is a good model for control design?". The capability of moving algebraically from model to controller is important in removing the usual opacity of the control design phase. However, the cost of this is to consider only model reference design with the specification embodied in the reference model. IMC also results in an algebraic control design, which can be recast as model reference with a special target transfer function. Our task here will be to explore this modification and to interpret the consequences for system identification in the windsurfer framework.

Section 2 is devoted to a brief development of the IMC controller design in the windsurfer method, especially in terms of it dependence on model parameters and design specification. Section 3 reviews the parametrization of the set of all models which yield a stable model reference controller for the actual plant. This is then extended to IMC designs. Section 4 develops the interpretation of the stabilization and performance conditions for model-based IMC in terms of identification requirements for models in the windsurfer method.

11.2 Internal Model Control for Stable Systems

Internal Model Control is amply treated by the book of Morari and Zafiriou [3] and is a design approach on the use of feedback to correct for plant-model mismatch, which includes the feedback of unknown disturbance signals. As such, it is interesting from a perspective of robustness analysis in addition to performance evaluation. Its appeal for adaptive windsurfer design is that, for stable plants, the performance specification of IMC is captured by a single parameter, λ, the closed-loop bandwidth. As successive designs are completed, tested and evaluated, this parameter may be increased until either the ultimate specification is reached or a decision is made that further improvement is impossible [1,2].

Denote the actual continuous-time scalar plant system by $P(s)$, where we assume the plant to be stable and strictly proper. Denote the i-th continuous-time plant model by $P_i(s)$, which we also assume to be stable and strictly proper with relative degree n_i. Define the (inner-outer) factorization

$$P_i(s) = [P_i]_a(s)[P_i]_m(s), \tag{11.1}$$

where $[P_i]_a$ is a stable all-pass factor, i.e. an inner function whose zeros are the k zeros of P_i in the closed right half-plane, whose poles are the complex conjugate of these zeros and which is bi-proper, and $[P_i]_m$ is the stable minimum-phase factor of P_i. That is, $[P_i]_m$ has poles at the poles of P_i (which we assumed stable), zeros at the zeros of P_i in the open left half-plane and at the complex conjugate of the zeros of P_i in the closed right half-plane, and n_i zeros at infinity. $[P_i]_m$ is an outer function.

Select the nominal closed-loop bandwidth $\lambda_i > 0$. Define the filter

$$F_i(s) = \left(\frac{\lambda_i}{s+\lambda_i}\right)^n. \tag{11.2}$$

The IMC structure is shown in Figure 11.2, where model error feedback

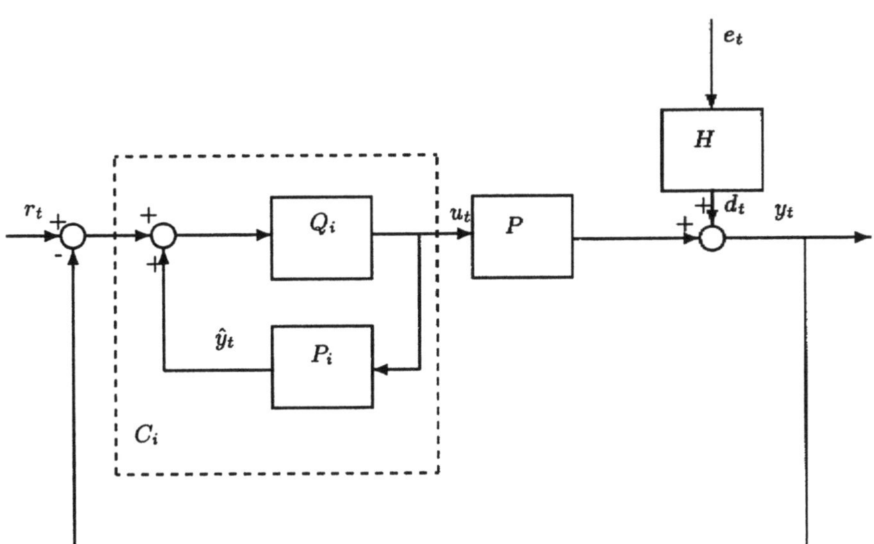

Fig. 11.1. Internal Model Control Structure.

controller

$$Q_i = [P_i]_m^{-1} F_i, \tag{11.3}$$

or, equivalently, output feedback controller

$$C_i = \frac{Q_i}{1 - P_i Q_i} = \frac{[P_i]_m^{-1} F_i}{1 - [P_i]_a F_i}, \tag{11.4}$$

which is stable, proper and minimum-phase by construction. The signals in Figure 11.2 are the externally applied reference r_t, the plant input u_t, the plant output y_t, and the additive output disturbance d_t, written as the output of a stable proper linear system $H(s)$ driven by white noise e_t.

This choice of controller has the effect of making the designed complementary sensitivity function,

$$\bar{T}_i(s) = \frac{P_i C_i}{1 - P_i C_i} = P_i Q_i = [P_i]_a F_i. \tag{11.5}$$

The designed sensitivity function is

$$\bar{S}_i(s) = 1 - \bar{T}_i = 1 - P_i Q_i. \tag{11.6}$$

Since $[P_i]_a$ is an all-pass transfer function, its magnitude is one for all frequencies and thus the designed complementary sensitivity function is flat in its passband and rolls off at frequency λ_i with attenuation -3n dB. The designed sensitivity function may be prevented from remaining close to zero throughout the closed-loop passband because of the phase properties of $[P_i]_a$. Provided Q_i is stable and proper, which it is by design, this controller internally stabilizes the loop in Figure 11.2 when $P = P_i$ [3]. As $\lambda_i \to 0$ the controller tends to zero and so the stability of the actual closed loop is ensured for λ_i sufficiently small. Hence, the connection between the IMC scheme and the tenets of the windsurfer method of progressively increasing design bandwidth until the model/plant-mismatch leads to discernible performance degradation, as measured by the achieved tracking error.

The signals of the achieved closed loop are given by

$$\begin{aligned} y_t &= \frac{H}{1 + PC_i} e_t + \frac{PC_i}{1 + PC_i} r_t, \qquad (11.7) \\ &= \frac{\bar{S}_i H}{\bar{S}_i + PQ_i} e_t + \frac{PQ_i}{\bar{S}_i + PQ_i} r_t, \\ &= \frac{(1 - P_i Q_i) H}{1 + (P - P_i) Q_i} e_t + \frac{PQ_i}{1 + (P - P_i) Q_i} r_t, \\ u_t &= \frac{-HC_i}{1 + PC_i} e_t + \frac{C_i}{1 + P_i C_i} r_t, \qquad (11.8) \\ &= \frac{-Q_i H}{1 + (P - P_i) Q_i} e_t + \frac{Q_i}{1 + (P - P_i) Q_i} r_t, \\ y_t - \hat{y}_t &= \frac{(P - P_i) Q_i}{1 + (P - P_i) Q_i} r_t + \frac{H}{1 + (P - P_i) Q_i} e_t. \qquad (11.9) \end{aligned}$$

Achieved closed-loop performance is measured via y_t, while the plant-model mismatch is discernible by correlating r_t with $y_t - \hat{y}_t$. Stability robustness is apparent for small λ_i.

We note from the formulæ for C_i and Q_i above and the respective degree properties, zero positions and pole positions of F_i and $[P_i]_m$ that, by construction, Q_i is stable, minimum-phase and is bi-proper, i.e. is proper with a proper inverse or, equivalently, has a finite, non-zero limit as $s \to \infty$. Similarly, C_i, is stable except for a single pole at $s = 0$, is minimum-phase and is bi-proper. These observations assist in some later interpretations of interpolation conditions.

The windsurfer method uses successive controller redesign using a fixed model and increasing the nominal closed-loop bandwidth λ_i. When the achieved performance exhibits poor behavior vis-à-vis the designed performance, an identification phase in introduced and the plant model refitted to reflect the currently demanded bandwidth. From our perspective, the attraction of this IMC method for the windsurfer method is that the controller is simply algebraically computed from the model P_i and the performance specification λ_i. In the next section we shall connect this to the class of acceptable models for IMC design as measured by robust stability and robust performance.

11.3 Models for Robust Stability and Performance

We consider the following questions.

Stability: *What is the complete class of stable models, $\mathcal{P} = \{P_i\}$, for which the IMC controller above is guaranteed internally stabilizing in closed loop with the true stable plant, P?*

Performance: *Within this class $\mathcal{P}$, what subset also is ensured of providing achieved complementary sensitivity T satisfying*

$$||T - \bar{T}|| < \epsilon, \tag{11.10}$$

for some positive ϵ and appropriate transfer function norm?

11.3.1 Unstable Zeros and Ones and Robustness

To tackle these questions, we draw on the work of Blondel *et al.* [11] dealing with model reference control design.

Theorem 1. *Consider the actual plant $P(s)$ and candidate model $\hat{P}(s)$. For each such model, consider the target designed complementary sensitivity function $\bar{T}(s)$ and its associated model-based controller $C(\hat{P}, \bar{T}, s)$ defined by*

$$\bar{T} = \frac{\hat{P}C}{1 + \hat{P}C}, \quad \text{or} \quad C = \frac{\bar{T}}{\hat{P}(1 - \bar{T})}, \tag{11.11}$$

where we assume that no zeros of the specified $\bar{T}$ are unstable poles of P and that no zeros of $\bar{S} = 1 - \bar{T}$ are unstable zeros of P.

The set, $\mathcal{P}$, of all models $\hat{P}$ which yield controllers C which internally stabilize $\hat{P}$ and internally stabilize P is given by

$$\mathcal{P} = \left\{ \hat{P} : \hat{P} = \frac{\bar{T}}{1-\bar{T}} \times \frac{1-R}{R} \times P \right\}, \tag{11.12}$$

where R is any proper stable transfer function satisfying the following interpolation constraints. For any point α in the closed right half-plane,

$$\begin{array}{l} (a)\ P(\alpha) = 0 \Rightarrow R(\alpha) = 0 \Rightarrow \{P(\alpha) = 0 \text{ or } \bar{T}(\alpha) = 0\}, \\ (b)\ P(\alpha) = \infty \Rightarrow R(\alpha) = 1 \Rightarrow \{P(\alpha) = \infty \text{ or } \bar{T}(\alpha) = 1\}. \end{array} \tag{11.13}$$

Subject to these conditions, the achieved complementary sensitivity function satisfies the following relationship

$$\frac{PC}{1+PC} = R, \tag{11.14}$$

whence

$$\left\| \frac{PC}{1+PC} - \bar{T} \right\| = \| R - \bar{T} \|. \tag{11.15}$$

Theorem 1 provides a full description of the class of models and target complementary sensitivity functions whose resultant controllers internally stabilize both P and $\hat{P}$. This class is fully parametrized by the stable transfer function R. The coda (11.15) shows that, for the class of models for which the additional condition $\|R - \bar{T}\| < \epsilon$ holds, we can further guarantee that the achieved complementary sensitivity satisfies

$$\left\| \frac{PC}{1+PC} - \bar{T} \right\| = \left\| \frac{1}{1+PC} - \bar{S} \right\| < \epsilon. \tag{11.16}$$

The conditions contained in the theorem specify interpolation requirements in the closed right half-plane. They restrict the permissible points in this domain at which the transfer functions may take on the values zero, one or ∞. These points are referred to as unstable zeros, ones and poles. The codicil concerning the permissible unstable zeros of $\bar{T}$ and $\bar{S}$ vis-à-vis the zeros and poles of P is a very mild genericity assumption on the choice of target complementary sensitivity function.

We shall now move from the consideration of these conditions in the general model reference control framework to their interpretation in IMC design. From (11.5-11.6) we see that the controller C_i is able to be viewed as a model reference design with a special dependence on the model properties.

11.3.2 Interpolation and Model Quality for IMC

Direct comparison between IMC and the specification of target complementary sensitivity function is given by (11.5),

$$\bar{T}_i(s) = \frac{P_i C_i}{1 - P_i C_i} = P_i Q_i = [P_i]_a F_i,$$

with IMC filter

$$F_i(s) = \left(\frac{\lambda_i}{s + \lambda_i}\right)^n.$$

Writing the inner-outer factorizations of model P_i and plant P as

$$P_i = [P_i]_a [P_i]_m \qquad \text{and} \qquad P = [P]_a [P]_m,$$

we see that the class $\mathcal{P}_{\text{IMC}}$ of Theorem 1 is given by

$$\mathcal{P}_{\text{IMC}} = \left\{ \hat{P} : \hat{P} = [P_i]_a [P_i]_m = \frac{[P_i]_a F_i}{1 - [P_i]_a F_i} \times \frac{1 - R}{R} \times [P]_a [P]_m \right\}, \tag{11.17}$$

subject to the interpolation constraints (11.13). Let us examine the implications of these interpolation constraints.

> **(11.13a) For α in the closed right half-plane**
> $P(\alpha) = 0 \Rightarrow R(\alpha) = 0 \Rightarrow \{P(\alpha) = 0 \text{ or } \bar{T}(\alpha) = 0\}$:

[I] By construction, the unstable zeros of $\bar{T}$ coincide with the unstable zeros of P_i only, i.e. the zeros of $[P_i]_a$ plus ∞.

[II] $R(\alpha) = 0$ for every unstable α such that $P(\alpha) = 0$ including ∞. Thus

$$R(s) = [P]_a(s) M(s),$$

for some stable M.

[III] $M(s)$ may have no unstable zeros except at ∞, where it has as many zeros as does P. This can also be seen from $R = \frac{PC_i}{1+PC_i}$ and the biproperness of C_i.

> **(11.13b) For α in the closed right half-plane**
> $P(\alpha) = \infty \Rightarrow R(\alpha) = 1 \Rightarrow \{P(\alpha) = \infty \text{ or } \bar{T}(\alpha) = 1\}$:

[IV] By construction,

$$\bar{T}(s) = [P_i]_a(s) F(s) = 1, \qquad \text{Re}(s) \geq 0,$$

has a single solution at $s = 0$. This may be established by the analyticity of $\bar{T}$ in the closed right half-plane and the maximum modulus principle. Thus, the only unstable one of $\bar{T}$ is at the origin.

[V] By assumption, P is stable and so $P(\alpha) = \infty$ at no point in the closed right half-plane. Thus, R may have a single unstable one at $s = 0$ and nowhere else. Since $\bar{T}(0) = 1$, the genericity assumption of Theorem 1 implies that $P(0) \neq 0$. Considering (11.12) evaluated at $s = 0$ and the stability of P_i, we have $R(0) = M(0) = 1$. We note that this latter condition does NOT imply that $P_i(0) = P(0)$. This is evident by taking any stable P and P_i with very small λ_i — closed-loop stability does not depend on matching the dc gains. (De L'Hôpital's rule must be used to evaluate the ratio of dc gains.)

[VI] The two functions M and F_i are analytic in the closed right half-plane, are minimum-phase and possess solitary unstable ones at $s = 0$. While the structure of F_i ensures that $1 - [P_i]_a F_i$ has no unstable zeros other than at $s = 0$, the same cannot be said of $1 - [P]_a M$, unless we enforce such a condition on M.

(11.16) $\left\| \frac{PC}{1+PC} - \bar{T} \right\| = \left\| \frac{1}{1+PC} - \bar{S} \right\| < \epsilon.$

[VII] From above, we have $T(s) = [P_i]_a(s) F_i(s)$ and $R(s) = [P]_a(s) M(s)$. Thus,

$$\|R - T\| < \epsilon \Leftrightarrow \|[P]_a M - [P_i]_a F_i\| < \epsilon \quad (11.18)$$

$$\Leftrightarrow \|M - [P]_a^{-1}[P_i]_a F_i\| < \epsilon$$

$$\Leftrightarrow \|[P_i]_a^{-1} M - [P]_a^{-1} F_i\| < \epsilon. \quad (11.19)$$

Here we have used the all-pass character of the $[\cdot]_a$ components to shuffle them in the frequency-domain norm.

Theorem 2. *For a given plant P and a given performance objective F_i, the model class $\mathcal{P}_{\text{IMC}}$ which achieves simultaneous internal stability plus closed-loop complementary sensitivity performance error less than ϵ is given by*

$$\mathcal{P}_{\text{IMC}} = \left\{ P_i(s) : P_i = [P_i]_a [P_i]_m = [P_i]_a \times \frac{F_i}{1 - [P_i]_a F_i} \times \frac{1 - [P]_a M}{M} \times [P]_m \right\},$$

where

i. $[P_i]_a$ *is an arbitrary, stable, all-pass, transfer function,*

ii. M *is an arbitrary stable, minimum-phase transfer function with* $M(0) = 1$ *and with* $1 - [P]_a M$ *possessing no other unstable zero than* $s = 0$.

iii. $[P_i]_a$ *and* M *satisfy*

$$\|[P_i]_a^{-1} M - [P]_a^{-1} F_i\| < \epsilon. \quad (11.20)$$

iv. $[P_i]_m$ *is determined by*

$$[P_i]_m = \frac{F_i}{1 - [P_i]_a F_i} \times \frac{1 - [P]_a M}{M} \times [P]_m, \quad (11.21)$$

which is stable and minimum-phase by construction.

11.3.3 Model Requirements in Windsurfer Design

Our task in this subsection is to interpret the conditions of Theorem 2 in terms of the system identification requirements for the model P_i of the plant P when the Windsurfer design is to be used. Recall the following two features: the Windsurfer design proceeds by extending the designed closed-loop bandwidth λ via (11.2), and the model class that we are considering, $\mathcal{P}_{\text{IMC}}$, is the complete set of all models yielding controllers which simultaneously stabilize P_i and P, while satisfying the complementary sensitivity performance robustness specification (11.15). The class of models is fully parametrized by $[P_i]_a$ and M — hence the form of (11.20) which distinguishes the specification side from the model determination side.

Stability of the achieved closed loop
Analyzing the achieved sensitivity function in (11.7-11.9) illustrates that the central issue for the simultaneous stability of the systems is that the zeros of $1+(P-P_i)Q_i$ be limited to stable values of s. Rewriting this in terms of the design parameters, we arrive at a modeling requirement

$$1+(P-P_i)Q_i \neq 0 \Leftrightarrow 1+(P-P_i)[P_i]_m^{-1}F_i \neq 0$$
$$\Leftarrow \left|\frac{P-P_i}{P_i}\right| |F_i| < 1. \tag{11.22}$$

That is, the relative model error between P and P_i should be bounded in the passband of the designed complementary sensitivity function magnitude F_i. Since the relative model error is exposed by the closed-loop data [9,12], it is natural to combine closed-loop identification with the Windsurfer design. The more revealing feature from (11.22) is that the quality of model fitting is more relaxed for less demanding (i.e. lower bandwidth specification) control objectives. This is well understood.

Unstable ones of $[P]_aM$
If we commence by ignoring, for the moment, the performance requirement (11.15) and concentrate only on the simultaneous stability properties, we may treat the specification on the minimum-phase part of the model embodied in (11.21) as constraining the factor M. By forcing $M(0) = 1$ we satisfy the need to have $[P]_aM$ with an unstable one at $s = 0$, since $[P]_a$ is all-pass. However, the need for this to be the *only* unstable one further restricts permissible Ms to those that are feasible for the achieved complementary sensitivity. (Recall that, subject to the satisfaction of the interpolation conditions, $R = [P]_aM$ is the achieved complementary sensitivity function and that C_i and P are stable by construction. The condition on unstable ones is fulfilled automatically in this case, as an unstable one would imply a point in the right half-plane where PC_i is unbounded.) We note that $[P]_aM$ may have no other unstable ones, even though this frequency response may exceed one in magnitude at many places.

Alternative expressions for M are

$$M = \frac{[P]_m Q_i}{1+(P-P_i)Q_i} = \frac{[P]_m [P_i]_m^{-1} F_i}{1+(P-P_i)Q_i} = \frac{[P]_m F_i}{[P_i]_m + (P-P_i)F_i}. \quad (11.23)$$

Requiring that M have only the sole unstable one imposes further matching conditions on the outer factor within the control bandwidth.

Roll-off and Zeros at Infinity

By construction of the controller, where the parameter n is selected to make C_i bi-proper, the relative degree of P_i (which is the same as that of $[P_i]_m$) is bound to that of F_i. This is reflected in (11.21). Similarly, the relative degrees of P, $[P]_m$ and M are identical. It is not until the passband of designed complementary sensitivity F_i exceeds the passband of P_i that the uncertainty of the model structure begins to dominate the picture in (11.23). That is, for high bandwidth designs, the model should match the roll-off of the true plant at least up to the design bandwidth. This will then ensure that the magnitude of M is bounded and rolls off properly.

Performance Considerations

We have concentrated on the stability conditions so far. Including the performance bound (11.15) further constrains the permissible models in a manner which builds on the limitations imposed by stability. The relation (11.23) may be substituted into the formula (11.20) as follows.

$$\begin{aligned}
\|[P_i]_a^{-1} M - [P]_a^{-1} F_i\| &= \|[P]_a [P_i]_a^{-1} M - F_i\|, \\
&= \left\| \frac{P F_i}{P_i + (P-P_i) F_i [P_i]_a} - F_i \right\|, \\
&= \left\| F_i \left(\frac{P}{P_i + (P-P_i)\bar{T}_i} - 1 \right) \right\|, \\
&= \left\| F_i (P-P_i) \left(\frac{\bar{S}}{P_i + (P-P_i)\bar{T}_i} \right) \right\|.
\end{aligned}$$

Ensuring this latter expression is less than ϵ in magnitude reflects, again, the need to find a model properly tuned to the designed closed-loop bandwidth λ_i. Further, the model needs to fit well where the designed sensitivity function $\bar{S}$ amplifies the gain in the passband of F_i. Thus, the performance requirement is reflected in the mix between model fit and control specification. It also reinforces the need evident in (11.20) that the in-band non-minimum-phase zeros need to be well captured in the model, since these are the source of $\bar{S}$ exceeding one in magnitude.

An alternative viewpoint to the model fit may be taken by looking directly to (11.20). Since both $[P]_a$ and $[P_i]_a$ are stable all-pass functions, their inverses are strictly unstable systems, while M and F_i are stable and minimum-phase. Regarded as a problem in the approximation of systems with mixed stable and unstable modes by a model with mixed modes, (11.20) indicates the importance of capturing the non-minimum-phase characteristics,

$[P]_a$, of the plant by the non-minimum-phase model characteristics, $[P_i]_a$, in the bandwidth of the control demand only. Again this makes good practical sense.

11.4 Conclusions

Our task here has been to refine to IMC design the description of the model class for which model reference control yields internally stabilizing controllers for the model and for the actual plant. This design of IMC controllers is found in the Windsurfer scheme of Adaptive Control, where it involves successive control design and model identification, as the control bandwidth is increased. We note that this is the complete class of all such models and that this class is rather simply parametrized. In developing these results, we have used the property that IMC design for stable plants and models is a direct variant of the model reference method, with a particular choice of the reference model. For such systems, IMC can also be viewed as the LQ optimal controller [3].

The model fitting principles generated here for the Windsurfer scheme broadly reflect the conditions understood by the progenitors of the method. Although here we have studied the complete class and have separated to some extent the fitting of the all-pass part and that of the minimum-phase component. It still remains to interpret the results via a specific identification methods, such as the Hansen scheme used in the Windsurfer approach. What is important here is the link established between the design specification (as embodied in λ_i) to the identification requirements.

Areas in which fruitful research might further be conducted could include: the study of IMC control for unstable systems, the incorporation of disturbance rejection performance beyond simply the specification of a designed complementary sensitivity function, understanding the role of model complexity in achieving closed-loop performance [13,14].

11.5 Coda

This is a presentation given to honour the work of Professor Brian Anderson, the inspirational scientist, engineer, leader, educator, polyglot and windsurfer. Much of Brian Anderson's work over a significant period of time has been devoted to the inclusion of the *closed loop* into many aspects of control design: system identification [15–20], model reduction [21], controller reduction [13,22–24], joint modeling and control design [25–27], discretization [28,29], cautious adaptation[30,31]. The Windsurfer Scheme and its (self)-anthropomorphism is an important part of his contribution to tying two of these closed-loop-dependent areas together and is yet another of his fundamental contributions to Adaptive Control.

References

1. Lee, W. S., Anderson, B. D. O., Kosut, R. L., Mareels, I. M. Y. (1993) "A new approach to adaptive robust control," *International Journal of Adaptive Control and Signal Processing,* vol. 7, pp. 183-211.
2. Lee, W. .S., Anderson, B. D. O., Mareels, I. M. Y., Kosut, R. L. (1995) "On some key issues in the windsurfer approach to adaptive robust control," *Automatica,* vol. 31, pp.1619-1636.
3. Morari, M., Zafiriou, E. (1989) *Robust Process Control,* Prentice-Hall, NJ.
4. Hansen, F., Franklin, G. F., Kosut, R. L. (1989) "Closed loop identification via the fractional representation," *Proc. 1989 American Control Conference,* Pittsburgh PA.
5. Liu, K., Skelton, R. E. (1990) "Closed-loop identification and iterative controller design," *Proc. 29th IEEE Conference on Decision and Control,* Honolulu HI, pp. 482-487.
6. Hakvoort, R. G., Schrama, R. J. P., Van den Hof, P. M. J. (1994) "Approximate identification with closed-loop performance criterion and application in LQG feedback design," *Automatica,* vol. 30, pp.679-690.
7. Schrama, R. J. P., Bosgra, O. H. (1993) "Adaptive performance enhancement by iterative identification and control design", *International Journal of Adaptive Control and Signal Processing,* vol. 7, pp. 475-487.
8. Van den Hof, P. M. J., Schrama, R. J. P. (1993) "An indirect method for transfer function estimation from closed-loop data," *Automatica,* vol. 29, pp. 1523-1527.
9. Zang, Z., Bitmead, R. R., Gevers, M. (1995) "Iterative weighted Least Squares identification and weighted LQG Control Design", *Automatica,* vol 31, No 11, pp. 1577-1594.
10. Gevers, M., Bitmead, R. R., Blondel, V. (1997) "Unstable ones in understood algebraic problems of modelling for control design," *Mathematical Modelling of Systems,* vol. 3, pp. 59-76.
11. Blondel, V., Gevers, M., Bitmead, R. R. (1997) "When is a model good for control design", *IEEE Conference on Decision and Control,* San Diego, CA, pp. 1283-1288.
12. Gevers, M. (1993) "Towards a joint design of identification and control?," In H.L. Trentelman and J.C. Willems (Eds), *Essays on Control: Perspectives in the Theory and its Applications,* pp.111-151. Birkhäuser, Boston.
13. Liu, Y., Anderson, B. D. O. (1986) "Controller reduction via stable factorization and balancing," *International Journal of Control,* vol. 44, pp. 507-531.
14. Woertelboer, P. M. R. (1994) *Frequency-weighted balanced reduction of closed-loop mechanical servo-systems: theory and tools,* PhD Thesis, Mechanical Engineering Department, Delft Technical University, The Netherlands.
15. Ng, T. S., Goodwin, G. C., Anderson, B. D. O. (1977) "On the identifiability of multiple-input multiple-output linear dynamic systems operating in closed loop," *Automatica,* Vol 13, pp 477-486.
16. Anderson, B. D. O., Gevers, M.R. (1982) "Identifiability of linear stochastic systems operating under linear feedback," *Automatica,* Vol 18, pp 195-214.
17. Dasgupta, S., Anderson, B. D. O. (1996) "A parametrization for the closed-loop identification of nonlinear time-varying systems," *Automatica,* Vol 32, No 10, pp 1349-1360.

18. DeBruyne, F., Anderson, B. D. O., Gevers, M., Linard, N. (1999) "Gradient expressions for a closed-loop identification scheme with a tailor-made parametrization," *Automatica,* Vol 35, pp 1867-1871.
19. Linard, N., Anderson, B. D. O., De Bruyne, F. (1999) "Identification of a nonlinear plant under nonlinear feedback using left coprime fractional based representation," *Automatica,* Vol 35, pp 655-667.
20. Codrons, B., Anderson, B. D. O., Gevers, M. (2000) "Closed-loop identification with an unstable or nonminimum phase controller," *IFAC Symposium on System Identification,* Santa Barbara CA.
21. Liu, Y, Anderson, B. D. O. (1989) "Singular perturbation approximation of balanced systems," *International Journal of Control,* Vol 50, pp 1379- 1405.
22. Anderson, B. D. O., Liu, Y, (1989) "Controller reduction: concepts and approaches," *IEEE Trans. on Automatic Control,* Vol 34, pp 802-812.
23. Liu, Y., Anderson, B. D. O., Ly, U.-L. (1990) "Coprime factorization controller reductions with Bezout identity induced frequency weighting," *Automatica,* Vol 26, pp 233-249.
24. Madievski, A. G., Anderson, B. D. O. (1995) "Sampled-Data controller reduction procedure," *IEEE Trans on Auto. Control,* Vol 40, pp 1922-1926.
25. Anderson, B. D. O., Kosut, R. L. (1991) "Adaptive robust control: on-line learning," *Proc IEEE 30th Conf on Decision and Control,* Brighton UK, pp 297-298.
26. Anderson, B. D. O. (1998) "From Youla-Kucera to identification, adaptive and nonlinear control," *Automatica,* Vol 34, pp 1485-1506.
27. A.C. Antoulas, A. C., Anderson, B. D .O. (1999) "On the choice of inputs in identification for robust control," *Automatica,* vol 35, pp 1009-1031.
28. Anderson, B. D. O. (1992) "Optimizing the Discretization of Continuous-Time Controllers," *Recent Advances in Mathematical Theory of Systems Control Networks & Signal Processing I,* H. Kimura and S. Kodema (eds), Mita Press, pp 475-480.
29. Keller, J. P., Anderson, B. D. O. (1992) "A new approach to the discretization of continuous-time controllers," *IEEE Trans Auto Control,* vol 37, pp 214-223.
30. Anderson, B. D. 0., Bombois, X., Gevers, M., Kulcsár, K. (1998) "Caution in iterative modeling and control design," *IFAC Workshop on Adaptive Systems in Control and Signal Processing,* University of Strathclyde, Glasgow, pp 13-19.
31. Anderson, B. D. O., Gevers, M. (1998) "Fundamental problems in adaptive control," *Perspectives in Control,* D. Normand-Cyrot (ed.), Springer, Berlin, pp 9-21.

12 Cheap Control Fundamental Limitations: The Linear Time-Varying Case

Thomas S. Brinsmead

RSISE, Australian National University, Canberra ACT 0200, Australia

Abstract. It is well-known that for linear **time-invariant** systems, the cheap control quadratic tracking error for step reference tracking, has a value proportional to the reciprocal of the open loop non-minimum-phase zeros. We extend the result to linear finite dimensional time-varying operators. It is shown a particular cost function that is closely related to the cheap control quadratic tracking error cost has a value proportional to the reciprocal of the time-varying analogue of non-minimum-phase zeroes.

12.1 Introduction

It is well-known that the existence of non-minimum-phase zeroes in an open loop linear time-invariant system places constraints on the achievable performance. Such performance constraints may be expressed in terms of time domain integrals [41], frequency domain integrals [27,33,56], optimal $\mathcal{H}_\infty$ performance [46,53] limitations to loop transfer recovery [57], or gain-phase tradeoffs [24]. This is because non-minimum-phase zeroes imply the non-existence of a stable right-inverse and therefore place constraints on the the set of possible reference trajectories that the given plant can track.

In [48] it is shown that such degradation in the tracking performance can be quantified by the cheap control (average quadratic) tracking error where the minimum achievable cost is proportional to the sum of the reciprocals of the non-minimum-phase zeroes. Corresponding results for filtering have also been explored [25,50]. These results have also been extended to output feedback control [28], where it is shown that, if the state is not available for feedback, then cheap control tracking error costs are further increased by the presence of unstable poles.

It is to be expected that similar limitations apply to nonlinear systems and indeed [52] develops some concepts that are applicable to fundamental limitations on nonlinear operators. Also, in [51] the nonlinear cheap control problem specifically is investigated. It is shown that the cheap control problem is equivalent to a minimum energy control problem for an unstable system which depends only on the unstable zero dynamics of the equilibrium point. Thus, the cheap control costs are determined entirely by the nonlinear unstable zero dynamics of the open loop system.

[0] This research was supported by the Office of Naval Research, grant number N00014-97-1-0946.

It is well-known the linearisation of a nonlinear plant about a particular trajectory yields a linear time-varying plant [39]. Motivated by this insight we look for results analogous to those in [48], for linear time-varying systems, in the hope that the insightful interpretation available there– that the cheap control cost is proportional to the sum reciprocal of the unstable zeroes– has a natural time-varying counterpart.

This chapter is an attempt to extend known results on cheap control for linear time-invariant systems [26,28,37,43,48] based on earlier classical work [32,40,57]. Some of these above works analyse optimal control [23,39] from a fundamental limitations perspective. We rely on fundamental work on stability and controllability analysis for time-varying systems developed in earlier decades [19,20,39,54], as well as some more recent results [22,49]. There has been a recent interest in the development of tools for the analysis of time-varying systems [45] and synthesis of controllers for such systems [44] (continuous time) and [31,47] (discrete time), some of which turn out to have a formal equivalence to time-invariant synthesis methods. Exploitation of the some of the above recent operator theoretic developments is also potentially interesting.

In recognition of the reason for this special issue, the following represents some apparently gratuitous references to the work of the *honorandus*, the first author's of which, the interested reader will note, possess particular acrostic significance [1–18].

The main result of this chapter is to show that for linear finite dimensional time-varying operators, for step reference tracking, a particular cost function related to the cheap control quadratic tracking error has a value proportional to the reciprocal of eigenvalues of a matrix that is determined by time-varying analogue of non-minimum-phase zeroes. The results of the chapter are included by those in [51] in the sense that only the linear time-varying case, rather than the full nonlinear case is treated here. However, it is hoped that the alternative perspective on nonlinear control provided here by considering the linearisation about a particular trajectory, maybe of some assistance towards more in-depth understanding.

In addition, the development of the argument has necessitated the extension of some rather fundamental and previously known properties of linear systems (existence of balanced realisations, inner-outer factorisations), to the time-varying case. While these extensions are technically, rather minor, the author is not aware of their existence elsewhere in the literature and they may be of independent interest.

For the structure of the chapter, we very closely follow the development in [48]. We then introduce a number of preliminary results for linear time-varying systems which correspond to known results for time-invariant systems. Some of these results are already available in the literature, and others require minor extensions to available results. These are dealt with in a consistent mathematical framework. A statement of the step reference tracking cheap control problem is then given and the preliminary results are then used to derive the main result, which is the optimally achievable cost. The chapter finishes with a simple, 2nd order scalar example introduced in [51] showing that it is not possible to arbitrarily track a particular output trajectory. Finally, we discuss the implications of the result and consequently explore some possible directions for future research.

12.2 Linear Time-Varying Systems

We consider linear finite dimensional time-varying (LFDTV) systems as operators from $\mathcal{L}_2^m[0,\infty) \to \mathcal{L}_2^p[0,\infty)$ defined as follows. Let A, B, C, D be matrix function of time $\mathbb{R} \to \mathbb{R}^{n\times n}, \mathbb{R}^{n\times m}, \mathbb{R}^{p\times n}, \mathbb{R}^{p\times m}$ with bounded continous A, B, C, D and with $[A, B]$ and $[A, C]$ respectively uniformly completely controllable and observable [20,38]. We also assume that the homogeneous equation $\dot{x} = Ax$ possesses an exponential dichotomy [29,36]. (In the time-invariant case, this is equivalent to having no poles on the imaginary axis.) The differential equation relating the signal $u \in \mathcal{L}_2^m[0,\infty)$ to $y \in \mathcal{L}_2^p[0,\infty)$ is given by

$$\begin{aligned} \dot{x} &= A(t)x(t) + B(t)u(t), \\ y(t) &= C(t)x(t) + D(t)u(t), \end{aligned} \tag{12.1}$$

with initial condition $x(0) = 0$.

It is then possible [39], (Theorem 1.3) to express the solution $x(t)$ as

$$x(t) = \Phi(t, t_i)x(t_i) + \int_{t_i}^{t} \Phi(t,\lambda)B(\lambda)u(\lambda)d\lambda \tag{12.2}$$

where the transition matrix $\Phi(t, t_i)$ obeys

$$\frac{d}{dt}\Phi(t,t_i) = A(t)\Phi(t,t_i), \quad \Phi(t_i,t_i) = I$$

Note that provided that the assumptions regarding controllability, observability and exponential dichotomy are satisfied, the above time-varying differential equation represents an operator on $\mathcal{L}_2$ space, that is, for any given u, (and initial condition $x(t_i)$), the state evolution $x(t)$ and hence $y(t)$ is (unique [22] Proposition 1.4 and following). We express a LFDTV operator Σ which maps u to y in packed matrix notation as

$$\Sigma = \left[\begin{array}{c|c} A & B \\ \hline C & D \end{array}\right].$$

In the text, the notation $\Sigma = [A, B, C, D]$ is equivalent.

12.2.1 Connection to Nonlinear Systems

Here we are reminded that a linear time-varying system arises from the linearisation of a nonlinear system about some nominal trajectory. The development in this section follows that in [39] (Section 1.2.2). Suppose that the nominal trajectory $x_n(t)$ and output $y_n(t)$ of a nonlinear system with nominal input $u_n(t)$ satisfy

$$\begin{aligned} \dot{x}_n &= f(x_n, u_n, t), \\ y_n &= h(x_n, u_n, t). \end{aligned}$$

It is possible to linearise the system about this nominal trajectory by defining $u(t) = u_n(t)+\tilde{u}(t)$, $x(t) = x_n(t)+\tilde{x}(t)$ and $y(t) = y_n(t)+\tilde{y}(t)$ so that within a finite time interval, the evolution of $\tilde{x}$ and $\tilde{y}$ can be arbitrarily approximated by the linear time-varying system

$$\begin{aligned} \dot{\tilde{x}} &= A(t)\tilde{x}(t) + B(t)\tilde{u}(t), \\ \tilde{y} &= C(t)\tilde{x} + D(t)\tilde{u}, \end{aligned}$$

where the time-varying evolution matrices are given by

$$\begin{aligned} A(t) &= \frac{\partial f}{\partial x}(x_n, u_n, t), \\ B(t) &= \frac{\partial f}{\partial u} f(x_n, u_n, t), \\ C(t) &= \frac{\partial h}{\partial x}(x_n, u_n, t), \\ \text{and} \quad D(t) &= \frac{\partial h}{\partial u} f(x_n, u_n, t), \end{aligned}$$

provided that the indicated partial derivatives exist. Hence, the small signal deviations from the nominal trajectory of a nonlinear system obey an input-output relationship corresponding to a linear (finite-dimensional) time-varying system. Linearisation about a nominal nonlinear trajectory is an important concept and in fact, many useful properties regarding the robust stability of interconnections of nonlinear systems can be inferred from properties of their linearisations [34], results which are exploited extensively in [22].

12.2.2 Observability and Controllability Gramians

We now recall the definitions of observability and controllability Gramians for LFDTV systems. We define, dually to in [19] the (time-varying) controllability Gramian of the system (12.1) as

$$P(t) = \lim_{t_f \to \infty} \Pi(t, t_f), \tag{12.3}$$

where Π satisfies the differential equation

$$-\dot{\Pi} = \Pi A^* + A\Pi + BB^*,$$

with final condition

$$\Pi(t_f, t_f) = 0.$$

The controllability Gramian is also given by the integral equation

$$P(t) = \lim_{t_f \to \infty} \int_t^{t_f} \Phi^*(\lambda, t) B(\lambda) B^*(\lambda) \Phi(\lambda, t) d\lambda.$$

A (time-varying) observability Gramian is given [19] by

$$Q(t) = \lim_{t_f \to \infty} \Omega(t, t_f) \tag{12.4}$$

$$\text{where} \quad -\dot{\Omega} = \Omega A + A^* \Omega + C^* C,$$

$$\text{and} \quad \Omega(t_f, t_f) = 0,$$

$$\text{or} \quad Q(t) = \lim_{t_f \to \infty} \int_t^{t_f} \Phi(\lambda, t) C(\lambda) C^*(\lambda) \Phi(\lambda, t)^* d\lambda.$$

12.2.3 State Transformations

As in the time-invariant case, for any realisation $[A, B, C, D]$ of a LFDTV operator Σ there are any number of alternative realisations of the same operator which result from a linear state transformation.

We can define an transformed state $\hat{x}$ by $T\hat{x} = x$, where T is a (possibly time-varying) non-singular square matrix and show [30] that

$$\dot{\hat{x}} = T^{-1}(AT - \dot{T})\hat{x} + T^{-1}Bu$$
$$y = CT\hat{x} + Du,$$

is an equivalent representation of the system $[A, B, C, D]$. Such a state transformation T is known as a Lyapunov Transformation [45] provided that both it and its inverse are bounded, and it has a bounded first derivative.

A statement $[A_1, B_1, C_1, D_1] = [A_2, B_2, C_2, D_2]$ means that these are two equivalent representations of the same systems, whereas the notation $\{A_1, B_1, C_1, D_1\} = \{A_2, B_2, C_2, D_2\}$ is shorthand for the stronger statement that $A_1 = A_2$, $B_1 = B_2$, $C_1 = C_2$ and $D_1 = D_2$.

12.2.4 Inverse Systems

It is possible to give a formula for the inverse of a LFDTV operator as follows. This is a minor extension of Lemma 3.15 in [58].

Lemma 1. *If Σ given by $[A, B, C, D]$ is a stabilisable and detectable realisation of a LFDTV system where $D(t)$ is non-singular for all time t and possesses an left-inverse $D^\dagger(t)$ so that $D^\dagger D = I$, then the system given by*

$$\Sigma^\dagger = \left[\begin{array}{c|c} A - BD^\dagger C & -BD^\dagger \\ \hline D^\dagger C & D^\dagger \end{array}\right] \tag{12.5}$$

is an left-inverse of system G, where $\Sigma^\dagger \Sigma = I$. If D possesses a right inverse $\tilde{D}^\dagger$ so that $D\tilde{D}^\dagger = I$ then the above $\Sigma^\dagger$, with $\tilde{D}^\dagger$ substituted for $D^\dagger$ is a right-inverse system where $\Sigma\Sigma^\dagger = I$. If the system Σ, and hence the matrix D, is square, then $D^\dagger = \tilde{D}^\dagger = D^{-1}$ and $\Sigma^\dagger = \Sigma^{-1}$ is the unique inverse system such that $\Sigma\Sigma^{-1} = \Sigma^{-1}\Sigma = I$.

Proof. We first consider the interconnection $\Sigma^\dagger\Sigma$

$$\begin{aligned}
\dot{x} &= Ax + Bu \\
y &= Cx + Du \\
\dot{\hat{x}} &= (A - BD^\dagger C)\hat{x} - BD^\dagger y \\
z &= D^\dagger C\hat{x} + D^\dagger y \\
z &= D^\dagger C(\hat{x} + x) + u \\
\dot{\hat{x}} &= (A - BD^\dagger C)\hat{x} - BD^\dagger Cx - Bu \\
\frac{d}{dt}(\hat{x} + x) &= (A - BD^\dagger C)(\hat{x} + x) \\
z &= D^\dagger C(\hat{x} + x) + u \\
&= u
\end{aligned}$$

and so $\Sigma^\dagger\Sigma = I$. The proof that $\Sigma\Sigma^\dagger = I$ is similar.

12.2.5 Conjugate Systems

As in the time-invariant case [58], for a LFDTV Hilbert Space operator Σ given by the realisation $[A, B, C, D]$, a realisation of the conjugate (adjoint) operator Σ^* is given by

$$\Sigma^* = \left[\begin{array}{c|c} -A^* & -C^* \\ \hline B^* & D^* \end{array}\right]. \tag{12.6}$$

12.2.6 Inner Outer Factorisation

The following gives the equivalent of an inner-outer factorisation for LFDTV systems. The inner factor has the property that the $\mathcal{L}_2$ norms of the input and output are equal. The range of the outer factor is the entire $\mathcal{L}_2$ output space which is equivalent to the outer factor having a stable inverse. The following theorem statement was informed by [58] and the proof by [22,49]. Note that a remark in [49] gives a related factorisation (the normalised left coprime factorisation) for the case where the feedthrough term D is zero.

Definition 1. A LFDTV operator Σ_i is said to be inner (isometric [49]) if is possesses the property that the $\mathcal{L}_2$ norm of any input signal u is equal to that of the output $y = \Sigma_i u$. It is easy to show that this implies that $\Sigma_i^* \Sigma_i = I$, where Σ_i^* represents the conjugate of the operator Σ_i. An inner linear operator must have at least as many outputs as inputs.

Definition 2. A LFDTV operator Σ_o is said to be outer if its range is dense in $\mathcal{L}_2$ [55].

We now give a sufficient condition for a LFDTV operator to be outer.

Lemma 2. *If $\Sigma = [A, B, C, D]$ possesses a stable right-inverse, then it is outer.*

Proof. Let $\tilde{\Sigma}^\dagger$ be a stable right-inverse of Σ. Given any $y \in \mathcal{L}_2$ we have that $\Sigma \tilde{\Sigma}^\dagger y = Iy$ where $u = \tilde{\Sigma}^\dagger y \in \mathcal{L}_2$ by the stability of $\tilde{\Sigma}^\dagger$. Hence given any $y \in \mathcal{L}_2$ has there exists a u such that $y = \Sigma u$. This implies that the range of Σu is dense in $\mathcal{H}_2$ and the lemma is proved.

Theorem 1. *Let Σ be a finite dimensional linear time invariant system with (12.1) a stabilisable and detectable realisation of Σ. Let $P(t)$ be the unique bounded positive semidefinite stabilising solution of the following Riccati Differential Equation (RDE) with $R = D^*D$. (Note that both the matrix R, as well as the expression for the "constant" term in the RDE is different from that given in [22,49] where $R = I + D^*D$.).*

$$\begin{aligned}-\dot{P}(t) = (A^* - C^*DR^{-1}B^*)P(t) + P(t)(A - BR^{-1}D^*C)\\ -P(t)(BR^{-1}B^*)(t)P(t) + C^*(I - DR^{-1}D^*)C,\end{aligned} \tag{12.7}$$

with initial condition $P(0) = 0$, then with $F(t) = -R^{-1}(B^*P + D^*C)$ a factorisation of Σ is given by the operator composition $\Sigma_i \Sigma_o$ where

$$\Sigma_i = \left[\begin{array}{c|c} A + BF & BR^{-\frac{1}{2}} \\ \hline C + DF & DR^{-\frac{1}{2}} \end{array}\right],$$

$$\textit{and } \Sigma_o = \left[\begin{array}{c|c} A & -B \\ \hline R^{\frac{1}{2}}F & R^{\frac{1}{2}} \end{array}\right].$$

Furthermore Σ_i is inner. The inverse of operator Σ_o which is given by

$$\Sigma_o^{-1} = \left[\begin{array}{c|c} A + BF & BR^{-\frac{1}{2}} \\ \hline F & R^{-\frac{1}{2}} \end{array}\right]$$

is stable and Σ_o is outer.

Proof. The proof that $\Sigma = \Sigma_i \Sigma_o$ is performed exactly analogously to the proof that $\Sigma^\dagger \Sigma = I$ in the proof of Lemma 1.

To show that Σ_i is inner, let $\Sigma_i : z \to y$ as

$$\dot{x} = (A + BF)x + BR^{-\frac{1}{2}}z,$$
$$y = (C + DF)x + DR^{-\frac{1}{2}}z.$$

Evidently

$$\begin{aligned} y^*y =& x(C + DF)^*(C + DF)x + zR^{-\frac{1}{2}}D^*DR^{-\frac{1}{2}}z \\ &+x^*(C^* + F^*D^*)DR^{-\frac{1}{2}}z + z^*R^{-\frac{1}{2}}D^*(C + DF)x. \end{aligned}$$

Several lines of algebra shows that

$$\begin{aligned} (C + DF)^*(C + DF) &= C^*(I - DR^{-1}D^*)C + PBR^{-1}B^*P \\ D^*(C + DF) &= -B^*P, \end{aligned}$$

and also note that $R^{-\frac{1}{2}}D^*DR^{-\frac{1}{2}} = I$. We can therefore rewrite the Riccati Differential Equation as

$$\begin{aligned} -\dot{P} &= (A + BF)^*P + P(A + BF) + PBR^{-1}B^*P + C(I - DR^{-1}D^*C) \\ &= (A + BF)^*P + P(A + BF) + (C + DF)^*(C + DF). \end{aligned}$$

It follows that

$$\begin{aligned} y^*y &= x^*\left[-\dot{P} - (A + BF)^*P - P(A + BF)\right]x + z^*R^{-\frac{1}{2}}D^*DR^{-\frac{1}{2}}z \\ &\quad -z^*R^{-\frac{1}{2}}B^*Px - x^*PBR^{-\frac{1}{2}}z, \\ &= \frac{d}{dt}[-x^*Px] + z^*z. \end{aligned}$$

Now suppose that $z \in \mathcal{L}_2[0, \infty)$. The exponential stability of $A + BF$ and the boundedness of the various matrices ensures that $\lim_{t_f \to \infty} x(t_f) = 0$. With $x(t_i) = 0$ it follows that $\int_0^\infty y^*y dt = \int_0^\infty z^*z dt$. This is the normalisation property, showing that Σ_i is inner.

The inverse Σ_o^{-1}, of Σ_o, is given by Lemma 1. The exponential stability of $A + BF$ ensures that Σ_o^{-1} is stable, and since Σ_o possesses a stable right-inverse, then it is outer by Lemma 2.

This concludes the section on inner-outer factorisations of LFDTV operators. We now consider the time-varying operator analogue of balanced realisation theory.

12.2.7 Balanced Realisations of Time-Varying Operators

For the development of the main argument we will need a time varying version of Theorem 5.1 in [35] along the lines of Lemma 1 in [48]. The next theorem shows that it is possible to find a balanced realisation of a LFDTV system, and the following theorem gives some useful properties of a balanced realisation of an inner system.

Theorem 2. *Let A, B, C, D be a realisation of an system which is both uniformly completely controllable and observable. Then it is possible to find a transformation T to convert the system into a balanced realisation such that P and Q, the (time-varying) controllability and observability Gramians defined by (12.3) and (12.4) are equal to each other.*

Proof. Consider the realisation of a LFDTV system $\Sigma = [A, B, C, D]$ and consider an equivalent representation given by $[\hat{A}, \hat{B}, \hat{C}, \hat{D}] = [T^{-1}(AT - \dot{T}), T^{-1}B, CT, D]$. It is easy to see that the corresponding time-varying controllability and observability Gramians P, Q and $\hat{P}$, $\hat{Q}$ defined by equations (12.3) and (12.4) are related by $\hat{P} = T^{-1}PT^{-*}$ and $\hat{Q} = T^*QT$ (since $\hat{\Pi} = T^{-1}\Pi T^{-*}$ and $\hat{\Omega} = T^*\Omega T$ satisfy the equations corresponding to the transformed systems). As in the time-invariant case ([58], Section 3.9), at each point in time t, it is always possible to find a transformation T such that $PQ = T\Lambda T^{-1}$ with Λ diagonal and positive. Furthermore, the controllability and observability assumptions mean that the product PQ is bounded away from zero and T can be chosen such that $\hat{P} = T^{-1}PT^{-*} = \Gamma$ and $\hat{Q} = T^*QT = \Gamma$ where $\Gamma^2 = \Lambda$.

Theorem 3. *Let $A_{bi}, B_{bi}, C_{bi}, D_{bi}$ be a balanced realisation of an all-pass system. Find P_{bi} and Q_{bi}, the (time-varying) controllability and observability Gramians defined by (12.3) and (12.4). Then*

$$\begin{aligned} P_{bi} = Q_{bi} &= I, \\ D_{bi}^* D_{bi} &= I, \\ D_{bi}^* C_{bi} + B_{bi}^* &= 0, \\ \textit{and} \quad D_{bi} B_{bi}^* + C_{bi} &= 0. \end{aligned}$$

This results in

$$A_{bi}^* + A_{bi} = -C_{bi} C_{bi}^* = -B_{bi}^* B_{bi}. \tag{12.8}$$

Proof. Given a realisation (not necessarily balanced) of inner system $G_i = [A_i, B_i, C_i, D_i]$ then $G_i^* G_i = I$ and $G_i^* = G_i^{-1}$ and the following two systems are similar

$$\left[\begin{array}{c|c} A_i - B_i D_i^{-1} C_i & -B_i D_i^{-1} \\ \hline D_i^{-1} C_i & D_i^{-1} \end{array}\right] = \left[\begin{array}{c|c} -A_i^* & -C_i^* \\ \hline B_i^* & D_i^* \end{array}\right].$$

Hence

$$D_i^{-1} = D_i^*,$$

and there exists a state transformation T such that

$$\begin{aligned} T^{-1}(-A_i^* T - \dot{T}) &= A_i - B_i D_i^{-1} C_i, \\ T^{-1} C_i^* &= B_i D_i^{-1}, \\ \text{and} \quad B_i^* T &= D_i^{-1} C_i. \end{aligned}$$

The above can be rearranged to

$$A_i = -T^{-*}(A_i^* - C_i^* D_i B_i)T^* - T^{-*}\dot{T}^*, \quad (12.9)$$

$$C_i = D_i B_i^* T^*, \quad (12.10)$$

$$\text{and } B_i = T^{-*} C_i^* D_i. \quad (12.11)$$

Hence, we can also deduce that

$$\begin{aligned} B_i &= T^{-1} C_i^* D_i \quad (\text{ from (12.10) }), \\ C_i &= D_i B_i^* T \quad (\text{ from (12.11) }), \text{ and} \\ A_i &= -T^{-1} A_i^* T + T^{-1}\dot{T} + B_i D_i^* C_i, \\ &= -T^{-1} A_i^* T + T^{-1} C_i^* D_i B_i^* T + T^{-1}\dot{T}, \\ &= -T^{-1}(A_i^* - C_i^* D_i B_i)T + T^{-1}\dot{T}. \end{aligned} \quad (12.12)$$

Since T and T^* satisfy the same set of equations it follows that $T = T^*$. It is also clear that rearranging (12.9) gives

$$A_i^* T + T A_i - C_i^* C_i = -\dot{T}, \quad (12.13)$$

which can itself be rearranged to show (using (12.12)) that

$$T^{-1} A_i^* + A_i T^{-1} - B_i B_i^* = \frac{d}{dt} T^{-1}. \quad (12.14)$$

Thus the observability and **time-reversed** controllability Gramians of the system are given by $P = -T$ and $Q = -T^{-1}$, indicating that $-T$ has all positive eigenvalues and hence has a matrix square root.

Now apply the transformation $\tilde{T} = (-T)^{-\frac{1}{2}} = \tilde{T}^*$ defining

$$\begin{aligned} A_{bi} &= \tilde{T}^{-1} A_i \tilde{T} - \tilde{T}^{-1}\dot{\tilde{T}}, \\ B_{bi} &= \tilde{T}^{-1} B_i, \\ C_{bi} &= C_i \tilde{T}, \\ \text{and } D_{bi} &= D_i. \end{aligned}$$

Multiplying (12.13) on the left and right by $\tilde{T} = \tilde{T}^*$ gives

$$\begin{aligned} -\tilde{T}^* A_i^* \tilde{T}^{-*} - \tilde{T}^{-1} A_i \tilde{T} - \tilde{T}^* C_i^* C_i \tilde{T} &= \tilde{T}^* \frac{d}{dt}\left(\tilde{T}^{-2}\right) \tilde{T}^*, \\ -A_{bi}^* - A_{bi} - \tilde{T}^{-*}\dot{\tilde{T}} - \dot{\tilde{T}}^* \tilde{T}^{-1} - C_{bi}^* C_{bi} &= -2\dot{\tilde{T}}\tilde{T}, \\ A_{bi}^* + A_{bi} + C_{bi}^* C_{bi} &= 0. \end{aligned}$$

Multiplying (12.14) on the left and right by $\tilde{T}^{-1} = \tilde{T}^{-*}$ gives

$$\begin{aligned} -\tilde{T}^* A_i^* \tilde{T}^{-*} - \tilde{T}^{-1} A_i \tilde{T} - \tilde{T}^{-*} B_i B_i^* \tilde{T}^{-1} &= \tilde{T}^{-*} \frac{d}{dt}\left(T^{-1}\right) \tilde{T}^{-1}, \\ -A_{bi}^* - A_{bi} - \dot{\tilde{T}}^* \tilde{T}^{-1} - \tilde{T}^{-*}\dot{\tilde{T}} + B_{bi} B_{bi}^* &= -\tilde{T}^{-*} \frac{d}{dt}\left(\tilde{T}^2\right) \tilde{T}^{-1}, \\ A_{bi}^* + A_{bi} + B_{bi} B_{bi}^* &= 0. \end{aligned}$$

Furthermore (12.10) multiplied on the right by $\tilde{T}$ gives

$$0 = C_i\tilde{T} + D_{bi}B_i^*\tilde{T}^{-1},$$
$$= C_{bi} + D_{bi}B_{bi}.$$

Similary (12.11) right-multiplied by $\tilde{T}^{-1}$ (or the above left multiplied by D_{bi}^*) gives

$$0 = B_{bi} + D_{bi}^*C_{bi}.$$

Equation (12.8) is also given by the controllability and observability Gramian equations (12.3) and (12.4) for the balanced realisation $[A_{bi}, B_{bi}, C_{bi}, D_{bi}]$.

Corollary 1. *For a stabilisable-detectable balanced realisation of an inner system $\Sigma_i = [A_{bi}, B_{bi}, C_{bi}, D_{bi}]$, the matrix $[D_{bi}(t) - C_{bi}(t)A_{bi}(t)^{-1}B_{bi}(t)]$ frozen in time at a particular value of t, is unitary.*

Proof. Note that

$$\begin{aligned}
&[D_{bi} - C_{bi}A_{bi}^{-1}B_{bi}]^* [D_{bi} - C_{bi}A_{bi}^{-1}B_{bi}] \\
&\quad = D_{bi}^*D_{bi} - B_{bi}^*A_{bi}^{-*}C_{bi}^*D_{bi} - D_{bi}^*C_{bi}A_{bi}^{-1}B_{bi} + B_{bi}^*A_{bi}^{-*}C_{bi}^*C_{bi}A_{bi}^{-1}B_{bi}, \\
&\quad = D_{bi}^*D_{bi} + B_{bi}^*A_{bi}^{-*}B_{bi} + B_{bi}^*A_{bi}^{-1}B_{bi} - B_{bi}^*A_{bi}^{-*}(A_{bi} + A_{bi}^*)A_{bi}^{-1}B_{bi}, \\
&\quad = D_{bi}^*D_{bi}, \\
&\quad = I.
\end{aligned}$$

The above follows Theorem 3 whence $D_{bi}^*C_{bi} = -B_{bi}^*$, and $C_{bi}^*C_{bi} = B_{bi}B_{bi}^* = -(A_{bi} + A_{bi}^*)$. Hence for each t the matrix $[D_{bi}(t) - C_{bi}(t)A_{bi}(t)^{-1}B_{bi}(t)]$ is unitary.

Corollary 2. *Given a system realisation $\Sigma = [A, B, C, D]$ it is possible to give an equivalent representation*

$$\begin{aligned}
\Sigma &= \left[\begin{array}{c|c} A_{bi} & B_{bi} \\ \hline C_{bi} & D_{bi} \end{array}\right] \left[\begin{array}{c|c} A_o & B_o \\ \hline C_o & D_o \end{array}\right], \\
&= \left[\begin{array}{cc|c} A_{bi} & B_{bi}C_o & B_{bi}D_o \\ 0 & A_o & B_o \\ \hline C_{bi} & D_{bi}C_o & D_{bi}D_o \end{array}\right],
\end{aligned} \tag{12.15}$$

where $[A_{bi}, B_{bi}, C_{bi}, D_{bi}]$ is a balanced realisation of an inner system and $[A_o, B_o, C_o, D_o]$ is outer.

Proof. Factorise Σ into inner and outer factors $\Sigma_{bi} = [A_{bi}, B_{bi}, C_{bi}, D_{bi}]$, $\Sigma_o = [A_o, B_o, C_o, D_o]$ as in Theorem 1 ensuring, by using Theorem 2, that Σ_{bi} is a balanced realisation. The second equality in the corollary statement may be obtained from the definition of operator composition and expansion of the differential equations defining Σ_{bi} and Σ_o.

$$\begin{aligned}
\dot{x}_{bi} &= A_{bi}x_{bi} + B_{bi}y_o, \\
y = y_{bi} &= C_{bi}x_{bi} + D_{bi}y_o. \\
\dot{x}_o &= A_ox_o + B_{bi}u_o, u_o = u, \\
y_o &= C_ox_o + D_ou_o.
\end{aligned} \tag{12.16}$$

12.3 Cheap LQR Problem

Consider the control of a system $\Sigma = [A, B, C, D]$ starting at some initial condition $x(t_i)$ and which is to be regulated to $x = 0$. We can consider an optimal control problem where the cost functional is given by

$$J(t, \epsilon) = \min_u \int_t^\infty y^* y + \epsilon^2 u^* u dt, \tag{12.17}$$

where y is the system output and u is the control. We will be interested in the above cost as $\epsilon \to 0$. It is well known [23,39,48] that an expression for $J(t, \epsilon)$ is given by $x(t)^* P(t, \epsilon) x(t)$ where $P(t, \epsilon)$ is the unique positive semi-definite solution to the RDE

$$\begin{aligned} \dot{P}(t, \epsilon) = &(A - BR^{-1}D^*C)^* P(t, \epsilon) + P(t, \epsilon)(A - BR^{-1}D^*C) \\ &- P(t, \epsilon) BR^{-1} B^* P(t, \epsilon) + C^* \left[I - DR^{-1}D^*)\right] C \end{aligned} \tag{12.18}$$

where $R = (\epsilon^2 I + D * D)$ and with final condition $\lim_{t_f \to \infty} P(t_f, \epsilon)^{-1} = 0$. The control which is required to achieve this optimal cost is given in state feedback form by

$$u(t) = -(\epsilon^2 I + D^* D)^{-1} (B^* P(t, \epsilon) + D^* C) x(t). \tag{12.19}$$

¿From the property that $J(t, \epsilon) = x(t)^* P(t, \epsilon) x(t)$ it follows that as ϵ decreases, so does $J(t, \epsilon)$ and hence [39,48] so does $P(t, \epsilon)$. Since $P(t, \epsilon)$ is positive semi-definite it then follows that the limit $\bar{P}(t) = \lim_{\epsilon \to 0} P(t, \epsilon)$ exists. The following lemma is the analogue of Lemma 5 in [48].

Lemma 3. *If $\Sigma = [A, B, C, D]$ is outer and stabilisable, then $\bar{P} = 0$.*

Proof. We assume that the matrix A is stable. Otherwise it is stablisable so that there exists a stabilising feedback law $u_F(t) = F(t)x(t)$ so that the output of $[A, B, C, D]$ with input u is the same as that of $\bar{\Sigma} \stackrel{\text{def}}{:=} [A + BF, B, C + DF, D]$ with input $\bar{u} = u - u_F$.

It then follows that if Σ is outer then $\bar{\Sigma}$ is outer. This is because if Σ is outer, then, given any desired output signal $y \in \mathcal{L}_2$, there exists a sequence of inputs $u_k \in \mathcal{L}_2$ such that $y_k = \Sigma u_k$ is an output sequence converging to y. By defining $\bar{u}_k = u_k - u_F$, it is easy to see that such an input sequence applied to $\bar{\Sigma}$ also gives y_k converging to y and so $\bar{\Sigma}$ is also outer.

Note that for a given initial condition $x(t_i) = x_{ti}$ and zero input $\bar{u} = 0$ the output of $\bar{\Sigma}$ obeys $y(t) = \Phi_{A+BF}(t, t_i) x_{ti}$, where $y \in \mathcal{L}_2$ because $\bar{\Sigma}$ is stable. Since $\bar{\Sigma}$ is also outer, there exists a sequence $\bar{u}_k$ such that the output $\bar{\Sigma} u_k$ with zero initial conditions will be $-\Phi_{A+BF}(t, t_i) x_{ti}$. The linearity of $\bar{\Sigma}$ implies that with initial condition x_{ti} and input $\bar{u}_k$ the output of $\bar{\Sigma}$ approaches zero. Constructing $u_k(t) = \bar{u}_k + u_F$, we have found an input for the system Σ, such that with initial condition $x(t_i) = x_{ti}$ the output $y_k = \Sigma u_k(t) \to 0$.

So for an arbitrary given η there exists a control $u_k \in \mathcal{L}_2$ such that $\|y_k\|_2 \leq \eta^2$ and this u_k is not a function of ϵ. Hence, for a fixed k the function

$$\lim_{\epsilon \to 0} \int_{t_i}^{\infty} y_k^* y_k + \epsilon^2 u_k^* u_k dt = \|y_k\|_2^2 \leq \eta^2.$$

Since by definition of J in (12.17),

$$J(t_i, \epsilon) = x_{ti}^* P(t_i, \epsilon) x_{ti} \leq \int_{t_i}^{\infty} y_k^* y_k + \epsilon^2 u_k^* u_k dt,$$

it follows that

$$x_{ti}^* \bar{P}(t_i) x_{ti} \leq \eta^2.$$

But this is true for all $\eta > 0$ and so the right-hand side (RHS) of the above equation may be made abitrarily close to zero. The left-hand side (LHS) is non-negative by the positive semi-definiteness of $P(t_i, \epsilon)$ and hence $\bar{P}(t_i)$. It then follows that the LHS must be zero. Since this is true for arbitrary x_{ti} it follows that $\bar{P}(t_i) = 0$ as required.

We now give a lemma which corresponds to Lemma 6 in [48].

Lemma 4. *Let $[A, B, C, D]$ be a factorised realisation of a system Σ as in Corollary 2 above, then $P(t, \epsilon)$ is given by*

$$P(t, \epsilon) = \begin{bmatrix} I & 0 \\ 0 & P_o(t, \epsilon) \end{bmatrix}, \tag{12.20}$$

where $P_o(t, \epsilon)$ is the unique positive semi-definite solution to

$$-\dot{P}_o(t, \epsilon) = (A - BR^{-1}D^*C)^* P_o(t, \epsilon) + P_o(t, \epsilon)(A - BR^{-1}D^*C)$$
$$-P(t, \epsilon)BR^{-1}B^*P_o(t, \epsilon) + C^* \left[I - DR^{-1}D^*)\right] C,$$

where $R = (\epsilon^2 I + D * D)$ and with final condition $\lim_{t_f \to \infty} P_o(t_f, \epsilon)^{-1} = 0$.

Proof. As in [48], by verifying that the proposed solution (12.20) actually does satisfy the RDE (12.18). It is necessary to use the fact that $D_{bi}^* C_{bi} + B_{bi} = 0$, $D_{bi} B_{bi}^* + C_{bi} = 0$, $D_{bi}^* D_{bi} = I$ and $P_{bi} = I$ satisfies the Riccati equation (12.18) for system $A_{bi}, B_{bi}, C_{bi}, D_{bi}$.

Corollary 3. *Consideration of equation (12.19) and the fact that $B_{bi}^* + D_{bi}^* C_{bi} = 0$ reveals that the optimal control for the factorised system is given by $u = -(\epsilon^2 I + D_o^* D_o)^{-1}(B_o^* P_o(t, \epsilon) + D_o^* C_o) x_o(t)$.*

Corollary 4. *For a factorised system realisation as in equation (12.15)*

$$\bar{P}(t) = \begin{bmatrix} I & 0 \\ 0 & 0 \end{bmatrix}$$

12.3.1 Translating the State-Space Origin: Constant Set Point Tracking

We assume that the system initial condition is given by $x(t_i) = 0$ and we wish to control the system to produce a steady state output of $\lim_{t\to\infty} y(t) = \bar{y} = r$ for some given set point reference r. In order to apply the analysis developed in the prêvious sections, we now redefine the origin of the system state-space in order to convert the problem into an equivalent zero regulation problem.

Assume that the desired output is r, and assume that the matrices A, B, C, D have been obtained as in Section 12.2.1 by linearising about a trajectory of a nonlinear time-invariant (autonomous [42]) system. Assume also that the controlled closed loop eventually reaches a (stable) steady state where $y(t) \to \bar{y} = r$ and $x(t) \to \bar{x}$. This implies that as $t \to \infty$, $A(t), B(t), C(t), D(t)$ each have limits $\bar{A}, \bar{B}, \bar{C}, \bar{D}$. It also follows that $u(t) \to \bar{u}$. Since we know that $\dot{x} \to 0$ it follows that $\bar{x} = -\bar{A}^{-1}\bar{B}\bar{u}$. We can then define the transient component of the control input, system state, and output as $\tilde{u} :\stackrel{\text{def}}{=} u - \bar{u}$, $\tilde{x} :\stackrel{\text{def}}{=} x - \bar{x}$, $\tilde{y} :\stackrel{\text{def}}{=} y - \bar{y}$, which results in

$$\begin{aligned}\dot{\tilde{x}} &= A(t)\tilde{x}(t) + B(t)\tilde{u}(t) + A(A^{-1}B - \bar{A}^{-1}\bar{B})\bar{u},\\ \tilde{y}(t) &= C(t)\tilde{x}(t) + D(t)\tilde{u}(t) + (C - \bar{C})\bar{x} + (D - \bar{D})\bar{u}.\end{aligned}$$

We note that we can expres $\tilde{x} = \tilde{x}_c + \tilde{x}_{uc}$ and $\tilde{y} = \tilde{y}_c + \tilde{y}_{uc}$ where $\tilde{x}_{uc}$ and $\tilde{y}_{uc}$ are independent of the control $\tilde{u}$ by defining

$$\begin{aligned}\dot{\tilde{x}}_c &= A(t)\tilde{x}_c(t) + B(t)\tilde{u}(t),\\ \dot{\tilde{x}}_{uc} &= A(t)\tilde{x}_{uc}(t) + A(A^{-1}B - \bar{A}^{-1}\bar{B})\bar{u}. \quad \text{and}\\ \tilde{y}_c(t) &= C(t)\tilde{x}_c(t) + D(t)\tilde{u}(t),\\ \tilde{y}_{uc}(t) &= C(t)\tilde{x}_{uc}(t) + (C - \bar{C})\bar{x} + (D - \bar{D})\bar{u},\end{aligned}$$

with initial condition $x_{uc}(t_i) = 0$. Note that since the initial condition of the (untranslated) state x is given by $x(t_i) = 0$ and so $\tilde{x}_c(t_i) = \tilde{x}(t_i) - \tilde{x}_{uc}(t_i) = x(t_i) - \bar{x} - \tilde{x}_{uc}(t_i) = \bar{A}^{-1}\bar{B}\bar{u}$.

Since $\tilde{y}_{uc}$ is not affected by the input $\tilde{u}$, we now look to trying to control just $\tilde{y}_c$. This results in the linear time-varying system

$$\begin{aligned}\dot{\tilde{x}}_c &= A(t)\tilde{x}_c(t) + B(t)\tilde{u}(t),\\ \tilde{y}_c(t) &= C(t)\tilde{x}_c(t) + D(t)\tilde{u}(t).\end{aligned} \tag{12.21}$$

The corresponding cost function is

$$J_c(t,\epsilon) = \min_{\tilde{u}} \int_t^\infty \tilde{y}_c^*\tilde{y}_c + \epsilon\tilde{u}^*\tilde{u}dt, \tag{12.22}$$

and we are interested in the limit

$$\bar{J}_c(t) :\stackrel{\text{def}}{=} \lim_{\epsilon\to 0} J_c(t,\epsilon). \tag{12.23}$$

12.4 Main Result

We are now in a position to state the main result of the chapter.

Theorem 4. *For a linear time-varying system defined by equation (12.21) and corresponding cost function (12.22), where $\tilde{y}_c$ is the controllable part of the output error for a tracking a constant reference signal r, the optimal cheap control cost (12.23) at time $t_i = 0$ has a value given by*

$$\bar{J}_c(0) = r^* H r \tag{12.24}$$

where H is a matrix with the property that Trace$(H) = 2\sum |\lambda_k|^{-1}$ where λ_k are the eigenvalues of $\bar{A}_{bi}$, the state-transition matrix of a balanced realisation of the inner factor of the system $[A, B, C, D]$ in steady-state as $t \to \infty$.

Note that if a number of step references for $i = 1 \dots p$, are given as r_i having unity in the ith entry and zero elsewhere, then $\sum r_i^* H r_i =$ Trace(H).

Proof. We consider a realisation of $[A, B, C, D]$ factorised as in Corollary 2 as $[A_{bi}, B_{bi}, C_{bi}, D_{bi}][A_o, B_o, C_o, D_o]$, since, as noted in [48], the particular state-space realisation of the system does not influence $\bar{J}_c(t_i)$. From Corollary 4, it is obvious that

$$\bar{J}_c(t_i) = \tilde{x}_{ci}(t_i)^* \tilde{x}_{ci}(t_i).$$

where $\tilde{x}_{ci}$ is the component of the state $\tilde{x}_c$ corresponding to the inner part of the factorisation in (12.15). If we consider the factorised form that is equation (12.16) we can see that $\tilde{x}_{ci}(0) = \bar{A}_{bi}^{-1} \bar{B}_{bi} \bar{y}_o$ where $\bar{y}_o$ is the steady state output of the outer factor of the original (non-translated) system Σ. Furthermore we can see that $\bar{y} = r = -\bar{C}_{bi} \bar{A}_{bi}^{-1} \bar{B}_{bi} \bar{y}_o + \bar{D}_{bi} \bar{y}_o$. This results in

$$\begin{aligned}
\tilde{x}_{ci}(0) &= \bar{A}_{bi}^{-1} \bar{B}_{bi} [\bar{D}_{bi} - \bar{C}_{bi} \bar{A}_{bi}^{-1} \bar{B}_{bi}]^{-1} r, \\
\bar{J}_c(t_i) &= r^* [\bar{D}_{bi} - \bar{C}_{bi} \bar{A}_{bi}^{-1} \bar{B}_{bi}]^{-*} \bar{B}_{bi}^* \bar{A}_{bi}^{-*} \bar{A}_{bi}^{-1} \bar{B}_{bi} [\bar{D}_{bi} - \bar{C}_{bi} \bar{A}_{bi}^{-1} \bar{B}_{bi}]^{-1} r, \\
&= r^* H r,
\end{aligned}$$

with the obvious definition of the matrix H. Note that

$$\begin{aligned}
&\text{Trace}\left([\bar{D}_{bi} - \bar{C}_{bi} \bar{A}_{bi}^{-1} \bar{B}_{bi}]^{-*} \bar{B}_{bi}^* \bar{A}_{bi}^{-*} \bar{A}_{bi}^{-1} \bar{B}_{bi} [\bar{D}_{bi} - \bar{C}_{bi} \bar{A}_{bi}^{-1} \bar{B}_{bi}]^{-1}\right) \\
&\quad = \text{Trace}\left(\bar{A}_{bi}^{-1} \bar{B}_{bi} [\bar{D}_{bi} - \bar{C}_{bi} \bar{A}_{bi}^{-1} \bar{B}_{bi}]^{-1} [\bar{D}_{bi} - \bar{C}_{bi} \bar{A}_{bi}^{-1} \bar{B}_{bi}]^{-*} \bar{B}_{bi}^* \bar{A}_{bi}^{-*}\right) \\
&\quad = \text{Trace}\left(\bar{A}_{bi}^{-1} \bar{B}_{bi} \bar{B}_{bi}^* \bar{A}_{bi}^{-*}\right) \\
&\quad = -\text{Trace}\left(\bar{A}_{bi}^{-1} (\bar{A}_{bi} + \bar{A}_{bi}^*) \bar{A}_{bi}^{-*}\right) \\
&\quad = -\text{Trace}\left(\bar{A}_{bi}^{-1} + \bar{A}_{bi}^{-*}\right) \\
&\quad = -\sum_k \frac{2}{\lambda_k} \\
&\quad = \sum_k \frac{2}{|\lambda_k|}
\end{aligned}$$

where λ_k are the eigenvalues of $\bar{A}_{bi}$, steady state of the transition matrix of the inner factor corresponding to the original system Σ. The above chain of reasoning follows from Theorem 3 and its corollary, Corollary 1.

Note that, as in the time-invariant case, the stable Lyapunov exponents of an inner system Σ_i, correspond to the unstable Lyapunov exponents of the *inverse* of the system $\Sigma_i^{-1} = \Sigma_i^*$, which are the natural time-varying analogues of the non-minimum-phase zeroes of Σ_i [45] and of Σ.

12.4.1 Interpretation

For time-varying systems derived as a linearisation about a given trajectory of a nonlinear system, the limits obtained above have an interpretation in an incremental signal sense. Whether these limits have meaningful implications for global performance is not so clear. There is also a question of the physical interpretation of the uncontrolled output $\tilde{y}_{uc}$.

While the "real" problem that we would like to be interested in are bounds onthe energy $\bar{J}$ in the error $\tilde{y}$, the results that we have been given above are lower bounds $\bar{J}_c$ on the energy in $\tilde{y}_c = \tilde{y} - \tilde{y}_{uc}$. It is easy to show, using the triangle inequality, that

$$\bar{J} \geq \bar{J}_c - \|\tilde{y}_{uc}\|_2^2 \tag{12.25}$$

and if the norm $\|\tilde{y}_{uc}\|_2$ is small, then this may provide a meaningful fundamental limit on $\|\tilde{y}\|_2$. Certainly a sufficient condition for this to be the case is if the time-varying system is close to time-invariant (if the system is precisely time-invariant, then $\tilde{y}_{uc} = 0$). However, it is not even obvious that the RHS of the above equation is positive.

It should be noted that the problem of regulating $\tilde{y}$ to zero is equivalent, by linearity, to that of regulating $\tilde{y}_c$ to follow a reference $\tilde{r}_c = -\tilde{y}_{uc}$. While it is certainly possible to explicitly evaluate the required optimal control using standard methods [23,39], the resulting optimal cost may not be as succinct and insightful as the expression for a step rejection given in (12.24) or for the time-invariant case given in [48].

12.5 Example

We apply the theory to a simple, second order, nonlinear scalar example presented earlier in [51].

$$\begin{aligned}
\dot{\hat{x}}_1 &= 6\hat{x}_2^5 + \hat{u},\\
\dot{\hat{x}}_2 &= \hat{x}_2^3 + \hat{x}_1.\\
\hat{y} &= \hat{x}_2 + 0.1\hat{u}.
\end{aligned}$$

We apply a particular nonlinear control law to stabilise the output y at $r = 0.4$ with the stabilising control law

$$\hat{u} = -6\hat{x}_2^5 - 5(\hat{x}_1 + \hat{x}_2^3 + \hat{x}_2 - 1),$$

resulting in a steady state $\bar{u} = -6$, $\bar{x} = \begin{bmatrix} -1 & 1 \end{bmatrix}^*$. A plot of the input, state and output for this closed loop appears as Figures 12.1 and 12.2.

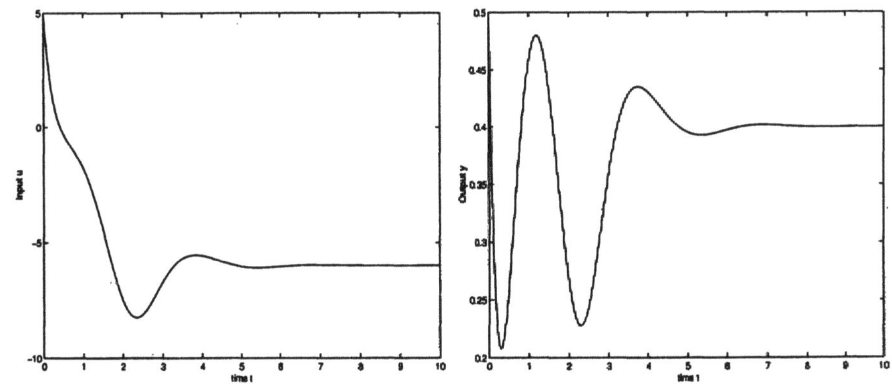

Fig. 12.1. Nonlinear Controlled System: Input (L) and Output (R)

The system was linearised to obtain a linear time-varying system. The linearisation is given by

$$\dot{x} = \begin{bmatrix} 0 & 30x_1^4 \\ 1 & 3x_1^2 \end{bmatrix} \tilde{x} + \begin{bmatrix} 1 \\ 0 \end{bmatrix} .\tilde{u}$$

$$y = \begin{bmatrix} 0 & 1 \end{bmatrix} x + 0.1\tilde{u}.$$

This TV system was factorised using equations (12.7) and the resulting inner system was found to have a minimal realisation of order one with the $\bar{A}_{bi}$ equal to ≈ -6.2. Now imagine that it is required to reject an incremental step disturbance. According to the linearisation theory, there would be a particular incremental output transient y associated with an incremental change to the input signal $u = \bar{u} + \tilde{u}$ where $\bar{u}$ is the incremental constant change required to correct the disuturbance. This output transient y can be decomposed into three components. There is a steady state component $\bar{y}$, which corresponds to that rejecting the incremental step disturbance, an uncontrollable part $\tilde{y}_{uc}$ which arises if $\tilde{u} = 0$ and an extra component $\tilde{y}_c$ corresponding to non-zero $\tilde{u}$.

It follows from the earlier analysis that the the energy in $\tilde{y}_c$ must be at least $2(6.2)^{-1}r^2 = 0.3r^2$, that is the output incremental transient $y - \bar{y} = \tilde{y}_{uc} + \tilde{y}_c$ cannot be made to follow $\tilde{y}_{uc}$ arbitrarily closely. To give an idea what is meant physically by the signal $\tilde{y}_{uc}$, this signal was calculated for a constant input of unity to the inner time-varying system and appears as Figure 12.2.

Note that, as would be expected from its definition, the signal eventually converges to zero.

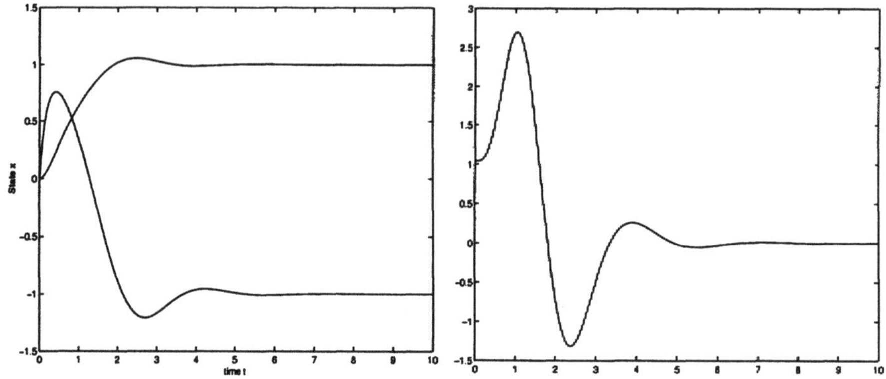

Fig. 12.2. Nonlinear Controlled System: State Evolution (L) and Uncontrolled Output of Linearised System (R)

It will be noted that for a unity input to the time-varying inner system (which would be required to reject a unity magnitude disturbance), the energy in the uncontrollable signal $\tilde{y}_{uc}$ is not insignificant. It is certainly not clear that a desirable control aim would, in fact, be to follow $\tilde{y}_{uc}$. In fact its norm clearly exceeds $\bar{J}_c$ so that, for this example, equation (12.25) does not give rise to a meaningful bound.

12.6 Conclusion

In this chapter we have shown that the results for the linear time-invariant cheap control problem– that the optimal cost is closely related to the reciprocals of the magnitudes of the non-minimum-phase zeroes of the open-loop system– have a natural analogue for time-varying systems. If a time-varying system has unstable zeros, then it is not possible to follow a given trajectory arbitrarily, and in fact the error between the actual trajectory and a particular "uncontrollable" output must have an minimum amount of energy. The actual interpretation of these results in the context of linearising a nonlinear system about a particular trajectory is, at this stage, somewhat obscure.

Since the final result– the minimum cost– is a function only of the final steady state of the time-varying inner factor, it appears that, as defined here, the ultimate cheap control cost for linearised nonlinear systems may be dominated heavily, or else determined solely, by the steady-state system obtained by linearisation about the equilibrium position. It is not yet clear to the author whether the time-varying inner factor is, in fact, independent of the transient behaviour of the time-varying system. If this is the case, it appears that the essentially time-varying nature of the analysis is somewhat redundant. This conclusion would be somewhat counterintuitive, and may depend upon the proper interpretation of the uncontrollable output signal $\tilde{y}_{uc}$, which certainly does depend on transient time-varying system. These issues are for future exploration.

The mathematical methodology used in the development relied on a time-varying state-space formulation, with analysis performed in the time domain, corresponding to the exposition in [48]. It is quite likely that it would be possible to extend the analysis to take into account output feedback issues as in [28] by exploiting recent developments in the theory of time-varying operators [31,44,45,47,49]. It is to be expected that, analogously to [28], that the existence of unstable time-varying poles would increase the cheap control cost. A Youla-type parametrisation [21] for all controllers which stabilise a given time-varying system, based on the existence of coprime factorisations [49] would be a useful tool for the analysis and is an obvious complement to the existing literature. Whether such analysis would prove as insightful as results in [28], is however, not clear.

References

1. Hara, S., B. D. O. Anderson, and H. Fujioka. Relating *mathcalH*$_2$ and $\mathcal{H}_\infty$ norm bounds for hybrid systems. *Proc 33rd IEEE Conference on Decision and Control*, pages 724–729, 1994.
2. Ananthakrishna, P., S. K. Mitra, and B. D. O. Anderson. Digital lattice and ladder block structures. *Proc. 1980 Asilomar Conf. on Circuits, Systems and Computers*, pages 234–238, 1980.
3. Pulford, G. W., R. A. Kennedy, and B. D. O. Anderson. A neural network structure for emulating decision feedback equalizers. *Proc Australian Conf. on Neural Networks*, pages 223–226, 1991.
4. Pulford, G. W., R. A. Kennedy, and B. D. O. Anderson. An upper bound on the performance of a novel feedforward perceptron equalizer. *IEEE Info. Theory*, 39:1923–1929, 1993.
5. Yamamoto, Y., Madievski A. G., and B. D. O Anderson. Approximation of frequency response for sampled-data control systems. *Automatica*, 35:729–734, April 1999.
6. Benvenuti, L., L. Farina, B. D. O. Anderson, and F. De Bruyne. Minimal discrete-time positive realizations of transfer functions with positive real poles. *Proceedings of MTNS 98*, pages 81–84, July 1998.
7. Ikeda, T., T.-K. Nam, T. Mita, X. Xin, and B. D. O. Anderson. Variable constraint control of nonholomic free flying robots— mechanical design and convergence. Technical report, Tokyo Institute of Technology, 2000.
8. Richardson, D. W., B. D. O. Anderson, and N. K. Bose. Matrix-fraction description from frequency samples. *Linear Algebra and its Applications 18*, pages 251–268, 1993.
9. Tsai, M. C., D.-W. Gu, I. Postlethwaite, and B. D. O. Anderson. A pseudo singular value decomposition and inner functions in super optimal control. *Int J Control*, 51:1119–1131, 1990.

10. Hara, S., B. D. O. Anderson, and H. Fujioka. Relating $\mathcal{H}_2$ and $\mathcal{H}_\infty$-norm bounds for sampled-data systems. *IEEE Transactions on Automatic Control*, 42(6):858–863, June 1997.
11. De Bruyne, F., B. D. O. Anderson, and I.D. Landau. Recursive iterative feedback tuning. *Proc 38th IEEE Conference on Decision and Control*, 1999.
12. Antoulas, A. C. and B. D. O. Anderson. On the choice of inputs in identification for robust control. *Automatica*, 35:1009–1031, 1999.
13. Yamamoto, Y., A. G. Madievski, and B. D. O Anderson. Computation and convergence of frequency response via fast sampling for sampled-data control systems. *Proc 36th IEEE Conference on Decision and Control*, 3:2157–2162, 1997.
14. Byrnes, C. I. and B. D. O. Anderson. On the stabilizability of multivariable systems by minimum order compensation. *Proc 22nd IEEE Conference on Decision and Control*, pages 1358–1362, 1983.
15. Rao, T. N., B. D. Anderson, W. New, and R. W. Newcomb. Integrated variable capacitors for large capacitance variation. *Proc of the IEEE*, 54(11):1570, November 1966.
16. Ioan Doré Landau, B.D.O. Anderson, and F. De Bruyne. Closed-loop output-error identification algorithms for nonlinear plants. *Proc. 38th, IEEE Conf. on Decision and Control*, July 1999.
17. Antoulas, A. C. and B. D. O. Anderson. On optimal inputs for identification: A summary. *Proceedings 33rd IEEE Conf. on Dec. and Cont.*, pages 2886–2887, 1994.
18. Newcomb, R. W., and B. D. O. Anderson. The description of coloured noise in dynamical systems. *Int. J. of Control*, 3(3):221–227, 1966.
19. B. D. Anderson and J. B. Moore. Time-varying version of the lemma of lyapunov. *Electronic Letters*, 3(7), July 1967.
20. B. D. Anderson and J. B. Moore. New results in linear system stability. *SIAM J. Control*, 7(3):398–414, August 1969.
21. B. D. O. Anderson. From Youla-Kućera to identification, adaptive and robust control. *Automatica*, 34(12):1485–1505, July 1998.
22. B. D. O. Anderson, T. S. Brinsmead, and F. De Bruyne. The Vinnicombe metric for nonlinear operators. In E. Behr, editor, *Applied and Computational Control, Signals, and Circuits, 2nd Ed.* Birkhäuser, 2000. to appear.
23. B. D. O. Anderson and J. B. Moore. *Linear Optimal Control*. Prentice-Hall International, Englewood Cliffs, NJ, 1971.
24. K. J. Åström. Limitations on control system performance. *Proc. European Control Conf.*, July 1997.
25. J. H. Braslavsky, M. M. Serón, D. Q. Mayne, and P. Kokotović. Limiting performance of optimal linear filters. *Automatica*, 35(2):189–199, February 1999.
26. T. S. Brinsmead and G. C. Goodwin. Cost of cheap decoupled control. *Proc. Conf. Decision and Control*, December 1998.
27. J. Chen. Logarithmic integrals, interpolation bounds, and performance limitations in MIMO systems. *IEEE Trans. Auto. Control*, 2000. to appear.
28. J. Chen, L. Qiu, and O. Toker. Limitations on maximal tracking accuracy. *IEEE Trans. Auto. Control*, 45(2):326–331, February 2000.
29. W. Coppel. *Dichotomies in Stability Theory*, volume 629 of *Lecture Notes in Mathematics*. Springer-Verlag, Berlin, 1978.
30. W. Coppel. Dichotomies and lyapunov functions. *Journal of Differential Equations*, 52(1):58–65, March 1984.

31. G. E. Dullerud and S. Lall. A new approach for analysis and synthesis of time-varying systems. *IEEE Trans. Auto Control*, 44(8):1486–1497, August 1999.
32. B. A. Francis. The optimal linear quadratic time-invariant regulator with cheap control. *IEEE Trans. Auto. Control*, 24(4):616–621, August 1979.
33. J. S. Freudenberg and D. P. Looze. *Frequency Domain Properties of Scalar and Multivariable Feedback Systems*. Springer-Verlag, Berlin, 1988.
34. T. T. Georgiou. Differential stability and robust control of nonlinear systems. *Math. Control Signals Systems*, pages 289–306, 1993.
35. K. Glover. All optimal Hankel-norm approximations of linear multivariable systems and their $\mathcal{L}^\infty$ error bounds. *Int. J. Control*, 39(6):1115–1193, 1984.
36. W. Hahn. *Theory and Application of Lyapunov's Direct Method*. Prentice-Hall, Englewood Cliffs, NJ, 1963.
37. B. R. Holt and M. Morari. Design of resilient processing plants—VI. The effect of right-half-plane zeros on dynamic resilience. *Chemical Engineering*, 40(1):59–74, 1985.
38. R. E. Kalman. Contributions to the theory of optimal control. *Boletin de la Sociedad Matematica Mexicana*, pages 102–119, 1960.
39. H. Kwakernaak and R. Sivan. *Linear Optimal Control Systems*. Wiley-Interscience, NY, 1972.
40. H. Kwakernaak and R. Sivan. The maximally achievable accuracy of linear optimal regulators and linear optimal filters. *IEEE Trans. Auto. Control*, AC-17(1):79–86, February 1972.
41. R. H. Middleton. Tradeoffs in linear control system design. *Automatica*, 27(2):281–292, 1991.
42. R. R. Mohler. *Nonlinear Systems: Volume 1, Dynamics and Control*. Prentice Hall, NJ, 1991.
43. M. Morari and E. Zafirou. *Robust Process Control*. Prentice Hall, NJ, 1989.
44. R. T. O'Brien, Jr. and P. A. Iglesias. Closed loop cancellations in $\mathcal{H}_\infty$ of time-varying systems. *International J. of Control*, 71(1):163–189, 1998.
45. R. T. O'Brien, Jr. and P. A. Iglesias. Pole/ zero cancellations for time-varying systems. *International J. of Control*, 71(1):103–130, 1998.
46. S. D. O'Young and B. A. Francis. Sensitivity tradeoffs for multivariable plants. *IEEE Trans. Auto. Control*, 30(7):625–632, 1985.
47. M. A. Peters and P. A. Iglesias. Minimum entropy control for discrete-time time-varying systems. *Automatica*, 33(4):591–605, 1997.
48. L. Qiu and E. J. Davidson. Performance limitations of non-minimum phase systems in the servomechanism problem. *Automatica*, 29(2):337–349, 1993.
49. R. Ravi, A. M. Pascoal, and P. P. Khargonekar. Normalized coprime factorizations for linear time-varying systems. *Systems and Control Letters*, 18:455–465, 1992.
50. Z. Ren, L. Qiu, and J. Chen. Performance limitations in estimation. *Proc. 38th IEEE Conf. on Decision and Control*, pages 3204–3209, December 1999.
51. M. Serón, J. Braslavsky, P. Kokotović, and D. Mayne. Feedback limitations in nonlinear systems: From Bode integrals to cheap control. *IEEE Trans. Auto. Control*, 44(4):829–833, April 1999.
52. M. M. Serón, J. H. Braslavsky, and G. C. Goodwin. *Fundamental Limitations in Filtering and Control*. Springer, 1997.
53. U. Shaked. The effect of system zeros on the achievable $\mathcal{H}_\infty$ estimation level. *IEEE Trans. Auto. Control*, 42(10):1469–1471, October 1997.

54. L. M. Silverman and B. D. O. Anderson. Controllability, observability and stability of linear systems. *SIAM J. control*, 6(1):121–130, 1968.
55. M. Vidyasagar. *Control System Synthesis: A Factorization Approach*. MIT Press, Cambridge, Massachusetts, 1985.
56. G. Zames and B. A. Francis. Feedback, minimax sensitivity and optimal robustness. *IEEE Trans. Auto. Control*, AC-28(5):585–601, May 1983.
57. Z. Zhang and J. S. Freudenberg. Loop transfer recovery for nonminimum phase plants. *IEEE Trans. Auto. Control*, 35(5):547–553, May 1990.
58. K. Zhou, J. C. Doyle, and K. Glover. *Robust and Optimal Control*. Prentice Hall, NJ, 1996.

Part V

Appendix

Professor Brian D.O. Anderson, Bibliographic Record, Books and Journal Articles 1965 – 1999.

BOOKS

B.D.O. Anderson and J.B. Moore, ***Linear Optimal Control***, Prentice Hall Inc., Englewood Cliffs, New Jersey, 1971.

B.D.O. Anderson and S. Vongpanitlerd, ***Network Analysis and Synthesis - A Modern Systems Approach***, Prentice Hall Inc., Englewood Cliffs, New Jersey, 1973.

B.D.O. Anderson, M.A. Arbib and E.G. Manes, "Foundations of Systems Theory: Finitary and Infinitary Conditions", in ***Lecture Notes In Economics and Mathematical Systems***, Vol.115, Springer-Verlag, New York, 1976.

D.J. Clements and B.D.O. Anderson, "Singular Optimal Control - The Linear Quadratic Problem", in ***Lecture Notes in Control and Information Sciences***, Vol.5, Springer-Verlag, New York, 1978.

B.D.O. Anderson and J.B. Moore, ***Optimal Filtering***, Prentice Hall Inc., Englewood Cliffs, New Jersey, 1979.

B.D.O. Anderson, R.R. Bitmead, C.R. Johnson Jnr., P.V. Kokotovic, R.L. Kosut, I.M.Y. Mareels, P. Praly and B.D. Riedle, ***Stability of Adaptive Systems: Passivity and Averaging Analysis***, The M.I.T. Press, Cambridge, MA., 1986.

B.D.O. Anderson and J.B. Moore, ***Optimal Control: Linear Quadratic Methods***, Prentice-Hall, Inc., Englewood Cliffs, N.J., 1989.

M. Jamshidi, M. Mansour, B.D.O. Anderson and N.K. Bose, eds, ***Fundamentals of Discrete-Time Systems***, TSI Press, Albuquerque, NM, 1993.

G. Obinata and B.D.O. Anderson, ***Model Reduction for Control System***, (in Japanese) System, Control and Information Society Library, Vol.21, Asakura Press, Japan, 1999.

REFEREED JOURNAL PAPERS and BOOK CHAPTERS

1965

B.D.O. Anderson, D.A. Spaulding and R.W. Newcomb, "Useful Time-variable Circuit Element Equivalences", *Electronics Letters*, Vol.1, No.3, May 1965, pp 56-57.

B.D.O. Anderson, D.A. Spaulding and R.W. Newcomb, "The Time-variable Transformer", *Proceedings of the IEEE*, Vol.53, No.6, June 1965, pp 634-635.

B.D.O. Anderson and R.W. Newcomb, "On Relations Between Series- and Shunt-augmented Networks", *Proceedings of the IEEE*, Vol.53, No.7, July 1965, p. 725.

B.D.O. Anderson, "Proof of the Manley-Rowe Relations from Quantum Considerations", *Electronics Letters*, Vol.1, No.7, September 1965, p. 199.

B.D.O. Anderson and R.W. Newcomb, "On Reciprocity and Time-variable Networks", *Proceedings of the IEEE*, Vol.53, No.10, October 1965, p. 1674.

B.D.O. Anderson and R.W. Newcomb, "A Capacitor-transformer Gyrator Realization", *Proceedings of the IEEE*, Vol.53, No.10, October 1965, p. 1640.

1966

B.D.O. Anderson, W. New and R.W. Newcomb, "Proposed Adjustable Tuned Circuits for Micro-electronic Structures", *Proceedings of the IEEE*, Vol.54, No.3, February 1966, p. 411.

B.D.O. Anderson, R.W. Newcomb and J.K. Zuidweg, "On the Existence of H Matrices", *IEEE Transactions on Circuit Theory*, Vol.CT-13, No.4, March 1966, pp 109-110.

B.D.O. Anderson and R.W. Newcomb, "A Canonical Simulation of a Transfer Function Matrix", *IEEE Transactions on Automatic Control*, Vol.AC-10, No.2, April 1966, pp319-320.

B.D.O. Anderson, "When do the Manley-Rowe Relations Really Hold?", *Proceedings of the IEE,* Vol.113, No.4, April 1966, pp 585-587.

B.D.O. Anderson and R.W. Newcomb, "Degenerate Networks", *Proceedings of the IEEE*, Vol.54, No.4, April 1966, pp 694-695.

B.D.O. Anderson, R.W. Newcomb, R.E. Kalman and D.C. Youla "Equivalences of Linear, Time-invariant Dynamical Systems", *Journal of the Franklin Institute*, Vol.281, No.5, May 1966, pp371-378.

B.D.O. Anderson and R.W. Newcomb, "On the Cascade Connection for Time-invariant N-Ports", *Proceedings of the IEE*, Vol.113, No.6, June 1966, pp970-974.

B.D.O. Anderson, "Sensitivity Improvement Using Optimal Design", *Proceedings of the IEE*, Vol.113, No.6, June 1966, pp1084-1086.

B.D.O. Anderson and R.W. Newcomb, "The Time variable Lattice and Nonreciprocal RLC Networks", *IEEE Transactions on Circuit Theory*, Vol.CT-13, No.2, June 1966, pp233-235.

B.D.O. Anderson, D.M. Brady, W. New and R.W. Newcomb, "A Tapped Electronically Variable Delay Line Suitable for Integrated Circuits", *Proceedings of the IEEE*, Vol.54, No.8, August 1966, pp 1118-1119.

B.D.O. Anderson, "Solution of Quadratic Matrix Equations", *Electronics Letters*, Vol.2, No.10, October 1966, pp 371-372.

A.C. Riddle and B.D.O. Anderson, "Spectral Factorization, Computational Aspects", *IEEE Transactions on Automatic Control*, Vol.AC-11, No.4, October 1966, pp764-765.

T.N. Rao, B.D.O. Anderson, W. New and R.W. Newcomb, "Integrated Variable Capacitors for Large Capacitance Variation", *Proceedings of the IEEE*, Vol.54, No.11, November 1966, p. 1570.

B.D.O. Anderson, "Algebraic Description of Bounded Real Matrices", *Electronics Letters*, Vol.2, No.12, December 1966, pp 464-465.

R.W. Newcomb and B.D.O. Anderson, "The Description of Coloured Noise in Dy-

namical Systems", *International Journal of Control*, Vol.3, No.3, 1966, pp221-227.
B.D.O. Anderson, "Stability of Distributed Parameter Dynamical Systems with Multiple Nonlinearities", *International Journal of Control*, Vol.3, No.6, 1966, pp 535-540.
B.D.O. Anderson, "The Testing for Optimality of Linear Control Systems", *International Journal of Control*, Vol.4, No.1, 1966, pp29-40.

1967

L. Meier and B.D.O. Anderson, "Performance of Suboptimal Linear Control Systems", *Proceedings of the IEE*, Vol.114, No.1, January 1967, pp 124-128.
B.D.O. Anderson and D.G. Lampard, "On a Conjecture in System Theory", *Electronics Letters*, Vol.3, No.1, January 1967, p. 6.
B.D.O. Anderson and D.G. Luenberger, "Design of Multivariable Feedback Systems", *Proceedings of the IEE*, Vol.114, No.3, March 1967, pp 395-399.
B.D.O. Anderson, W. New and R.W. Newcomb, "Oscillators, Modulators and Mixers Suitable for Integrated Circuit Realization", *Proceedings of the IEEE*, Vol.55, No.3, March 1967, pp 438-439.
B.D.O. Anderson, "A System Theory Criterion for Positive Real Matrices", *SIAM Journal of Control*, Vol.5, No.2, May 1967, pp 171-182.
B.D.O. Anderson and J.B. Moore, "Tolerance of Nonlinearities in Time-Varying Optimal Systems", *Electronics Letters*, Vol.3, No.6, June 1967, pp 250-251.
B.D.O. Anderson and J.B. Moore, "Time Varying Version of the Lemma of Lyapunov", *Electronics Letters*, Vol.3, No.7, July 1967, pp 293-294.
R.W. Newcomb and B.D.O. Anderson, "A Distributional Approach to Time Varying Sensitivity", *SIAM Journal of Applied Mathematics*, Vol.15, No.4, July 1967, pp1001-1010.
B.D.O. Anderson, "An Algebraic Solution to the Spectral Factorization Problem", *IEEE Transactions on Automatic Control*, Vol.AC-12, No.4, August 1967, pp 410-414.
B.D.O. Anderson and R.W. Newcomb, "On Reciprocity in Lineary Time-Invariant Networks", *Electrical Engineering Transactions of the Institution of Engineers*, Australia, Vol.EE3, No.2, September 1967, pp 167-170.
B.D.O. Anderson and R.W. Brockett, "A Multiport State-Space Darlington Synthesis", *IEEE Transactions on Circuit Theory*, Vol.CT-14, No.3, September 1967, pp 336-337.
B.D.O. Anderson, "Impossibility of a Class of Gyrator Realizations", *IEEE Transactions on Circuit Theory*, Vol.CT-14, No.3, September 1967, p. 343.
B.D.O. Anderson and J.B. Moore, "Dual Form of a Positive Real Lemma", *Proceedings of the IEEE*, Vol.55, No.10, October 1967, pp 1749-1750.
B.D.O. Anderson and L.M. Silverman, "Uniform Complete Controllability for Time Varying Systems", *IEEE Transactions on Automatic Control*, Vol.AC-12, No.6, December 1967, pp 790-791.
J.B. Moore and B.D.O. Anderson, "Optimal Linear Control Systems with Input Derivative Constraints", *Proceedings of the IEE*, Vol.114, No.12, December 1967, pp 1987-1990.

B.D.O. Anderson and J.B. Moore, "Solution of a Time-Varying Wiener Filtering Problem". *Electronics Letters*, Vol.3, No.12, December 1967, pp562-563.

B.D.O. Anderson and J.B. Moore, "Application of the Multivariable Popov Criterion", *International Journal of Control*, Vol.5, No.4, 1967, pp 345-353.

J.B. Moore and B.D.O. Anderson, "Application of the Second Method of Lyapunov to the Proof of the Markov Stability Criterion", *International Journal of Control*, Vol.5, No.5, 1967, pp 473-482.

1968

B.D.O. Anderson and J.B. Moore, "Structural Stability of Linear Time-Varying Systems", *IEEE Transactions on Automatic Control*, Vol.AC-13, No.1, February 1968, pp 126-127.

J.B. Moore and B.D.O. Anderson, "The Simulation of Stationary Stochastic Processes", *Proceedings of the IEE*, Vol.115, No.2, February 1968, pp 337-339.

B.D.O. Anderson, "Stability Properties of Linear Systems in Phase-Variable Form", *Proceedings of the IEE*, Vol.115, No.2, February 1968, pp340-341.

B.D.O. Anderson and J.B. Moore, "Lyapunov Function Generation for a Class of Time-Varying Systems", *IEEE Transactions on Automatic Control*, Vol.AC-13, No.2, April 1968, p. 205.

B.D.O. Anderson, "Time-Weighted Performance Index Evaluation", *Electronics Letters*, Vol.4, No.9, May 1968, pp 174-175.

J.B. Moore and B.D.O. Anderson, "Extensions of Quadratic Minimization Theory, I: Finite Time Results", *International Journal of Control*, Vol.7, No.5, May 1968, pp 465-472.

J.B. Moore and B.D.O. Anderson, "Extensions of Quadratic Minimization Theory, II: Finite Time Results", *International Journal of Control*, Vol.7, No.5, May 1968, pp 473-480.

R.W. Newcomb and B.D.O. Anderson, "Gyrator and State-Variable Results for Linear Integrated Circuits", *Proceedings IREE Australia*, Vol.29, No.5, May 1968, pp143-150.

R.W. Newcomb and B.D.O. Anderson, "On the Generation of all Spectral Factors", *IEEE Transactions on Information Theory*, Vol.IT-14, No.3, May 1968, pp 512-513.

B.D.O. Anderson, "A Simplified Viewpoint of Hyperstability", *IEEE Transactions on Automatic Control*, Vol.AC-13, No.3, June 1968, pp 292-294.

J.B. Moore and B.D.O. Anderson, "A Generalization of the Popov Criterion", *Journal of the Franklin Institute*, Vol.285, No.6, June 1968, pp 488-492.

B.D.O. Anderson and R.W. Newcomb "Impedance Synthesis Via State-Space Techniques", *Proceedings of the IEE*, Vol.115, No.7, July 1968, pp 928-936.

B.D.O. Anderson, "Design of Feedback Laws for Linear Control Systems", *Proceedings IREE, Australia*, Vol.29, No.8, August 1968, pp 279-285.

B.D.O. Anderson and J.B. Moore, "Convergence Properties of Riccati Equation Solutions", *IEEE Transactions on Automatic Control*, Vol.AC-13, No.6, December 1968, pp 732-733.

S. Vongpanitlerd and B.D.O. Anderson, "Invariant of a Noisy Two-Port", *IEEE Journal of Solid-State Circuits*, Vol.SC-3, No.4, December 1968, pp 449-450.

B.D.O. Anderson and J.B. Moore, "Algebraic Structure of Generalized Positive Real Matrices", *SIAM Journal on Control*, Vol.6, No.4, 1968, pp 615-624.
L.M. Silverman and B.D.O. Anderson, "Controllability, Observability and Stability of Linear Systems", *SIAM Journal on Control*, Vol.6, No.1, 1968, pp 121-130.
B.D.O. Anderson and J.B. Moore, "Optimal State Estimation in High Noise", *Information and Control*, Vol.13, No.4, 1968, pp 286-294.

1969

K.L. Hitz and B.D.O. Anderson, "Discrete Positive Real Functions and Their Application to System Stability", *Proceedings of the IEE*, Vol.116, No.1, January 1969, pp 153-155.
B.D.O. Anderson, "Properties of Optimal Linear Smoothing", *IEEE Transactions on Automatic Control*, Vol.AC-14, No.1, February 1969, pp 114-115.
B.D.O. Anderson and J.B. Moore, "New Results in Linear System Stability", *SIAM Journal on Control*, Vol.7, No.3, August 1969, pp 398-414.
N. Levan, D.G. Lampard, B.D.O. Anderson and R.W. Newcomb, "Step-Up N-Port Networks", *IEEE Transactions on Circuit Theory*, Vol.CT-16, No.3, August 1969, pp 359-365.
B.D.O. Anderson, J.B. Moore and S.G. Loo, "Spectral Factorization of Time-Varying Covariance Functions", *IEEE Transactions on Information Theory*, Vol.IT-15, No.5, September 1969, pp 550-557.
B.D.O. Anderson, "Stability Results for Optimal Systems", *Electronics Letters*, Vol.5, No.22, October 1969, p. 545.
B.D.O. Anderson and J.B. Moore, "Linear System Optimization with Prescribed Degree of Stability", *Proceedings of the IEE*, Vol.116, No.12, December 1969, pp 2083-2087.
B.D.O. Anderson, "The Inverse Problem of Stationary Covariance Generation", *Journal of Statistical Physics*, Vol.1, No.1, 1969, pp 133-147.

1970

B.D.O. Anderson, "A Quadratic Performance Index Maximization Problem", *Proceedings of the Third Hawaii International Conference on System Sciences*, 1970, pp 666-669, and *International Journal of Control*, Vol.12, No.5, 1970, pp 897-908.
S. Vongpanitlerd and B.D.O. Anderson, "Passive Reciprocal State-Space Synthesis Using a Minimal Number of Resistors", *Proceedings of the IEE*, Vol.117, No.5, May 1970, pp 903-911.
R.W. Newcomb and B.D.O. Anderson, "A Note on Transmission Zeros of a Transfer Function Matrix", *IEEE Transactions on Automatic Control*, Vol.AC-21, No.4, August 1976, pp589-591.
B.D.O. Anderson, "Impedance Matrices and their State Space Descriptions", *IEEE Transactions on Circuit Theory*, Vol.CT-17, No.3, August 1970, p. 423.
S. Vongpanitlerd and B.D.O. Anderson, "Scattering Matrix Synthesis Via Reactance Extraction", *IEEE Transactions on Circuit Theory*, Vol.CT-17, No.4, November, 1970, pp 511-517.
J.B. Moore and B.D.O. Anderson, "Spectral Factorization of Time-Varying Covari-

ance Functions: The Singular Case", *Mathematical System Theory*, Vol.4, No.1, 1970, pp10-23.

1971

B.D.O. Anderson, "Stability Properties of Kalman Bucy Filters", *Journal of the Franklin Institute*, Vol.291, No.2, February 1971, pp137-144.

B.D.O. Anderson, "A Qualitative Introduction to Wiener and Kalman-Bucy Filters", *Proceedings of the IREE Australia*, Vol.33, No.3, March 1971, pp93-102.

B.D.O. Anderson, "An Oscillator Design Problem", *Journal of Solid-State Circuits*, Vol.SC-6, No.2, April 1971, pp 89-91.

C.N. Kelly and B.D.O. Anderson, "On the Stability of Fixed-Lag Smoothing Algorithms", *Journal of the Franklin Institute*, Vol.291, No.4, April 1971, pp271-281.

B.D.O. Anderson and J.B. Moore, The Kalman-Bucy Filter as a True Time-Varying Wiener Filter", *IEEE Transactions on Systems, Man and Cybernetics*, Vol.SMC-1, No.2, April 1971, pp 119-127.

T. Kailath and B.D.O. Anderson, "Some Integral Equations with Nonsymmetric Separable Kernels", *SIAM Journal Applied Mathematics,* Vol.20, No.4, June 1971, pp659-669.

P.J. Moylan and B.D.O. Anderson, "Nonlinear Regulator Theory and an Inverse Optimal Control Problem", *IEEE Transactions. on Automatic Control,* Vol.AC-18, No.5, October 1973, pp 460-464, and Proceedings of Joint Automatic Control Conference, Washington University, August 1971, pp 462-467.

B.D.O. Anderson and T. Kailath, "The Choice of Signal-Process Models in Kalman-Bucy Filtering", *Journal of Mathematical Analysis and Applications*, Vol.35, No.3, September 1971, pp 659-668.

B.D.O. Anderson and J.B. Moore, "Formulas for Multidimensional Optimal Linear Filtering in White Noise", *Electronics Letters,* Vol.7, No.18, September 1971, p.515.

B.D.O. Anderson and S. Chiraratananon, "Smoothing as an Improvement on Filtering: A Universal Bound", *Electronics Letters*, Vol.7, No.18, September 1971, p.524.

M.G. Wood, J.B. Moore and B.D.O. Anderson, "Study of an Integral Equation Arising in Detection Theory", *IEEE Transactions on Information Theory*, Vol.IT-17, No.6, November 1971, pp 677-686.

S. Chiraratananon, and B.D.O. Anderson, "The Fixed-Lag Smoother as a Stable Finite-Dimensional Linear System", *Automatica*, Vol.7, No.6, November 1971, pp657-669.

B.D.O. Anderson, "External and Internal Stability of Linear Systems - A New Connection", Proceedings of Joint Automatic Control Conference, Washington University, August 1971, pp 53-58 and *IEEE Transactions on Automatic Control,* Vol.AC-17, No.1, February 1972, pp 107-111.

1972

B.D.O. Anderson, "On the Computation of the Cauchy Index", *Quarterly of Applied Mathematics,* Vol.29, No.4, January 1972, pp577-582.

B.D.O. Anderson, K.L. Hitz and T.E. Fortmann, "A Note on Bounds on Solutions of the Riccati Equation", *IEEE Transactions on Automatic Control,* Vol.AC-17,

No.1, February 1972, p.178.
B.D.O. Anderson, "The Small Gain Theorem, The Passivity Theorem and their Equivalence", *Journal of the Franklin Institute,* Vol.293, No.2, February 1972, pp 105-115.
B.D.O. Anderson and S. Chiraratananon, "New Linear Smoothing Formulas", *IEEE Transactions on Automatic Control,* Vol.AC-17, No.1, February 1972, pp 160-161.
B.D.O. Anderson, "Fixed Interval Smoothing for Nonlinear Continuous Time Systems", *Information and Control,* Vol.20, No.3, April 1972, pp 1-7.
Moore, J.B. and B.D.O. Anderson, "Construction of Lyapunov Functions for Time-varying Systems Containing Memoryless Nonlinearities", *Automation and Remote Control*, Vol.33, No.5, May 1972, pp 716-722.
S.Chiraratananon and Anderson, B.D.O., "Outline Designs for Stable, Continuous-Time Fixed-Lag Smoothers", *Electronics Letters*, Vol.8, No.10, May 1972. pp 263-264.
B.D.O. Anderson, "Continuously Equivalent State Variable Realizations", *IEEE Transactions on Circuit Theory,* Vol.CT-19, No.2, May 1972, pp 286-287.
E.I. Jury and B.D.O. Anderson, "Some Remarks on Simplified Stability Criteria for Continuous Linear Systems", *IEEE Transactions on Automatic Control,* Vol.AC-17, No.3, June 1972, pp 371-372.
K.L. Kitz and B.D.O. Anderson, "An Iterative Method of Computing the Limiting Solution of the Matrix Riccati Differential Equation", *Proceedings of the IEE,* Vol.119, No.9, September 1972, pp 1402-1406.
B.D.O. Anderson, "Hurwitz Polynomials, LC and RC Positive Real Functions, and the Hermite Matrix", *IEEE Transactions on Circuit Theory*, Vol.CT-19, No.5, September 1972, pp 495-496.
B.D.O. Anderson, "The Reduced Order Hermite Criterion with Application Proof of the Lienard-Chipart Criterion", *IEEE Transactions on Automatic Control,* Vol.AC-17, No.5, October 1972, pp 669-672.
B.D.O. Anderson, "Problems of Optimal Control Theory: An Introduction", in ***Optimization***, Eds. R.S. Anderssen, L.S. Jennings and D.M. Ryan, University of Queensland Press, 1972, pp 187-203.

1973

B.D.O. Anderson, "Minimal Order Gyrator Lossless Synthesis", *IEEE Transactions on Circuit Theory,* Vol.CT-20, No.1, January 1973, pp 11-15.
L.H. Son and B.D.O. Anderson, "Design of Kalman Filters Using Signal Model Output Statistics", *Proceedings IEE,* Vol.120, No.2, February 1973, pp 312-318.
B.D.O. Anderson and E.I. Jury, "A Simplified Schur-Cohn Test", *IEEE Transactions on Automatic Control*, Vol.AC-18, No.2, April 1973, pp 157-163.
T.E. Fortmann and B.D.O. Anderson, "On the Approximation of Optimal Realizable Linear Filters Using a Karhunen-Loeve Expansion", *IEEE Transactions on Information Theory,* Vol.IT-19, No.4, July 1973, pp 561-564.
B.D.O. Anderson, "Partially Singular Linear-Quadratic Control Problems", *IEEE Transactions on Automatic Control,* Vol.AC-18, No.4, August 1973, pp407-409.
D.J. Clements and B.D.O. Anderson, "Well-Behaved Ito Equations with Simula-

tions that Always Misbehave", *IEEE Transactions on Automatic. Control,* Vol.AC-18, No.6, December 1973, pp 676-677.

B.D.O. Anderson, "Exponential Data Weighting in the Kalman-Bucy Filter", *Information Sciences,* Vol.5, 1973, pp 217-230.

B.D.O. Anderson, "Algebraic Properties of Minimal Degree Spectral Factors", *Automatica,* Vol.9, No.4, July 1973, pp 491-500. Also Vol.11, No.3, May 1975, pp 321-322.

1974

S. Chiraratananon and B.D.O. Anderson, "Stable Fixed Lag Smoothing of Continuous Time Processes", *IEEE Transactions on Information Theory,* Vol.IT-20, No.1, January 1974, pp 25-36.

B.D.O. Anderson and E.I. Jury, "Stability of Multidimensional Digital Filters", *IEEE Transactions on Circuit Theory,* Vol.CT-21, No.2, March 1974, pp 300-304.

B.D.O. Anderson, N.K. Bose and E.I. Jury, "A Simple Test for Zeros of a Complex Polynomial in a Sector", *IEEE Transactions on Automatic. Control,* Vol.AC-19, No.4, August 1974, pp 437-438.

B.D.O. Anderson and P.J. Moylan, "Synthesis of Linear Time Varying Passive Networks", *IEEE Transactions on Circuits and Systems,* Vol.CAS-21, No.5, September 1974, pp 678-687.

B.D.O. Anderson, K.L. Hitz and N. Diem, "Recursive Algorithms for Spectral Factorization", *IEEE Transactions on Circuits and Systems,* Vol.CAS-21, No.6, November 1974, pp 742-750.

B.D.O. Anderson and P.J. Moylan, "Spectral Factorization of a Finite-Dimensional, Nonstationary Matrix Covariance", *IEEE Transactions on Automatic Control,* Vol.AC-19, No.6, December 1974, pp 680-692.

B.D.O. Anderson, "An Orthogonal Decomposition Defined by a Pair of Skew-Symmetric Forms", *Linear Algebra and its Applications,* Vol.8, 1974, pp91-93.

E.I. Jury and B.D.O. Anderson, "On the Reduced Hermite and Reduced Schur-Cohn Matrix Relationships", *International Journal of Control,* Vol.19, No.5, 1974, pp 877-890.

1975

B.D.O. Anderson, N.K. Bose and E.I. Jury, "Output Feedback Stabilization and Related Problems - Solution via Decision Methods", *IEEE Transactions on Automatic Control,* Vol.AC-20, No.1, February 1975, pp 53-66. Also Proceedings 1975 Joint Automatic Cont. Conference.

B.D.O. Anderson, E.I. Jury and F.L.C. Chapparo, "Relations Between Real and Complex Polynomials for Stability and Aperiodicity Conditions", *IEEE Transactions on Automatic Control,* Vol.AC-20, No.2, April 1975, pp 244-246.

B.D.O. Anderson and P.J. Moylan, "The Brune Synthesis in State-Space Terms", *International Journal of Circuit Theory and Applications,* Vol.3, No.2, June 1975, pp 193-199.

B.D.O. Anderson, N.K. Bose and E.I. Jury, "On Eigenvalues of Complex Matrices in a Sector", *IEEE Transactions on Automatic Control,* Vol.AC-20, No.3, June

1975, p. 433.
D.J. Clements, and B.D.O. Anderson, "A Nonlinear Fixed Lag Smoother for Finite-State Markov Processes", *IEEE Transactions on Information Theory,* Vol.IT-21, No.4, July 1975, pp 446-452.

1976

B.D.O. Anderson and R.W. Newcomb, "Linear Passive Networks, Functional Theory", *Proceedings of the IEEE*, Vol.64, No.1, January 1976, pp 72-88.
B.D.O. Anderson, M.A. Arbib and E.G. Manes, "Foundations of Systems Theory: Multidecomposable Systems", *Journal of Franklin Institute*, Vol.301, No.6, June 1976, pp 497-508.
B.D.O. Anderson and E.I. Jury, "Inverse Lienard-Chipart Problem", *IEEE Transactions on Automatic Control,* Vol.AC-21, No.3, June 1976, p. 426.
D.J. Clements and B.D.O. Anderson, "Polynomial Factorization via the Riccati Equation", *SIAM Journal of Applied Mathematics,* Vol.31, No.1, July 1976, pp 179-205.
B.D.O. Anderson and E.I. Jury, "A Note on the Youla-Bongiorno-Lu Condition", *Automatica*, Vol.12, No.4, July 1976, pp 387-388.
B.D.O. Anderson and E.I. Jury, "Generalized Bezoutian and Sylvester Matrices in Multivariable Linear Control", *IEEE Transactions on Automatic Control,* Vol.AC-21, No.4, August 1976, pp 551-556.
R.W. Scott and B.D.O. Anderson, "Conditions for a Feedback Transfer Matrix to be Proper", *IEEE Transactions on Automatic Control,* Vol.AC-21, No.4, August 1976, pp632-634.
B.D.O. Anderson and E.I. Jury, "Proof of a Special Case of Shanks' Conjecture", *IEEE Transactions on Acoustics, Speech and Signal Processing*, Vol.ASSP-24, No.6, December 1976, pp 574-575.
N.D. Diem and B.D.O. Anderson, "New Results for Linear-Quadratic Discrete-Time Games", *International Journal of Control*, Vol.23, No.1, 1976, pp 17-38.
B.D.O. Anderson, "On the Root Distribution of a Real Polynomial with Respect to the Unit Circle", *Regelungstechnik* No.3, 1976, pp 101-102.
B.D.O. Anderson, E.I. Jury, and M. Mansour, "Schwarz Matrix Properties for Continuous and Discrete Time Systems", *International Journal of Control*, Vol.23, No.1, 1976, pp 1-16.
B.D.O. Anderson and J.B. Moore, "A Matrix Kronecker Lemma", *Linear Algebra and its Applications,* Vol.15, 1976, pp 227-234.

1977

B.D.O. Anderson and E.I. Jury, "A Simplest Possible Property of the Generalized Routh-Hurwitz Conditions", *SIAM Journal of Applied Mathematics,* Vol.15, No.1, January 1977, pp 177-183.
D.J. Clements and B.D.O. Anderson, "Transformational Solution of Singular Linear Quadratic Control Problems", *IEEE Transactions Automatic Control,* Vol.AC-22, No.1, February 1977, pp 57-60.
D.J. Clements, B.D.O. Anderson and P.J. Moylan, "Matrix Inequality Solution to

Linear-Quadratic Singular Control Problems", *IEEE Transactions on Automatic Control*, Vol.AC-22, No.1, February 1977, pp 55-57.

R.W. Scott and B.D.O. Anderson, "Parametric Solution to the Stable Exact Model Matching Problem", *IEEE Transactions on Automatic Control*, Vol.AC-22, No.1, February 1977, pp 137-138.

B.D.O. Anderson, "Exponential Stability of Linear Equations Arising in Adaptive Identification", *IEEE Transactions on Automatic Control*, Vol.AC-22, No.1, February 1977, pp 83-88.

D.J. and B.D.O. Anderson, "Extremal Solutions of Riemann-Stieltjes Inequalities of Linear Optimal Control", *IEEE Transactions on Automatic Control*, Vol.AC-22, No.1, February 1977, pp 139-140.

D.J. Clements and B.D.O. Anderson, "Linear-Quadratic Discrete-Time Control and Constant Directions", *Automatica*, Vol.13, No.3, May 1977, pp 255-264.

B.D.O. Anderson and R.W. Scott, "Output Feedback Stabilization - Solution by Algebraic Geometry Methods", *Proceedings of the IEEE*, Vol.65, No.6, June 1977, pp 849-860.

E.I. Jury, V.R. Kolavennu and B.D.O. Anderson, "Stabilization of Certain Two-Dimensional Recursive Filters", *Proceedings of the IEEE*, Vol.65, No.6, June 1977, pp 887-892.

B.D.O. Anderson, "An Approach to Multivariable System Identification", *Automatica*, Vol.13, No.4, July 1977, pp 401-408.

B.D.O. Anderson and R.R. Bitmead, "Stability of Matrix Polynomials", *International Journal of Control*, Vol.26, No.2, August 1977, pp 235-248.

T.S. Ng, G.C. Goodwin and B.D.O. Anderson, "On the Identifiability of Multiple-Input Multiple-Output Linear Dynamic Systems Operating in Closed Loop", *Automatica*, Vol.13, No.5, September 1977, pp 477-486.

R.R. Bitmead and B.D.O. Anderson, "Matrix Fraction Description of the Lossless Positive Real Property", *IEEE Transactions on Circuits and Systems*, Vol.CAS-24, No.10, October 1977, pp 546-550.

B.D.O. Anderson and H.B. Doan, "Design Procedure for Stable Suboptimal Fixed Lag Smoothers", *IEEE Transactions on Automatic Control*, Vol.AC-22, No.6, December 1977, pp 949-952.

R.R. Bitmead and B.D.O. Anderson, "The Matrix Cauchy Index: Properties and Applications, *SIAM Journal of Applied Mathematics*, Vol.33, No.4, December 1977, pp 655-672.

1978

B.D.O. Anderson, "Second Order Convergence Algorithms for the Steady-State Riccati Equation", *International Journal of Control*, Vol.28, No.2, 1978, pp 295-306.

B.D.O. Anderson and D.J. Clements, "Robust Linear-Quadratic Minimization", *JMAA*, Vol.62, No.3, March 1978, pp 620-628.

B.D.O. Anderson, "Covariance Factorization via Newton-Raphson Iteration", *IEEE Transactions on Information Theory*, Vol.IT-24, No.2, March 1978, pp 183-187.

R.W. Scott and B.D.O. Anderson, "Least Order Stable Solution of the Exact Model

Matching Problem", *Automatica*, Vol.14. No.5, September 1978, pp 481-492.
B.D.O. Anderson, J.B. Moore and R.M. Hawkes, "Model Approximation via Prediction Error Identification", *Automatica*, Vol.14, No.5, November 1978, pp 615-622.
R.R., Bitmead S-Y. Kung, B.D.O. Anderson and T. Kailalth, "Greatest Common Divisor via Generalized Sylvester and Bezout Matrices", *IEEE Transactions on Automatic Control*, Vol.AC-23, No.6, December 1978, pp 1043-1047.
R.R. Bitmead and B.D.O. Anderson, "A Dynamical Interpretation of Column and Row Properness of Matrix Fraction Descriptions", *IEEE Transactions on Automatic Control*, Vol.AC-33, No.6, December 1978, pp 1101-1102.
B.D.O. Anderson, "Structure of a Class of Unimodular Matrices", *Linear Algebra and Its Applications*, Vol.22, 1978, pp 167-174.
B.D.O. Anderson and N.T. Hung, "Multivariable Design Problem Reduction to Scalar Design Problems", ***Alternatives for Linear Multivariable Control***, Eds. M.K. Sain, J.L. Peczkowski and J.L. Melsa, National Engineering Consortium, Chicago, 1978, pp 88-95.
B.D.O. Anderson and T. Kailath, "Fast Algorithms for the Integral Equations of the Inverse Scattering Problem", *Journal of Integral Equations and Operator Theory*, Vol.1, No.1, 1978, pp 132-136.
B.D.O. Anderson and T. Kailath, "Forwards and Backwards Models for Finite-State Markov Processes", *Advances in Applied Probability*, Vol.11, 1978, pp 118-133.

1979

B.D.O. Anderson and T. Kailath, "Forward, Backward and Dynamically Reversible Models for Second Order Processes", *IEEE Transactions on Circuits and Systems*, Vol.CAS-26, No.11, November 1979, pp 956-965
N.T. Hung and B.D.O. Anderson, "Triangularization Technique for the Design of Multivariable Control Systems", *IEEE Transactions on Automatic Control*, Vol.AC-24, No.3, June 1979, pp 455-460.
J.M. Potter, A.S. Morse and B.D.O. Anderson, "Single Channel Control of a Two-Channel System", *IEEE Transactions on Automatic Control*, Vol.AC-24, No.3, June 1979, pp 491-492.
B.D.O. Anderson and T. Kailath, "Passive Network Synthesis via Dual Spectral Factorizations", *IEEE Transactions on Circuits and Systems*, Vol.CAS-26, No.10, October 1979, pp 866-873.
B.D.O. Anderson, "Practical System Inversion Without Impulse Response Approximation", *IEEE Transactions on Circuits and Systems*, Vol.CAS-26, No.11, November 1979, pp965-967.
K.K. Saluja and B.D.O. Anderson, "t-fault t/2 - step Sequentially Diagnosable Systems", *IEEE Proceedings Letters*, Vol.67, No.12, December 1979, pp 1678-1679.

1980

B.D.O. Anderson, "Comrade Matrix and Systems Excited by Colored Noise", *IEEE Transactions on Automatic Control*, Vol.AC-25, No.1, February 1980, pp 119-120.
J.M. Potter and B.D.O. Anderson "Partial Prior Information and Decision-Making" *IEEE Transactions on Systems, Man and Cybernetics*, Vol.SMC-10, No.3, March

1980, pp 125-133.
N.T. Hung and B.D.O. Anderson, "Analysis and Synthesis of Nonlinear Reciprocal Networks Containing Two Element Types and Transformers", *IEEE Transactions on Circuits and Systems*, Vol.CAS-27, No.4, April, 1980, pp 263-276.
R.M. Lewis and B.D.O. Anderson, "Insensitivity of a Class of Nonlinear Compartmental Systems to the Introduction of Arbitrary Delays", *IEEE Transactions Circuits and Systems*, Vol.CAS-27, No.7, July 1980, pp 604-612.
R.M. Lewis and B.D.O. Anderson, "Necessary and Sufficient Conditions for Delay Independent Stability of Linear Autonomous Systems", *IEEE Transactions on Automatic Control*, Vol.AC-25, No.4, August 1980, pp 735-739.
R.R. Bitmead and B.D.O. Anderson, "Lyapunov Techniques for the Exponential Stability of Linear Difference Equations with Random Coefficients", *IEEE Transactions on Automatic Control*, Vol.AC-25, No.4, August 1980, pp 782-788.
R.R. Bitmead and B.D.O. Anderson, "Performance of Adaptive Estimation Algorithms in Dependent Random Environments", *IEEE Transactions on Automatic Control*, Vol.AC-25, No.4, August 1980, pp 788-794.
B.D.O. Anderson, "Nonlinear Networks and Onsager-Casimir Reversibility", *IEEE Transactions Circuits and Systems*, Vol.CAS-27, No.11, November 1980, pp 1051-1058.
R.J. Evans, A. Cantoni and B.D.O. Anderson, "Theoretical Aspects of Optimal Autocorrelation Synthesis", *Information Sciences*, Vol.22, 1980, pp 171-199.
J.B. Moore and B.D.O. Anderson, "Coping with Singular Transition Matrices in Estimation and Control Stability Theory", *International Journal of Control*, Vol.31, No.3, 1980, pp 571-586.
Saluja, K.K. and Anderson, B.D.O., "Fault Diagnosis in Loop-connected Systems", *Information Sciences*, Vol.21, No.1, pp 75-92, 1980.
R.R. Bitmead and B.D.O. Anderson, "Asymptotically Fast Solution of Toeplitz and Related Systems of Linear Equations", *Linear Algebra and its Applications*, Vol.34, 1980, pp 103-116.

1981

R.R. Bitmead and B.D.O Anderson, "Adaptive Frequency Sampling Filters", *IEEE Transactions on Circuits and Systems, Special Issue on Adaptive Signal Processing*, Vol.CAS-29, No.6, June 1981, pp 499-510.
B.D.O. Anderson and J.B. Moore, "Time-varying Feedback Laws for Decentralized Control", *IEEE Transactions on Automatic Control*, Vol.AC-26, No.5, October 1981, pp1133-1139.
B.D.O. Anderson and J.B. Moore, "Detectability and Stabilizability of Discrete-time Linear Systems", *SIAM Journal on Control and Optimization*, Vol.19, No.1, January 1981, pp 20-32.
J.M. Potter and B.D.O. Anderson, "Time-Limited Time-invariant Wiener Filtering", *Information Sciences*, Vol.23, No.3, April 1981, pp 153-158.
E.I. Jury and B.D.O. Anderson, "A Note on the "Reduced Schur-Cohn Criterion", *IEEE Transactions on Automatic Control*, Vol.AC-26, No.2, April 1981, pp 612-614.

C.R. Johnson Jnr., M.G. Larimore, J.R. Treichler and B.D.O. Anderson, "Sharf Convergence Properties", *IEEE Transactions on Acoustics, Speech and Signal Processing*, Vol.ASSP-29, No.3, June 1981.

Anderson, B.D.O. and M.R. Gevers, "On Multivariable Pole-Zero Cancellations and the Stability of Feedback Systems", *IEEE Transactions on Circuits and Systems*, Vol.CAS-28, No.8, August 1981, pp830-833.

B.D.O. Anderson and D.J. Clements, "Algebraic Characterization of Fixed Modes in Decentralised Control", *Automatica*, Vol.17, No.5, September 1981, pp 703-712.

B.D.O. Anderson and D.J. Clements, "Algebraic Characterization of Fixed Modes in Decentralised Control", *Automatica*, Vol.17, No.5, September 1981, pp 703-712.

C.R. Johnson Jnr. and B.D.O. Anderson, "Sufficient Excitation and Stable Reduced Order Adaptive IIR Filtering", *IEEE Transactions Acoustics, Speech and Signal Processing*, Vol.ASSP-29, No.6, December 1981, pp 1212-1215.

M.R. Gevers and B.D.O. Anderson, "Representations of Jointly Stationary Stochastic Feedback Processes", *International Journal of Control*, Vol.33, No.5, 1981, pp 777-809.

1982

B.D.O.Anderson, "Transfer Function Matrix Description of Decentralized Fixed Modes", *IEEE Transactions on Automatic Control*, Vol.AC-27, No.6, December 1982.

B.D.O. Anderson and A.C. Tsoi, "Connecting Forward and Backward Autoregressive Models", *IEEE Transactions on Automatic Control*, Vol.AC-29, No.10, October.

B.D.O. Anderson and C.R. Johnson Jnr., "On Reduced Order Adaptive Output Error Identification and Adaptive IIR Filtering", *IEEE Transactions on Automatic Control*, Vol.AC-27, No.4, August 1982, pp927-933.

B.D.O. Anderson and C.R. Johnson Jnr., "Exponential Convergence of Adaptive Identification and Control Algorithms", *Automatica*, Vol.18, No.1, January 1982, pp 1-13..

B.D.O. Anderson and M.R. Gevers, "Identifiability of Linear Stochastic Systems Operating Under Linear Feedback", *Automatica*, Vol.18, No.2, March 1982, pp 195-214.

M.R. Gevers and B.D.O. Anderson, "On Jointly Stationary Feedback Free Stochastic Processes", *IEEE Transactions on Automatic Control*, Vol.AC-27, No.2, April 1982, pp 431-436.

B.D.O. Anderson, "Internal and External Stability of Linear Time-varying Systems", *SIAM Journal of Control and Optimization*, Vol.20, No.3, May 1982, pp 408-413.

J. Fadavi-Ardekani, S.K. Mitra and B.D.O. Anderson, "Extended State-Space Model of Discrete-Time Dynamical Systems", *IEEE Transactions on Circuits and Systems*, Vol.CAS-29, No.8, August 1982, pp 547-556.

R.M. Johnstone, C.R. Johnson Jnr., R.R. Bitmead and B.D.O. Anderson, "Exponential Convergence of Recursive Least Squares with Exponential Forgetting Factor", *Systems and Control Letters*, Vol.2, No.2, August 1982, pp 77-82.

R.M. Johnstone and B.D.O. Anderson, "Exponential Convergence of Recursive Least Squares with Exponential Forgetting Factor - Adaptive Control", *Systems and Control Letters*, Vol.2, No.2, August 1982, pp 69-76.
B.D.O. Anderson, "Reverse-Time Diffusion Equation Models", *Stochastic Processes and Their Applications*, Vol.12, No.31, 1982, pp 313-326.
B.D.O. Anderson, "Port Properties of Nonlinear Reciprocal Networks", *Circuits, Systems and Signal Processing,* Vol.1, No.1, 1982, pp77-92.
B.D.O. Anderson and H-M Hong, "Structural Controllability and Matrix Nets", *International Journal of Control*, Vol.35, No.3, 1982, pp 397-416.
B.D.O. Anderson, "Developments in System Theory", *Golden Jubilee Publication of IREE Australia*, 1982, pp 24-25.

1983

B.D.O. Anderson and R.M. Johnstone, "Global Adaptive Pole Placement: Detailed Analysis of a First Order System", *IEEE Transactions on Automatic Control*, Vol.AC-28, No.8, August 1983, pp 852-855.
A.S. Morse, W.A. Wolovich and B.D.O. Anderson, "Generic Pole Assignment: Preliminary Results", *IEEE Transactions on Automatic Control*, Vol.AC-28, No.4, April 1983, pp 503-506.
N. Hamada and B.D.O. Anderson, "Simplified Stability Tests for Delay Differential Systems", *IEEE Transactions on Circuits and Systems*, Vol.CAS-30, No.7, July 1983, pp488-492.
P. Ananthakrishna, B.D.O. Anderson and S.K. Mitra, "Realization of Digital Transfer Functions Using Cascaded Lattice and Ladder Block Structures", *Journal of the Franklin Institute*, Vol.316, No.1, July 1983, pp 93-124.
J.M. Potter and B.D.O. Anderson, "Statistical Inference with Partial Prior Information", *IEEE Transactions on Information Theory*, Vol.IT-29, No.5, September 1983, pp 688-695.
B.D.O. Anderson, "A Physical Basis for Krein's Prediction Formulas", *Stochastic Processes and Their Applications*, Vol.15, No.1, 1983, pp 133-154.
B.D.O. Anderson and I.B. Rhodes, "Smoothing Algorithms for Nonlinear Finite-Dimensional Systems", *Stochastics*, Vol.9, 1983, pp 139-165.
B.D.O. Anderson and M.R. Gevers, "On the Minimality of Feedback Realizations", *International Journal of Control*, Vol.37, No.1, 1983, pp 145-158.
B.D.O. Anderson and R.M. Johnstone, "Adaptive Systems and Time-Varying Plants", *International Journal of Control*, Vol.37, No.2, 1983, pp 367-377.

1984

N. Hamada and B.D.O. Anderson, "Markov Parameter Characterization of the Strict Positive Real Property", *IEEE Transactions Circuits and Systems*, Vol.CAS-31, No.9, September 1984, pp 814-819.
B.D.O. Anderson and M. Deistler, "Identifiability in Dynamic Errors-in-Variables Models", *Journal of Time Series Analysis*, Vol.5, No.1, 1984 pp 1-13.
C.I. Byrnes and B.D.O. Anderson, "Output Feedback and Generic Stabilizability", *SIAM Journal of Control and Optimization*, Vol.22, No.3, May 1984, pp 362-380.

A.C. Tsoi and B.D.O. Anderson, "D-step Lattice-ladder Predictor in Terms of 1-step Lattice-ladder Predictors", *IEEE Transactions on Automatic Control*, Vol.AC-29, No.5, May 1984, pp 464-465.

B.D.O. Anderson, "Behaviour of the Output Error Identification Algorithm for Small Stepsize Gains", *Systems and Control Letters*, Vol.4, No.3, May 1984, pp 119-124.

N. Hamada and B.D.O. Anderson, "The Unit Circle Cauchy Index: Definition, Characterization and Linear Systems Applications", *SIAM Journal on Applied Mathematics*, Vol.44, No.4, August 1984, pp 803-818.

B.D.O. Anderson and A. Linemann, "Spreading the Control Complexity in Decentralised Control of Interconnected Systems", *Systems and Control Letters*, Vol.5, No.1, October 1984, pp 1-8.

1985

B.D.O. Anderson and R.M. Johnstone, "Global Adaptive Pole Positioning", *IEEE Transactions on Automatic Control*, Vol.AC-30, No.1, January 1985, pp 11-21.

G.A. Latham and B.D.O. Anderson, "Frequency-Weighted Optimal Hankel Norm Approximation of Stable Transfer Functions", *System and Control Letters*, No.4, February 1985, pp 229-236.

B.D.O. Anderson, S. Dasgupta and A-C. Tsoi, "On the Convergence of a Model Reference Adaptive Control Algorithm with Unknown High Frequency Gain", *System and Control Letters*, No.5, April 1985, pp 303-307.

B.D.O. Anderson and D.L. Mingori, "Use of Frequency Dependence in Linear-Quadratic Control Problems to Frequency-Shape Robustness", *Journal Guidance and Control*, Vol.8, No.3, May-June 1985, pp 397-401.

B.D.O. Anderson, "Adaptive Systems, Lack of Persistency of Excitation and Bursting Phenomena", *Automatica*, Vol.21, No.3, May 1985, pp 247-258.

B.D.O. Anderson and P.C. Parks, "Lumped Approximation of Distributed Systems and Controllability Questions", *Proceedings of the IEEE*, Vol.132, PtD, No.3, May 1985.

S.K. Mitra, P.P. Vaidyanathan and B.D.O. Anderson, "A General Theory and Synthesis Procedure for Low Sensitivity Active RC Filters", *IEEE Transactions Circuits and. Systems*, Vol.CAS-32, No.7, July 1985, pp 687-699.

B.D.O. Anderson, "Identification of Scalar Errors-in-Variables Models with Dynamics", *Automatica*, Vol.21, No.6, 1985, pp 709-716.

B.D.O. Anderson, J.A. Gibson and H.R. Sirisina, "Phase Lag and Lead Weighting in Linear Quadratic Control", *Optimal Control Applications and Methods*, Vol.6,No.3, July/September 1985, pp 249-263.

B.D.O. Anderson, "Continuity of the Spectral Factorization Operation", *Matematica Aplicada E Computacional*, Vol.4, No.2, December 1985, pp 139-156.

B.D.O. Anderson and J.B. Moore, "Decentralized Control Using Time Varying Feedback Laws", ***in Control and Dynamic Systems: Advances in Theory and Applications***, Ed. C.T. Leondes, pp 85-115, Academic Press, Orlando, Florida, 1985.

R.J. Elliot and B.D.O. Anderson, "Reverse Time Diffusions", *Stochastic Proceed-*

ings and Applications, Vol.19, 1985, pp 327-339.
B.D.O. Anderson, W.A. Coppel and D.J. Cullen, "Strong System Equivalence", *Journal of the Australian Mathematical Society, (Series B)*, Vol.27, 1985, pp194-222.
B.D.O. Anderson, "Wavenumber Support Regions for Arbitrary Arrays", *Circuits, Systems and Signal Processing*, Vol.4, No.4, 1985, pp439-457.
B.D.O. Anderson and V. Kucera, "Matrix Fraction Construction of Linear Compensators", *IEEE Transactions on Automatic Control*, Vol.AC-30, No.11, 1985, pp1112-1114.

1986

M. Green and B.D.O. Anderson, "Identification of Multivariable Errors-In-Variables Models with Dynamics", *IEEE Transactions on Automatic Control*, Vol.AC-31, No.5, 1986, pp467-471.
R.L. Kosut and B.D.O. Anderson, "A Local Stability Analysis for a Class of Adaptive Systems", *IEEE Transactions on Automatic Control*, Vol.AC-31, No.1, January 1986, pp86-89.
G. Kreisselmeier and B.D.O. Anderson, "Robust Model Reference Adaptive Control", *IEEE Transactions on Automatic Control*, Vol.AC-31, No.2, February 1986, pp 127-133.
B.D.O. Anderson, P. Agathoklis, E.I. Jury and M. Mansour, "Stability and the Matrix Lyapunov Equation of Discrete 2-Dimensional Systems", *IEEE Transactions on Circuits and Systems*, Vol.CAS-33, No.3, March 1986, pp261-267.
M. Bodson, S. Sastry, B.D.O. Anderson, I.M.Y. Mareels and R.R.. Bitmead, "Nonlinear Averaging Theorems, and the Determination of Parameter Convergence Rates in Adaptive Control", *Systems and Control Letters*, Vol.7, No.3, June 1986, pp 145-158.
A.C. Antoulas and B.D.O. Anderson, "On the Scalar Rational Interpolation Problem: The Scalar Case", *IMAJ of Mathemastics Control and Information*, Vol.3, 1986, pp 61-88.
B.D.O. Anderson, "Weighted Hankel Approximation: Calculation of Bounds", *Systems and Control Letters*, Vol.7, No.4, July 1986, pp 247-255.
S. Dasgupta, B.D.O. Anderson, and R.J. Kaye, "Output Error Identification Methods for Partly Known Systems", *International Journal of Control*, Vol.43, No.1, 1986, pp 171-191.
E.I. Jury, B.D.O. Anderson and M. Mansour, "On Model Reduction of Discrete Time Systems", *Automatica*, Vol.22, No.6, 1986, pp717-721.
S. Dasgupta, B.D.O. Anderson and A.C. Tsoi, "Exponential Convergence of a Model Reference Adaptive Controller for Plants with Known High Frequency Gain", *System and Control Letters*, Vol.7, 1986, pp343-350.
Y. Liu and B.D.O. Anderson, "Controller Reduction Via Stable Factorization and Balancing", *International Journal of Control*, Vol.44, No.2, 1986, pp 507-531.
R.E. Skelton and B.D.O. Anderson, "q-Markov Covariance Equivalent Realizations", *International Journal of Control*, Vol.44, No.5, 1986, pp 1477-1490.

1987

B.D.O. Anderson and A. Linemann, "Control of Decentralized Systems with Distributed Controller Complexity", *IEEE Transactions on Automatic Control*, Vol.AC-32, No.7, July 1987, pp 625-629.

R.L. Kosut, B.D.O. Anderson and I.M.Y. Mareels, "Stability Theory for Adaptive Systems, Methods of Averaging and Persistency of Excitation", *IEEE Transactions on Automatic Control*, Vol.AC-32, No.1, January 1987, pp26-34.

S. Dasgupta and B.D.O. Anderson, "Physically Based Parametrization for Designing Adaptive Algorithms", *Automatica*, Vol.23, No.4, July 1987, pp 469-477.

R. Kennedy and B.D.O. Anderson, "Error Recovery of Decision Feedback Equalizers on Exponential Impulse Response Channels", *IEEE Transactions on Communications*, Vol.COM-35, August 1987, pp846-848.

B.D.O. Anderson, E.I. Jury and M. Mansour, "On Robust Hurwitz Polynomials", *IEEE Transactions on Automatic Control*, Vol.AC-32, No.10, October 1987, pp 909-913.

R.A. Kennedy and B.D.O. Anderson, "Recovery Times of Decision Feedback Equalizers - 1. Noiseless Case", *IEEE Transactions on Communications*, Vol.COM-35, No.10, October 1987, pp 1012-1021.

R.A. Kennedy, B.D.O. Anderson and R.R. Bitmead, "Tight Bounds on the Error Probabilities of Decision Feedback Equalizers", *IEEE Transactions on Communications*, Vol.COM-35, No.10, October 1987, pp 1022-1028.

M. Green and B.D.O. Anderson, "On the Continuity of the Wiener-Hopf Factorization", *J.A.M.S., Series B*, Vol.28, 1987, pp443-461.

B.D.O. Anderson and P.V. Kokotovic, "Optimal Control Problems for Large Time Intervals", *Automatica*, Vol.23, No.3, 1987, pp355-363.

M. Green and B.D.O. Anderson, "Factorization of All-Pass Matrix Functions", *International Journal of Control*, Vol.45, No.1, 1987, pp215-235.

M. Green and B.D.O. Anderson, "State-Space Formulas for the Stable Minimum Phase Factors of All-Pass Matrix Functions", *International Journal of Control*, Vol.45, No.5, 1987, pp1575-1602.

Liu, Y. and Anderson, B.D.O., "Model Reduction with Time Delay", *Proceedings of the IEEE, Part D*, Vol.134, No.6, 1987, pp 349-367.

D.A. Gray, B.D.O. Anderson and P.K. Sim, "The Estimation of Structured Covariances for Use in Array Beamforming", *Circuits, Systems, and Signal Processing*, Vol.6, No.4, 1987, pp 421-447.

P.J. Parker and B.D.O. Anderson, "Unstable Rational Function Approximation", *International Journal on Control*, Vol.46, No.5, 1987, pp 1783-1801.

B.D.O. Anderson and M. Deistler, "Dynamic Error-in-Variables Systems with Three Variables", *Automatica*, Vol.23, No.5, 1987, pp 611-616.

B.D.O. Anderson and P.V. Kokotovic, "Optimal Control Problems for Large Time Intervals", *Automatica*, Vol.23, No.3, 1987, pp 355-363.

1988

F.J. Kraus, B.D.O. Anderson, E.I. Jury and M. Mansour, "On Robustness of Low-Order Schur Polynomials", *IEEE Transactions on Circuits and Systems*, Vol.CAS-

35, No.5, May 1988, pp 570-577.
B.D.O. Anderson and M. Green, "Hilbert Transform and Gain/Phase Error Bounds for Rational Functions", *IEEE Transactions on Circuits and Systems*, Vol.CAS-35, No.5, May 1988 pp 528-535.
W. Sethares, I.M.Y. Mareels, B.D.O. Anderson, C.R. Johnson Jnr. and R.R. Bitmead, "Excitation Conditions for Signed Regressor Least Means Squares Adaptation", *IEEE Transactions on Circuits and Systems*, Vol.CAS-35, No.6, June 1988, pp 613-624.
S. Dasgupta, B.D.O. Anderson and R.J. Kaye, "Identification of Physical Parameters in Structured Systems", *Automatica*, Vol.23, No.2, 1988, pp 217-225.
D.J.N. Limebeer and B.D.O. Anderson, "An Interpolation Theory Approach to H^8 Controller Degree Bounds", *Linear Algebra and Its Applications*, Vol.98, 1988, pp 347-386.

1989

R.A. Kennedy, B.D.O. Anderson and R.R. Bitmead, "Channels Leading to Rapid Error Recovery for Decision Feedback Equalizers: Passivity Analysis", *IEEE Transactions on Communication*, Vol.COM-37, No.11, 1989, pp1126-1135
B.D.O. Anderson and M. Green, "Model Reduction by Phase Matching", *Math. of Control., Signals and Systems*, Vol.2, No.3, 1989, pp 221-263.
B.D.O. Anderson, F.J. Kraus and M. Mansour, "Robust Stability of Polynomials with Multilinear Parameter Dependence", *International Journal of Control*, Vol.50, 1989, pp 1745-1762, and in ***Robustness in Identification and Control***, Eds. M Milanese, R Tempo and A Vicino, Plenum, New York, 1989, pp 263-280.
R.A. Kennedy, G. Pulford, B.D.O. Anderson and R.R. Bitmead, "When has a Decision-Directed Equalizer Converged?", *IEEE Transactions on Communications*, Vol.COM-37, No.8, August 1989, pp 879-884.
A.C. Antoulas and B.D.O. Anderson, "On the Stable Rational Interpolation Problem", *Journal of Linear Algebra and Applications*, Vol.122-124, 1989, pp 301-329.
O. Shentov, S.K. Mitra and B.D.O. Anderson, "Stability Testing of 2-D Recursive Digital Filters Based on a Circuit-Theoretic Approach", *IEEE Transactions on Circuits and Systems*, Vol.CAS-36, No.3, March 1989, pp 387-392.
A. Linemann, I. Postlethwaite and B.D.O. Anderson, "Almost Disturbance Decoupling with Stabilization by Measurement Feedback", *Systems and Control Letters*, Vol.12, 1989, pp 225-234.
M.R. Frater, R.A. Kennedy and B.D.O. Anderson, "Reverse-Time Modeling Optimal Control and Large Deviations", *Systems and Control Letters*, Vol.12, 1989, pp351-356.
M. Vidyasagar and B.D.O. Anderson, "Approximation and Stabilization of Distributed Systems by Lumped Systems", *Systems and Control Letters*, Vol.12, No.2, Feb 1989, pp 95-101.
M.R. Frater and B.D.O. Anderson, "Fast Estimation of the Statistics of Excessive Backlogs in Tandem Networks of Queues", *Aust. Telecommunications Research*,Vol.23, No.1, June 1989, pp49-55.
W.A. Sethares, B.D.O. Anderson and C.R. Johnson Jnr, "Adaptive Algorithm with

Filtered Regressor and Error", *Math. of Control, Signals and Systems*, Vol.2, 1989, pp 381-403.
E. Yan and B.D.O. Anderson, "The Simultaneous Optimization Problem for Sensitivity and Gain Margin", *IEEE Transactions on Automatic Control*, Vol.AC-35, 1990, pp 558-563.
Y. Liu and B.D.O. Anderson, "Singular Perturbation Approximation of Balanced Systems", *International Journal of Control*, Vol.50, 1989, pp 1379-1405.
R.E. Skelton and B.D.O. Anderson, "Weighted q-Markov Covers", *International Journal of Control*, Vol.49, No.5, 1989, pp 1755-1771.

1990

M.R. Frater, R.R. Bitmead, R.A. Kennedy and B.D.O. Anderson, "Fast Simulation of Rare Events Using Reverse-time Models", *Computer Networks and ISDN Systems*, Vol.20, 1990, pp 315-321.
B.D.O. Anderson, S. Dasgupta, P. Khargonekar, F.J. Kraus and M. Mansour, "Robust Strict Positive Realness: Characterization and Construction", *IEEE Transactions on Circuits and Systems*, Vol.CAS-37, 1990, pp 869-876.
Y. Liu, B.D.O. Anderson and Uy-Loi Ly, "Coprime Factorization Controller Reductions with Bezout Identity Induced Frequency Weighting", *Automatica*, Vol.26, No.2, 1990, pp 233-249.
P.J. Parker and B.D.O. Anderson, "Frequency Tracking of Nonsinusoidal Periodic Signals in Noise", *Signal Processing*, Vol.20, 1990, pp 127-152.
B.D.O. Anderson and Y. Liu, "Controller Reduction: Concepts and Approaches", *Systems, Control and Information*, (in Japanese) Vol.34, No.9, 1990
Y. Liu and B.D.O. Anderson, "Frequency Weighted Controller Reduction Methods and Loop Transfer Recovery", *Automatica*, Vol.26, No.3, 1990, pp 487-497.
M.C. Tsoi, D-W. Gu, I. Postlethwaite and B.D.O. Anderson, "A Pseudo Singular Value Decomposition and Inner Functions in Super Optimal Control", *International Journal of Control*, Vol.51, 1990, pp1119-1131.
M. Matsumoto, B.D.O. Anderson and M. Mansour, "Sufficient Condition for State Space Representation of N-D Discrete Time Lossless Bounded Real Matrix and N-D Stability of Mansour Matrix", *IEEE Transactions Circuits and Systems*, Vol.CAS-37, 1990, pp 1151-1157.
B.D.O. Anderson and A.C. Antoulas, "Rational Interpolation and State Variable Realizations", *Linear Algebra and Its Applications*, Vol.137/138, 1990, pp 479-509.
A.C. Antoulas and B.D.O. Anderson, "State-space and Polynomial Approaches to Rational Interpolations", in ***Realization and Modelling in System Theory***, Birkhauser, 1990, pp 73-82.
M. Mansour and B.D.O. Anderson, "On the Markov Stability Criterion for Discrete Systems", *IEEE Transactions Circuits and Systems*, Vol.CAS-37, 1990, pp 1576-1578.
M.R. Frater, J. Walrand and B.D.O. Anderson, "Optimally Efficient Simulation of Buffer Overflows in Queues with Deterministic Service Times", *Aust. Telecommunication Research*, Vol.24, 1990, pp 1-8.
W. Yan, R.R. Bitmead and B.D.O. Anderson, "Design Problems for Sensitivity and

Complementary Sensitivity", *Systems and Control Letters*, Vol.15, No.2, 1990, pp 137-146.

S. Dasgupta, B.D.O. Anderson and A.C. Tsoi, "Input Conditions for Continuous-time Adaptive Systems Problems", *IEEE Transactions on Automatic Control*, Vol. AC-35, No.1, 1990, pp 78-82.

B.D.O. Anderson, S. Dasgupta, P. Khargonekar, F.J. Kraus and M. Mansour, "Robust Strict Positive Realness: Characterization and Construction", *IEEE Transactions on Circuits and Systems*, Vol.CAS-37, 1990, pp 869-876.

1991

J. Keller and B.D.O. Anderson, "H^8 Optimal Controller Discretization", *International Journal on Robust and Nonlinear Control*, Vol.1, 1991, pp 125-137.

G.A. Williamson, B.D.O. Anderson and C.R. Johnson Jnr., "On the Local Stability Properties of Adaptive Parameter Estimators with Composite Errors and Split Algorithms", *IEEE Transactions on Automatic Control*, Vol.AC-36, 1991, pp 463-473.

B.D.O. Anderson and J.B. Moore, "Kalman Filtering: Whence, What and Whither" *Kalman Festschrift Volume*, pp 41-54, 1991.

M. Deistler and B.D.O. Anderson, "Identification of Dynamic Systems from Noisy Data: The Case m*=1", *Kalman Festschrift Volume*. pp 423-436, 1991.

Z. Ding, R.A. Kennedy, B.D.O. Anderson and C.R. Johnson Jnr., "Ill-convergence of Godard Blind Equalizers in Data Communications", *IEEE Transactions on Communications*, Vol.COM-39, Sept 1991, pp 1313-1327.

J. Lam and B.D.O. Anderson, "L_1 Impulse response error bound for balanced truncation", *Systems and Control Letters*, Vol.18, No.1, 1991, pp129-137.

M.R. Frater, T.M. Lennon and B.D.O. Anderson, "Optimally Efficient Estimation of the Statistics of Rare Events in Queueing Networks", *IEEE Transactions on Automatic Control Special Issue on Queueing Networks*, Vol.AC-36, 1991, pp 1395-1405.

B.D.O. Anderson and M. Mansour, "New Relations Between the Schur-Cohn-Fujiwara and Nour Eldin Stability Criteria", *IEEE Transactions on Automatic Control*, Vol.AC-36, No.9 Sept 91, pp 1117-1120.

I.R. Petersen, B.D.O. Anderson, and E.A. Jonckheere, "A First Principles Solution to the Non-singular H8 Control Problem", *International Journal of Robust and Nonlinear Control*, Vol.1, 1991, pp 171-186.

J. Keller and B.D.O. Anderson, "H8 Optimal Controller Discretization", *International Journal of Robust and Nonlinear Control*, Vol.1, 1991, pp 125-137.

M-J. Er and B.D.O. Anderson, "Practical Issues in Multirate Output Controllers", *International Journal of Control*, Vol.53 No.5, 1991, pp 1005-1020.

G.A. Williamson, C.R. Johnson Jnr. and B.D.O. Anderson, "Locally Robust Identification of Linear Systems Containing Unknown Gain Elements with Application to Adapted IIR Lattice Models", *Automatica*, Vol.27, September 1991, pp 783-798.

S. Dasgupta, P.J. Parker, B.D.O. Anderson, F. Kraus and M. Mansour, "Frequency Domain Conditions for the Robust Stability of Linear and Nonlinear Dynamical Systems", *IEEE Trans Circuits and Systems*, Vol. CAS-38, 1991, pp 389-397.

1992

J.P. Keller and B.D.O. Anderson, "A new approach to the Discretization of Continuous Time Controllers", *IEEE Transactions on Automatic Control*, Vol.AC-37, 1992, pp 214-223.

W. Yan, B.D.O. Anderson and R.R. Bitmead, "The Combined Sensitivity and Phase Margin Problem", *Automatica*, Vol.28, No.2, 1992, pp 417-421.

L. White and B.D.O. Anderson, "Stationary Discrete-Time Covariance Factorization by Use of Newton Raphson Iteration", *Journal Mathematics of Control, Signals and Systems*, Vol.5, No.3. 1992.

D.J.N. Limebeer, B.D.O. Anderson, P.P. Khargonekar and M. Green, "A Game Theoretic Approach to H^8 Control for Time Varying Systems", *SIAM Journal on Control and Optimization*, Vol.30, No 2, 1992, pp262-283.

J.P. Keller and B.D.O. Anderson, " A New Approach to the Discretization of Continuous-time Controllers", *IEEE Transactions on Automatic Control*, Vol.AC-37, 1992, pp 214-223.

B.D.O. Anderson and J.B. Moore, "Comments on 'Stabilizability and Detectability of Discrete-time Time-varying Systems", *IEEE Transactions on Automatic Control*, Vol.AC-37 No.3, 1992, pp 409-412.

M.J. Er and B.D.O. Anderson, "Performance Study of Multi-rate Output Controllers Under Noise Disturbances", *International Journal of Control*, Vol.15, No.3, 1992, pp 531-545.

B.D.O. Anderson and S. Dasgupta, "Connections Between Real Schur Polynomials and Half Order Complex Schur Polynomials", *Systems and Control Letters*, Vol.18, 1992, pp 237-244.

D.J.N. Limebeer, B.D.O. Anderson, P.P. Khargonekar and M. Green, "A Game Theoretic Approach to H^8 Control for Time Varying Systems", *SIAM Journal on Control and Optimization*, Vol.30 No.2, 1992, pp 262-283.

B.D.O. Anderson, F. Kraus, M. Mansour and S. Dasgupta, "Easily Testable Sufficient Conditions for the Robust Stability of Systems with Multilinear Parameter Dependence", in ***Robustness of Dynamic Systems with Parameter Uncertainties***, Eds. M Mansour et al, Birkhauser, 1992, pp 81-92.

** F. Kraus, M. Mansour, W. Truöl and B.D.O. Anderson, "Robust Stability of Control Systems: Extreme Point Results for the Stability of Edges", *International Journal of Control*, Vol.55, 1992, pp 1039-1049.

L. Gang, B.D.O. Anderson, M. Gevers and J. Perkins, "Optimal FWL Design of State-space Digital Systems with Weighted Sensitivity Minimization and Sparseness consideration", in ***Recent Advances in Mathematical Theory of Systems Control Networks and Signal Processing***, Eds H. Kimura, S Kodema, Mita Press 1992, pp 549-554, *and IEEE Transactions on Circuits and Systems - I: Fundamental Theory and Applications*, (abbreviated version) Vol.39, 1992, pp365-377.

B.D.O. Anderson and S. Dasgupta, "Connections Between Real Schur Polynomials and Half Order Vomplex Schur Polynomials", *Systems and Control Letters*, Vol.18, 1992, pp 237-244.

B.D.O. Anderson, W-Y. Yan and R.R. Bitmead, "Trade-offs Among Conflicting

Objectives in Robust Control Design", *Advances in Control and Dynamic Systems*, Vol.50, 1992, pp 1-53.

J. Lam and B.D.O. Anderson, "L1 Impulse Response Error Bound for Balanced Truncation", *Systems and Control Letters*, Vol.18, No.1, 1992,pp 129-137.

B.D.O. Anderson, "Optimizing the Discretization of Continuous-time Controllers", in ***Recent Advances in Mathematical Theory of Systems Control Networks and Signal Processing I***, Eds. H. Kimura, S Kodema, Mita Press, 1992, pp475-480.

U. Helmke and B.D.O. Anderson, "Hermitian Pencils and Output Feedback Stabilization", *International Journal of Control*, Vol.56, No.4, 1992, pp 857-876.

1993

J.P. Keller and B.D.O. Anderson, "H8-optimierung Abgetasteter Regelsysteme", *Automatisierungstechnik* Vol.41, 1993, pp 114-123.

D.W. Richardson, B.D.O. Anderson and N.K. Bose, "Matrix-fraction Description From Frequency Samples", *Linear Algebra and its Applications*, Vol.181, 1993, pp 251-268, 1993.

S. Verdu, B.D.O. Anderson and R.A. Kennedy, "Blind Equalization Without Gain Identification", *IEEE Transactions on Information Theory*, Vol.IT-39, No.1, January 1993, pp 292-297.

Z. Ding, R.A. Kennedy, B.D.O. Anderson and C.R. Johnson Jnr., "Local Convergence of the Sato Blind Equalizer and Generalizations Under Practical Constraints", *IEEE Transactions on Information Theory*, Vol.IT-39, No.1, January 1993, pp 129-144.

B.D.O. Anderson, M. Gevers and G. Li, "Optimal FWL Design of a Digital Controller", in ***Fundamentals of Discrete-Time Systems***, Chicago Eds,
M. Jamshidi, M. Mansour, B.D.O. Anderson and N.K. Bose, FSI Press, Albuquerque 1993.

M-J. Er and B.D.O. Anderson, "Design of Reduced-order Multirate Input Compensators for Output Injection Feedback Laws", *International Journal of Control*, Vol.58, No.1, 1993 pp 169-185.

B.D.O. Anderson and M. Deistler, "Identification of Dynamic Systems From Noisy Data; Single Factor Case", *Journal Mathematics of Control, Signals, and Systems*, Vol.6, 1993, pp 1029.

M. Mansour and B.D.O. Anderson, "Kharitonov's Theorem and the Second Method of Lyapunov", *Systems and Control Letters*, Vol.20, 1993, pp 39-47, and in ***Robustness of Dynamic Systems with Parameter Uncertainties***, Eds, M Mansour et al, Birkhauser 1992, pp 3-11.

W-S. Lee, B.D.O. Anderson, R.L. Kosut and I.M.Y. Mareels, "A New Approach to Adaptive Robust Control", *International Journal of Adaptive Control and Signal Processing*, Vol.7, 1993, pp 183-211.

B.D.O. Anderson, "Controller Design: Moving from Theory to Practice", *IEEE Control Systems Magazine*, Vol.CAS-13, No.4 , August 1993, pp 16-25.

B.D.O. Anderson, "Control Engineering as an Integrating Discipline From the 17th To the 21st Century", *International Journal Soc. Materials Engineering for Re-*

sources, Vol.1, No.1, 1991, pp 10-26.
S. Dasgupta, B.D.O. Anderson and R.J. Kaye, "Adaptive Control of Systems with Unknown Physical Element Values", *International Journal of Adaptive Control and Signal Processing*, Vol.7, 1993, pp 137-150.
R.A. Kennedy, B.D.O. Anderson and R..R. Bitmead, "Blind Adaption of Decision Feedback Equalizers: Gross Convergence Properties", *International Journal of Adaptive Control and Signal Processing,* Vol.7, 1993, pp 497-523.
G.W. Pulford, R.A. Kennedy and B.D.O. Anderson, "An Upper Bound on the Performance of a Novel Feedforward Perception Equalizer", *IEEE Transactions on Information Theory*, Vol.IT-39, 1993, pp 1923-1929.
D.A. Gray, B.D.O. Anderson and R.R. Bitmead, "Towed Array Shape Estimation Using Kalman Filters Part I - Theoretical Model", *Journal of Oceanic Engineering*, Vol.18, 1993, pp 543-556.

1994

B.D.O. Anderson and J. Keller, "Discretization Techniques in Control Systems", *Control and Dynamic Systems Series*, Vol.66, 1994, pp 47-112.
M-J. Er and B.D.O. Anderson, "Discrete-time Loop Transfer Recovery via Generalized Sampled-data Hold Function Based Compensation", *International Journal of Robust and Nonlinear Control*, Vol.4, No.6, 1994, pp 741-756.
M. Karan, R.C. Williamson and B.D.O. Anderson, "Performance of the Maximum Likelihood Constant Frequency Estimator for Frequency Tracking", *IEEE Transactions on Signal Processing*, Vol.SP-42, No.10, 1994, pp 2749-2757.
M. Campi, W.S. Lee and B.D.O. Anderson, "New Filters for Internal Model Control Design", *International Journal of Robust and Nonlinear Control*, Vol.4, No.6, November-December 1994, pp 757-775.
B. James, B.D.O. Anderson and R.C. Williamson, "Conditional Mean and Maximum Likelihood Approaches to Multiharmonic Frequency Estimation",
IEEE Transactions on Signal Processing, June 1994, pp 1366-1375.
M.R. Frater and B.D.O. Anderson, "Fast Simulation of Buffer Overflows in Tandem Networks of GI/GI/1 Queues", *Annals of Operations Research*, Vol.49, 1994, pp 207 - 220.
** B.D.O. Anderson and I.D. Landau, "Least Squares Identification and the Robust Strict Positive Real Property", *IEEE Transactions on Circuits and Systems*, Vol.CAS-41, 1994, pp 601-607.
W. Yan, B.D.O. Anderson and R.R. Bitmead, "On the Gain Margin Improvement using Dynamic Compensation Based on Generalized Sampled-data Hold Functions", *IEEE Transactions on Automatic Control*, Vol.AC-39, 1994, pp 2347-2354.
S. Dasgupta, B.D.O. Anderson and R.J. Kaye, "Persistence of Excitation Conditions for Partially Known Systems", *Automatica*, Vol.30, 1994, pp 547-550.
D.J.N. Limebeer, B.D.O. Anderson and B. Hendel, "A Nash Game Approach to Mixed H_2/H8 Control", *IEEE Transactions on Automatic Control*, Vol.AC-39, 1994, pp 69-82.
R.C. Williamson, B. James, B.D.O. Anderson and J.P. Kootsookos, "Threshold Effects in Multiharmonic Maximum Likelihood Frequency Estimation", *Journal of*

Signal Processing, Vol.37, 1994, pp 309-331.
M-J. Er, B.D.O. Anderson and W-Y. Yan, "Gain Margin Improvement Using Generalised Sampled-data Hold Function Based Multirate Output Compensator", *Automatica*, Vol.3, 1994, pp 461-470.
S. Dasgupta, G. Chockalingham, B.D.O. Anderson and M. Fu, "Lyapunov Functions for Uncertain Systems with Applications to the Stability of Time Varying Systems", *IEEE Transactions on Circuits and Systems*, Vol.CAS-41, No.2, 1994, pp 93-106.

1995

B.D.O. Anderson, F. Kraus, M. Mansour, and S. Dasgupta, "Easily Testable Sufficient Conditions for the Robust Stability of Systems with Multilinear Parameter Dependence", *Automatica*, Vol.31, 1995, pp 25-40.
M-J. Er, and B.D.O. Anderson, "Design of Reduced Order Multirate Output Linear Functional Observer-based Compensators", *Automatica*, Vol.31, No.2, February 1995, pp 237-242.
B. James, B.D.O. Anderson and R.C. Williamson, "Characterization of Threshold for Single Tone Maximum Likelihood Frequency Estimation", *IEEE Transactions on Signal Processing*, Vol.SP-43, April 1995, pp 817-821.
A.G. Madievski, B.D.O. Anderson and M.R. Gevers, "Optimum Realization of Sampled-data Controllers for FWL Sensitivity Minimization", *Automatica*, Vol.31, 1995, pp 367-379.
F. De Bruyne, B.D.O. Anderson and M. Gevers, "Relating H_2 and H8 Bounds for Finite Dimensional Systems", *Systems and Control Letters*, Vol.24, 1995, pp 173-181.
S.W. Kim, B.D.O. Anderson and A.G. Madievski, "Error Bound for Transfer Function Order Reduction Using Frequency Weighted Balanced Reduction", *Systems and Control Letters*, No.4, 1995, pp 183-192.
Anderson, B.D.O., Mansour, M. and Kraus, F., "A New Test for Strict Positive Realness", *IEEE Transactions on Circuits and Systems, Part I*, Vol.CAS-42, 1995, pp 226-229.
S.W. Kim, B.D.O. Anderson and A.G. Madievski, "Multiplicative Approximation of Transfer Functions with Frequency Weighting", *Systems and Control Letters*, Vol.25, June 1995, pp 199-204.
W.S. Lee, B.D.O. Anderson, I.M.Y. Mareels and Kosut, R.L., "On Some Key Issues in the Windsurfer Approach to Adaptive Robust Control", *Automatica*, Vol.31, No.11, November 1995, pp 1619-1636.
M. Karan, B.D.O. Anderson and R C. Williamson, "An Efficient Calculation of the Moments of Matched and Mismatched Hidden Markov Models", *IEEE Transactions on Signal Processing*, Vol.SP-43, No.10, October 1995, pp 2422-2425.
A.G. Madievski, and B.D.O. Anderson, "Sampled-data Controller Reduction Procedure", *IEEE Transactions on Automatic Control*, Vol.AC-40, 1995, pp 1922 - 1926.
C.R Johnson Jnr. and B.D.O. Anderson, "Godard Blind Equalizer Error Surface Characteristics: White, Zero-mean, Binary Source Case", *International Journal of*

Adaptive Control and Signal Processing, Vol.9, No.4, July-Aug 1995, pp 301-324.
J.B. Matson, B.D.O. Anderson, A.J. Laub and D.J. Clements, "Riccati Difference Equations for Discrete Time Spectral Factorization with Unit Circle Zeros", ***Recent Trends in Optimization Theory and Applications, WSIAA***, Vol.5, Ed. R.P. Agarwal, pp 311-326.
B.D.O. Anderson, "Digital Control: What Do We Want and What Can We Now Achieve?", *International Journal of the Society of Materials Engineering for Resources*, Vol.3, No.1, 1995.

1996

S. Dasgupta and B.D.O. Anderson, "The Square Root of Linear Time-varying Systems with Applications", *IEEE Transactions on Circuits and Systems*, Vol.CAS-43, No.12, December 1996, pp 973-986.
B.D.O. Anderson, M. Deistler, L. Farina and L. Benevuti, "Nonnegative Realization of a Linear System with Nonnegative Impulse Response", *IEEE Transactions on Circuits and Systems*, Vol.CAS-43, No.2, February 1996, pp 134-142.
T.J. Endres, B.D.O. Anderson, C.R. Johnson Jnr. and L. Tong, "On the Robustness of FIR Channel Identification from Fractionally-spaced Received Signal Second-order Statistics", *Signal Processing Letters*, Vol.3, May 1996, pp 153-155.
M. Mansour and B.D.O. Anderson, "On the Robust Stability of Time-varying Linear Systems", International Series of Numerical Mathematics, Hurwitz Centenary Conference, ***Stability Theory***, Vol.121, pp 135 - 149, 1996, ISBN 3-7643-5474-7 and ISBN 0-8176-5474-7.
S. Dasgupta and B.D.O. Anderson, "A Parametrization for the Closed-loop Identification of Nonlinear Time-varying Systems", *Automatica*, Vol.32, No.10, 1996, pp 1349-1360.
B.D.O. Anderson, M. Deistler and W. Scherrer, "Solution Set Properties for Static Errors-in-variables Problems", *Automatica*, Vol.32 No.7, 1996, pp1031-1035.
B.D.O. Anderson, and S. Dasgupta, "Multiplier Theory and Operator Square Roots: Application to Robust and Time-varying Stability", International Series of Numerical Mathematics, Hurwitz Centenary Conference ***Stability Theory***, Vol.121, 1996, pp 113-124, ISBN:3-7643-5474-7 and ISBN 0-8176-5474-7.

1997

D.J. Clements, B.D.O. Anderson, A.J. Laub, and J.E. Matson, "Spectral Factorization with Imaginary Axis Zeros", *Linear Algebra and Applications*, Vol.250, 1997, pp 225-252.
S. Hara, B.D.O. Anderson and H. Fujioka, "Relating H_2 and H_8 Norm Bounds for Sampled-data Systems", *IEEE Transactions on Automatic Control*", Vol.AC-42, No.6, June 1997, pp 858-863.
T. Mita, X. Xin, K. Tomiyama and B.D.O. Anderson, "Extended H_8 Control - Unified Solutions to H8 Servo and H8 Estimation Problems", *Transactions SICE*, Vol.33, 1997, pp 656-664 (in Japanese).
B.D.O. Anderson, "New Developments in the Theory of Positive Systems", in ***Systems and Control in the 21st Century***, Eds. C.I. Byrnes, B.N. Datta, D.S.

Gilliam, C.F. Martin, (ISBN: 0-8176-3881-4/3-7643-3881-4) Birkhauser, Boston, 1997, pp17-36.
B.D.O. Anderson, "Building Special Linear System Realizations of Special Transfer Functions", in ***Communications, Computation, Control and Signal Processing***, Eds. A. Paulraj, V. Roychowdhary and C.D. Schaper, Kluwer, Boston, 1997, pp 333 – 341.

1998

T. Mita, J.B. Matson and B.D.O. Anderson, "A Complete and Simple Parametrization of Controllers for a Nonstandard H_8 Control Problem", *Automatica*, Vol.34, No.3, 1998, pp 369-374.
B.D.O. Anderson, "From Youla-Kucera to Identification, Adaptive and Nonlinear Control", *Automatica*, Vol.34 No.12, 1998, pp 39-59.
L. Shue, B.D.O. Anderson and S. Dey, "Exponential Stability of Filters and Smoothers for Hidden Markov Models", *IEEE Transactions on Signal Processing*, Vol.SP-46 No.8, 1998, pp 2180-2194.
F. DeBruyne, B.D.O. Anderson and M. Gevers, "On Plant and LQG Controller Continuity Questions", *Automatica*, Vol.34, No.5, 1998, pp 631-635.
N. Linard, B.D.O. Anderson and F. De Bruyne, "Coprimeness Properties of Nonlinear Fractional System Realizations", *Systems and Control Letters*, Vol.34, 1998, pp 265-271.
B.D.O. Anderson and M. Gevers, "Fundamental Problems in Adaptive Control", in ***Perspectives in Control***, Ed. D. Normand-Cyrot, Springer, Berlin, 1998, pp 9-21.
B.D.O. Anderson, M.R. James and D.J.N. Limebeer, "Robust Stabilization of Nonlinear Systems Via Normalized Coprime Factor Representation", *Automatica*, Vol.34, No.12, 1998, pp 1593-1599.
B.D.O. Anderson, "Positive System Realizations", Chapter 2 in ***Open Problems in Mathematical Systems and Control Theory***, Eds. Blondel, V.D., Sontag, E.D., Vidyasagar M., and Willems, J.C., Springer-Verlag, 1998.

1999

B.D.O. Anderson, "Riccati Equations, Network Theory and Brune Synthesis: Old Solutions for Contemporary Problems", in ***Dynamical Systems, Control, Coding, Computer Vision***, Eds. G. Picci and D.S. Gilliam, Birkhaüser, 1999. *Progress Systems Control Theory*, Vol.25, 1999, pp.1-25.
B.D.O. Anderson, "From Wiener to hidden Markov models: Common Threads of Filtering and Smoothing", *IEEE Control Systems Magazine*, 1999, pp 41-51.
B.D.O. Anderson , "The Realization Problem for Hidden Markov Models", *Mathematics of Control, Signals and Systems*, Vol.12, 1999, pp 80-120.
A.C. Antoulas and B.D.O. Anderson, "On the Choice of Inputs in Identification for Robust Control", *Automatica*, Vol.35, No.6, 1999, pp 1009-1031.
F. DeBruyne, B.D.O. Anderson, M. Gevers and Linard, N., "Gradient Expressions for a Closed-loop Identification Scheme with a Tailor-made Parametrization", *Automatica*, Vol.35, No.11, 1999, pp 1867-1871.

Y. Yamamoto, A.G. Madievski and B.D.O. Anderson, "Approximation of Frequency Response for Sampled-data Control Systems", *Automatica*, Vol.35, 1999, pp 729-734.

B.D.O. Anderson, K. Chongsrid, D.J.N. Limebeer and S. Hara, "Direct Reduced Order Discretization of Continuous-time Controller", *International Journal of Robust and Nonlinear Control*, Vol.9, No.3, 1999 pp 143-155.

N. Linard, B.D.O. Anderson and F. De Bruyne, "Identification of a Nonlinear Plant Under Nonlinear Feedback Using Left Coprime Fractional Based Representation", *Automatica*, Vol.35, 1999, pp 655-667.

T. Endres, B.D.O. Anderson, C.R. Johnson, Jnr. and M. Green, "Robustness to Fractionally-Spaced Equalizer Length Using the Constant Modulus Criterion", *IEEE Transactions on Signal Processing*, Vol.SP-47, No.2, 1999, pp 544-548.

MIX
Papier aus verantwortungsvollen Quellen
Paper from responsible sources
FSC® C105338

If you have any concerns about our products,
you can contact us on
ProductSafety@springernature.com

In case Publisher is established outside the EU,
the EU authorized representative is:
Springer Nature Customer Service Center GmbH
Europaplatz 3, 69115 Heidelberg, Germany

Printed by Libri Plureos GmbH
in Hamburg, Germany